Developments in
Biophysical
Research

Developments in
Biophysical Research

Edited by

Antonio Borsellino
University of Genoa
Genoa, Italy

Pietro Omodeo
University of Padua
Padua, Italy

Roberto Strom
University of Rome
Rome, Italy

Arnaldo Vecli
University of Parma
Parma, Italy

and

Enzo Wanke
CNR Laboratory of Cybernetics and Biophysics
Camogli (Genoa), Italy

PLENUM PRESS · NEW YORK AND LONDON

Library of Congress Cataloging in Publication Data

Congress on Developments in Biophysical Methods, Parma, 1979.
  Developments in biophysical research.

  Proceedings of the congress sponsored by the Italian Society for Pure and Applied
Biophysics, held Oct. 8-11, 1979, in Parma, Italy.
  Includes index.
  1. Biophysics—Methodology—Congresses. I. Borsellino, Antonio. II. Italian Society
for Pure and Applied Biophysics. III. Title.
QH505.C62 1979   574.19'1   80-25985
ISBN-13: 978-1-4684-1079-2   e-ISBN-13: 978-1-4684-1077-8
DOI: 10.1007/978-1-4684-1077-8

Proceedings of a Congress on Developments in Biophysical Methods
sponsored by the Italian Society for Pure and Applied Biophysics
and held October 8-11, 1979, in Parma, Italy.

© 1980 Plenum Press, New York
Softcover reprint of the hardcover 1st edition 1980
A Division of Plenum Publishing Corporation
227 West 17th Street, New York, N.Y. 10011

PREFACE

This volume derives from papers presented at the 4th biennial meeting of the Italian Biophysical Society, held in Parma in October 1979. It includes review lectures presented by guest scientists (R.H. Adrian, E. Neher, S. Ottolenghi, and G. Zaccai); the remaining reviews and papers present some of the problems currently under study in our country.

One can see that biophysical problems are studied under different academic roofs, i.e., at physiological or biochemical departments. We consider this a strength and a weakness at the same time.

The "Italian Bioenergetics Group" contributed to the success of the meeting, as much as the groups working in various laboratories of the Consiglio Nazionale delle Ricerche, in particular those of the National Group of Cybernetics and Biophysics. To them, and to the University of Parma, which contributed financial and organizational support, the Editors of this volume wish to express their appreciation.

Particular thanks are to Drs. E. Carbone and V. Lenci, who spent time and effort for the meeting and in collecting the papers. The special care in the typing goes to the credit of Miss Rampello, to whom the Editors express their deepest gratitude.

A. Borsellino
P. Omodeo
R. Strom
A. Vecli
E. Wanke

CONTENTS

EXCITABLE MEMBRANES

NONEXCITABLE MEMBRANES AND ARTIFICIAL SYSTEMS

PHOTOBIOLOGY

BIOMACROMOLECULES: PROTEIN STRUCTURE AND FUNCTION

EXCITABLE MEMBRANES

CHANNEL NOISE MEASUREMENTS IN NERVE

F. Conti

Laboratorio di Cibernetica e Biofisica, CNR

16032 Camogli, Italy

SUMMARY

The fluctuations of a macrovariable, such as the current flowing across a nerve membrane at constant membrane potential, can provide information about the elementary events which contribute to its total instantaneous value.

One type of information which can only be obtained from fluctuation analysis concerns the size of the elementary events. Under the assumption that ionic channels in nerve membranes have only one non-zero conductance value, this value can be estimated from measurements of the variance of current fluctuations associated with the random opening and closing of channels. The conductance of single sodium or potassium channels, measured with this method both in myelinated and unmyelinated nerve fibers, ranges between 4 and 12 pS.

The spectral analysis of fluctuations yields also valuable information, and it can be used to test the validity of kinetic schemes describing the transitions between different channel states. In Ranvier nodes the component of the sodium current fluctuations which corresponds to the sodium inactivation process observed in voltage-clamp experiments is much larger than expected from a simple Hodgkin-Huxley scheme with statistically independent activation and inactivation processes. This finding provides a strong argument in favour of the hypothesis that the inactivation process is at least partially sequential to the activation process.

Recordings of the elementary contributions to the total ionic current in squid giant axons due to single potassium channels have been recently obtained. These measurements confirm the validity of

the theory used to analyze macroscopic fluctuations and appear to
be the most fruitful future approach to the study of ionic channels
in nerve membranes.

INTRODUCTION

Fluctuations of macroscopic variables in thermodynamic systems
at equilibrium or in steady-state conditions have long been understood (1). Fluctuations are the macroscopic manifestation of the
discrete nature of matter and can be exploited to gain quantitative
information about the elementary components of large systems. A
classical example is the measurement by Perrin of Boltzmann's
constant (and Avogadro's number) from the application of Einstein's
theory of the Brownian motion of colloidal particles (2). Another
equally classical example is the evaluation of the charge of the
electron from the measurement of shot noise in vacuum tubes (3).

The use of fluctuation analysis has allowed in the last decade
a major step in the understanding of ionic permeation through excitable membranes. Previous measurements of electrical noise across
biological membranes had been used by Hagins (4) to evaluate single
photon responses of photoreceptor cells, and by Verveen and Derksen
(5) to study the stochastic nature of the responses of neural net-
works. However, it was only in 1970 that Katz & Miledi (6) reported
the first measurements of membrane noise associated with the random
opening of ionic channels which were used to estimate the size the
single channel currents. Since then, estimates of single channel
conductances have been obtained from noise analysis in many excitable
membranes. These measurements have answered two fundamental
questions. First, they demonstrated the existence of physically
distinct semimacroscopic patways for ion permeation. Second, the
large value of the estimated conductances, implying ionic fluxes of
up to 10^7 ions per second, ruled out the possibility that such pathways utilize a carrier mechanism, which has a much too slow turnover
(7).

This lecture is not aimed at presenting a comprehensive review
of fluctuation studies of nerve membranes, but rather a digest of
the experimental work which I consider to be most representative of
the present state of knowledge in this field. Only measurements of
current fluctuations in nerve membranes kept under voltage-clamp
conditions will be considered, reminding that equivalent information
can also be extracted from voltage fluctuations measurements provided
the membrane impedance locus is known (8). A brief review of the
basic theoretical concepts of fluctuations analysis is first
presented.

THEORETICAL BACKGROUND

Two kinds of basic information can be derived from fluctuation analysis. One concerns the amplitude of the elementary events contributing to the mean macroscopic behaviour and to the macroscopic fluctuations of the system under study. This information is derived from measurements of the mean square amplitude of the fluctuations together with the mean value of the parameter under study. The second type of information concerns the kinetics of the transitions between the different states of the elementary components of the system. This information is obtained from the statistical correlation existing between fluctuations measured at different times.

Amplitude of current fluctuations. Two very simple properties of random variables allow to relate the amplitude of macroscopic and microscopic fluctuations. Let the membrane current, I, be the sum of the contributions arising from N identical but independent ionic channels:

$$I = \sum_{j=1}^{N} i_j$$

The mean value of I, μ_I, and the mean square deviation of I from μ_I, σ_I^2, are then given by:

$$\mu_I = N \mu_i, \tag{1}$$

and:

$$\sigma_I^2 = N \sigma_i^2, \tag{2}$$

where μ_i and σ_i^2 are the mean and the variance of the current flowing through any individual channel. Combining eqns. (1) and (2) in different ways one obtains:

$$\sigma_I/\mu_I = \sigma_i/\mu_i \ \sqrt{N} \tag{3}$$

or:

$$\sigma_I^2/\mu_I^2 = \sigma_i^2/\mu_i. \tag{4}$$

Eqn. (3) stresses the well known fact that the ratio of the standard deviation to the mean of any macrovariable is inversely proportional to the number of its elementary contributions. As stressed by Neher & Stevens (9) the success of fluctuation analysis in the study of biological membranes is basically due to the fact that these systems have two-dimensional structures and contain, therefore, a relatively small number of components. Eqn. (4) shows that the ratio σ_I^2/μ_I is independent of N. From the definitions of σ_I^2 and μ_i it is also clear why their ratio can yield an estimate of the "size" of single channel currents: thus, if the size of the single channel currents is doubled, σ_I^2/μ_i is also doubled. The actual information contained in σ_I^2/μ_i depends on the particular model chosen to describe individual channels. Luckily enough, ionic channels in nerve membranes seem to have a very simple behaviour, that is: their conductance has only one non-zero value (10,11). In this case, eqn. (4) can be rewritten in a very simple form:

$$\sigma_I^2/\mu_I = (1-p)i, \tag{5}$$

where i is the current through an open channel and p is the open channel probability.

   Spectral analysis of current fluctuations. The time course of current fluctuations contains information about the kinetics of the transitions of ionic channels between their various states. This can easily be understood from the following simple example. Suppose that ionic channels are undergoing thermal fluctuations between two equiprobable open and closed configurations at an average rate of 10^5 transitions per second. The "memory" of each elementary event is then of the order of 10 µs and a large current fluctuation measured at any time will die out after few tens of microseconds. On the other hand a much slower on-off kinetics (say 10 transitions per second) would imply that fluctuations are fairly constant within 10 ms because a very small percentage of channels, on the average, will change their state during that fraction of time.

   The quantity which yields the best characterization of the above intuitive property is the mean product of the fluctuations at any two times separated by the interval, t, called the autocovariance function, $C_I(t)$. Alternatively, the poxer spectrum, $S_I(f)$, obtained either directly from the Fourier analysis of the fluctuations or by taking the Fourier transform of $C_I(t)$, yields a completely equivalent characterization. For ionic channels which can exist in n different states and udergo markoffian transitions between these states, $C_I(t)$ is the sum of (n-1) decaying exponentials (9,12):

$$C_I(t) = \sigma_I^2 \sum_{j=1}^{n-1} c_j \exp(-t/\tau_j) \ , \ (\sum_{j=1}^{n-1} c_j=1); \tag{6}$$

where the time constants, τ_j, are solely determined by the matrix of the rate constants of the transition between all possible channel states, and the amplitudes, c_j, depend both on the transition rates and on the relative conductance of the various channel states. The equivalent of eqn. (6) for the power spectral density, $S_I(f)$, is:

$$S_I(t) = \sigma_I^2 \sum_{j=1}^{n-1} c_j L(f,\tau_j) \tag{7}$$

where the <u>Lorentzian spectrum</u>, $L(f,\tau)$, is defined as:

$$L(f,\tau) = \frac{4\tau}{1+(2\pi\tau f)^2} \tag{8}$$

and its cut-off (half power) frequency is $(1/2\pi\tau)$.

The practical relevance of fluctuation spectroscopy for studying the kinetics of state-transitions is enhanced by complementary information obtainable from relaxation (voltage-clamp) measurements. Under the same assumptions leading to eqns. (6) and (7), the time course of the macroscopic current, $I(t)$, following a step in membrane potential, is given by:

$$I(t) = \sum_{j=1}^{n-1} I_j \exp(-t/\tau_j) + I(\infty) \quad , \tag{9}$$

where the exponentials have the same time constants as those appearing in the autocovariance function, but different relative amplitudes. Combined measurements of fluctuations and macroscopic relaxations can be analyzed according to eqns. (7) and (9) in two steps. First one verifies that the time constants needed to fit both types of data are in good quantitative agreement. Then, fluctuation spectra are fitted according to eqn. (7) using the time constants obtained from relaxation data and allowing the amplitudes, c_j, to vary. In this way, fluctuation measurements can provide $(n-2)$ additional quantities which depend on the kinetics of state transitions and may allow to discriminate between different schemes which are equally good to describe solely relaxation data.

THE CONDUCTANCE OF SODIUM AND POTASSIUM CHANNELS

According to what outlined above, the current through an open sodium (potassium) channel can be directly estimated from the variance of sodium (potassium) current fluctuations in nerve membranes kept under voltage clamp condition and in the presence of tetraethylammonium (TEA)(tetrodotoxin (TTX)) to block potassium (sodium) currents. The single channel current, $i_{Na}(i_K)$, can be then converted into open channel conductance, $\gamma_{Na}(\gamma_K)$, either by assuming Ohm's law (10,13) or a constant field current-voltage characteristic (14,15), or by using an experimental instantaneous I-V relation (11).

Following such straightforward approach, estimates of 4 pS for
γ_K(10) and of 7.7 pS for γ_{Na}(11) were obtained in frog nodes. A
most important additional result of these studies was the verification that the σ_I^2/μ_I versus p relation was in fairly good agreement with eqn.(5) over a wide range of p values, providing a very
strong support to the hypothesis that ionic channels in nerve
membranes can have only one non-zero conductance value.

A less direct estimate of γ_K and γ_{Na} is obtained from the amplitude of the power spectra of current fluctuations. Basically, power
spectra contain more information than the simple variance, as shown
by eqn.(7). The additional information is very useful to ascertain
to what extent the measured current fluctuations can be attributed
to the flickering of ion specific channels between open and closed
states rather than to other noise sources. This control was particularly desirable in the early studies of nerve membrane noise, which
revealed the presence of large 1/f spectral components of still un-
clear origin (16). Indeed, the first unequivocal characterizations
of sodium and potassium channel noise in the squid axon membrane (13)
and of sodium channel noise in frog nodes (14,15) were obtained from
the fitting of measured spectra with the superposition of Lorentzian-
like spectra plus 1/f components. From the low frequency asymptote
and the cut-off frequency of the Lorentzian components estimates of
γ_K=12 pS and γ_{Na}=4 pS were obtained for the ionic channels of squid
giant axons in normal physiological conditions (13). Similar measurements yielded γ_{Na}=7.9 pS in normal frog nodes (14) and γ_{Na}=6.4 pS
in frog nodes with sodium inactivation modified (decreased) by
various drugs (15). The estimate of γ_{Na}=2-3 pS, obtained by Van den
Berg et al. (17) in frog nodes, is likely affected by larger errors
because it was based on voltage fluctuation measurements (8) and it
was calculated using "standard" Hodgkin-Huxley (HH) parameters (18)
rather than macroscopic relaxation data from the same nodes.
Potassium current fluctuation spectra in frog nodes (19) yield
γ_K=2.9 pS in close agreement with the value obtained from single
variance measurements (10). Table 1 summarizes estimates of γ_{Na}
and γ_K obtained from fluctuation analysis.

TABLE I. Conductance of normal sodium and potassium channels,
estimated from fluctuation analysis.

γ_{Na}(pS)	Preparation	Type of measurement	Ref.
4	Squid axon	Current noise spectrum	(13)
7.9	Frog node	" " "	(14)
7.7	" "	" " variance	(11)
2.5	" "	Voltage noise spectrum	(17)
8.85	" "	Current noise variance corrected for bandwidth	(20)

$\gamma_K(pS)$	Preparation	Type of measurement	Ref.
12	Squid axon	Current noise spectrum	(13)
4	Frog node	" " variance	(10)
2.9	" "	" " spectrum	(19)

Although they are equivalent in principle, the two methods used to obtain the γ estimates reported above are affected by different errors.

The variance of the recorded fluctuations is equal to the integral of the power spectrum over the recording bandwidth and it is measured with much greater accuracy than the spectral distribution However, undesired noise contributions cannot be easily subtracted from the total variance without the help of the spectral analysis. Furthermore, the measured variance is an underestimation of the actual variance, due to bandwidth limitations (the integral of Lorentzian spectrum up to its cut-off frequency yields only 50% of the variance).

Power spectra show more clearly the presence of stray noise contributions, and the theoretical variance of Lorentzian components can be estimated from their low frequency asymptotes and their cut-off frequencies; but theoretical schemes of channel kinetics are needed to fit the measured spectra.

A compromise between the two methods has been recently used to evaluate γ_{Na} in frog nodes (20). Fluctuation spectra, obtained after subtracting TTX insensitive components, were analyzed and shown to fit a simple sum of relaxation spectra. The noise variance was simultaneously and independently measured as the mean square of the fluctuations. It was then corrected for bandwidth limitations according to the extrapolation of the spectra to infinite frequency, and inserted in eqn.(5), to yield i_{Na}, after subtraction of the TTX insensitive contribution. γ_{Na}, estimated in this way, was found to be fairly independent of membrane voltage and it had an average value of 8.85 pS. This seems at present the best estimate of γ_{Na} in frog nodes. It is in excellent agreement with the value of 7.7 pS obtained by Sigworth with a completely different method, taking into account that the latter estimate was not corrected for the limited bandwidth of the measurements.

FLUCTUATION SPECTRA AND CHANNEL KINETICS

In principle, as discussed in the theoretical section, power spectra can yield important information about the kinetics of the transitions of ionic channels between their various states. In practice the detailed theoretical information implied by eqns.(6) and (7) is not easy to obtain. It is common experience that the

dissection of the sum several exponentials into its single components
is feasible only if the time constants involved differ by large
factors. The same is obviously true for the dissection of single
Lorentzians out of eqn.(7).

Unfortunately, voltage-clamp experiments show that the time
constants describing the relaxation of sodium (or potassium) currents
do not fullfill this requirement (with the exception of the inactivation time constant, discussed later). According to the Hodgkin-
Huxley equations (18), the relaxation of potassium currents is
described by four time constants with values in the ratio of 1:2:3:4.
Likewise, the rising phase of sodium currents contains approximately
three exponentials, with time constants in the ratio of 1:2:3.
Even if the HH kinetic schemes may be considered as simple phenomenological descriptions of the data it is obvious that the actual
channel kinetics must imply time constants which are roughly in the
above ratios. Such small excursion in the values of the time
constants have made so far impossible to acquire detailed information
from potassium current noise spectra and from the high frequency
relaxation spectra of sodium current fluctuations. Indeed, the
superposition of Lorentzians expected from the HH kinetic schemes
are practically indistinguishable from one simple Lorentzian within
the accuracy of the data obtained until now.

In squid giant axons potassium and sodium current noise spectra
were simply fitted with the sum of a Lorentzian spectrum plus a 1/f
component (13) and their analysis was confined to the verification
that the cut-off frequencies of the Lorentzians were in qualitative
agreement with what expected from a simple HH kinetics.

The relaxation of the sodium current contains also a slow
inactivation phase with a time constant which is roughly one order
of magnitude larger than those involved in the activation phase.
It should then be possible to dissect the spectral contribution of
this slower relaxation if the spectral analysis extends to sufficiently low frequencies. The measurements of Conti et al. (14) covered
the frequency range of 6 to 5.000 Hz and could in principle allow
such dissection, since the cut-off frequency of the slow relaxation
noise associated with sodium inactivation is expected to range
between 15 and 100 Hz. However, as it was recognized later in the
analysis of the data, these were found to be affected by artifacts
below 300 Hz. Again, the only kinetic information which could be
obtained, was confined to the recognition of a general qualitative
agreement between the spectrum at higher frequencies and the expec-
tations of the HH model. A 1/f component, as well as the theoretical
HH inactivation noise, were assumed to yield substantial contributions to the total noise spectrum at low frequencies, but no confi-
dence could be placed on their fitted amplitudes.

Very recently, more reliable sodium current fluctuation spectra have been obtained in frog nodes in the frequency range of 3 to 5.000 Hz by avoiding the low frequency artifacts which were recognized to originate from slow systematic drifts in the mean current (20). True random fluctuations with zero mean were extracted from the measured signals by subtracting the systematic drifts. The power spectra of the resulting records are identical to those obtained in previous works (14,15) above 100 Hz, but they provide reliable information also in the range of 3 to 100 Hz were they show the tendency to reach the low frequency plateau expected from a simple Lorentzian behaviour. In this case no significant $1/f$ noise contributions were observed, and the measured spectra could be fitted by the superposition of a simple Lorentzian noise, $c_h \sigma_I^2 L(f, \tau_h)$ associated with the time constant of sodium inactivation, τ_h, and a pseudo-Lorentzian noise, $(1-c_h)\sigma_I^2 s_m(f)$, associated with the superposition of Lorentzians expected from the HH kinetics of the sodium activation process. The fitting of the data was done by varying the amplitude c_h and the sodium activation time constant, τ_m, with the remaining HH parameters (18), m_∞, h_∞ and τ_h, fixed according to the values obtained from the HH analysis of voltage-clamp currents in the same nodes. For the consistency of the HH scheme the best fit of the noise data should have been obtained for τ_m values equal to those measured from voltage-clamp data. The agreement was found to be reasonable, but the τ_m values obtained from noise spectra were systematically higher, beyond the experimental errors. The value of c_h expected according to the HH scheme, c_h^{HH}, can be derived from the parameters m_∞ and h_∞ according to the simple formula (14):

$$c_h^{HH} = \frac{m^3(1-h_\infty)}{1-m_\infty^3 \; h_\infty} \tag{9}$$

The best fit of the current noise spectra for membrane potentials around −50 mV was obtained with c_h values 3 to 5 times larger than expected according to eqn.(9). The disagreement decreased at more depolarized membrane potentials, vanishing around −30 mV.

From these results it was concluded that the assumption contained in the HH scheme, that the inactivation gate acts independently from the activation process, was not tenable. Indeed, such assumption has never been supported by direct evidence and it had already been challenged on the basis of the dependence of sodium gating currents upon the state of sodium inactivation (21). The above analysis of fluctuation data provides perhaps the most direct evidence against the hypothesis of independence, since it involves only the measurement of normal ionic currents. Interestingly enough, schemes having the sodium inactivation sequentially coupled to activation, as suggested by the results of gating current measurements (21), were found to yield c_h values which fit slightly better also the noise data.

THE ACTUAL SHAPE OF SINGLE CHANNEL CURRENTS

The analysis of current fluctuations in nerve membranes according to eqns. (5)-(8) is based upon the fundamental assumption that
ionic channels have a finite number of statistically significant
states in the time scale of usual recordings (say 1 μs). In other
words, it is assumed that the large set of all possible microscopic
states that an ionic channel can assume can be divided into a small
number of subsets in such a way that: 1. the random transitions
between the states of each subset occur at extremely fast rates;
2. the transitions from one subset to another occur much less
frequently, yielding an average lifetime for each subset which is
well within the range of our recording instruments. Under these
assumptions, the transitions between the subsets (which we identify
as the "channel states") will follow a simple markoffian process (1),
and eqns. (6)-(8) will follow.

The utlimate proof of the validity of the above picture can
only derive from direct observations of single channel events.
Indeed, the macroscopic noise expected from the random and sudden
opening of ionic channels which close gradually, with an exponential
time course of time constant, τ, has the same spectral characteristics of the fluctuations produced by ionic channels which jump
infinitely fast from the fully closed to the fully open state, and
viceversa, with rate constants, α and β, such that $1/(\alpha+\beta)=\tau$. The
first analysis of the acetylcholine (Ach) noise in the postsynaptic
membrane (6) was indeed based upon the first of the above two
pictures.

In 1976, Neher & Sackman (22) reported the first observations
of ionic currents associated with the opening of single channels in
denervated muscle fibers. These recordings showed that the opening
of Ach-channels is indeed all or nothing, confirming the markoffian
character of channel fluctuations.

Recordings of single channel events are much more difficult in
nerve membranes, due to the comparatively smaller conductance and
shorter lifetime of the electrically gated nerve channels. However,
it has been possible very recently to develop a technique for recording elementary current events associated with the functioning of
single potassium channels of the squid axon membrane (23). These
recordings were obtained from the voltage-clamp of small patches of
membrane (1 to 2 μm^2) under a reversed potassium gradient. The
membrane was approached from the intracellular face in axons intracellularly perfused with a low ionic strength solution. These
measurements demonstrate that single potassium channels open and
close in an all or nothing fashion and confirm the channel con-

ductance estimates obtained from previous measurements of macroscopic noise(13). Furthermore they seem to reveal fine structures of the potassium channel kinetics which would be very difficult to assess with less direct observations.

CONCLUSIONS

The progress made by fluctuation analysis of nerve membranes in the last decade is rather impressive. It has brought one of the most convincing and direct evidence that nerve excitation is ultimately related to the simple all or nothing behaviour of physically distinct, semimacroscopic pathways, called the sodium and potassium channels. Initial discrepancies between the results obtained by different groups died out rather fast and there is now a general agreement between the estimates of channel conductances obtained independently by a number of different authors. These values give the most convincing evidence that in their open state the ionic channels of nerve membranes have a pore-like structure spanning across the membrane.

With the present degree of mastery of the experimental techniques it is possible that much more accurate data can be collected in the near future, providing more information on detailed channel kinetics. However, major advances in our understanding of the physics of ionic channels in nerve membranes are more likely to be obtained from the analysis of single channel events. This promises to be the most fruitful approach of the forthcoming years.

REFERENCES

1. LAX,M. (1960). Fluctuations from the nonequilibrium state. Rev.mod.Phys. 32:25-64
2. KAC,M. (1947). Random walk and the theory of Brownian motion. Am.Math.monthly 54: 369-91. Also reprinted in Selected Papers on Noise and Stochastic Processes (ed. N.Wax). New York: Dover Publ.Inc., 1954.
3. VAN DER ZIEL,A. (1970). Noise: Sources, Characterization, Measurements. Engelwood Cliffs, N.Y.: Prentice-Hall, Inc.
4. HAGINS,W.A. (1965). Electrical Signs of Information Flox in Photoreceptors. Cold Spring Harbor Symposia on Quantitative Biology, vol. XXX, 403-418.
5. VERVEEN,A.A. & DERKSEN,H.H. (1965). Fluctuations in membrane potential of axons and the problem of coding, Kybernetic 2: 152-60.
6. KATZ,B. & MILEDI,R. (1970). Membrane noise produced by acetylcholine. Nature, Lond., 225:962-3.
7. LAÜGER,P. (1972). Carrier mediated ion transport. Science, N.Y. 178:24-30.

8. WANKE,E., DE FELICE,L.J. & CONTI,F. (1974). Voltage noise and
   current noise in space clamped squid giant axon. Pflügers
   Arch. 347:63-74.
9. NEHER,E. & STEVENS,C.F. (1977). Conductance fluctuations and
   ionic pores in Membranes. Ann.Rev.Biophys.Bioeng. 6:345-381.
10. BEGENISICH,T. & STEVENS,C.F. (1975). How many conductance states
    do potassium channels have? Biophys.J. 15:843-46.
11. SIGWORTH,F.J. (1977). Sodium channels in nerve apparently have
    two conductance states. Nature, Lond. 270:265-267.
12. CONTI,F. & WANKE,E. (1975). Channel noise in nerve membranes
    and liquid bilayers. Q.Rev.Bioph. 8:451-506.
13. CONTI,F., DE FELICE,L.J. & WANKE,E. (1975). Potassium and sodium
    ion current noise in the membrane of the squid giant axon.
    J. Physiol., 248:45-82.
14. CONTI,F. HILLE,B., NEUMCKE,B., NONNER,W. & STÄMPFLI,R. (1976a).
    Measurement of the conductance of the sodium channel from the
    current fluctuations at the node of Ranvier. J. Physiol. 262:
    699-727.
15. CONTI,F., HILLE,B., NEUMCKE,B., NONNER,W. & STÄMPFLI,R. (1976b).
    Conductance of the sodium channel in myelinated nerve fibres
    with modified sodium inactivation. J.Physiol. 262:729-742.
16. VERVEEN,A.A. & DE FELICE,L.J. (1974). Membrane noise. Prog.
    Biophys.molec.Biol. 28:189-265.
17. VAN DEN BERG,R.J., DE GOEDE,J., VERVEEN,A.A. (1975). Conductance
    fluctuations in Ranvier nodes. Pflügers Arch. 360:17-23.
18. HODGKIN,A.L. & HUXLEY,A.F. (1952). A quantitative description
    of membrane current and its application to conduction and
    excitation in nerve. J. Physiol. 117: 500-544.
19. VAN DEN BERG,R.J., SIEBENGA,E. & DE BRUIN,G. (1977). Potassium
    ion noise currents and inactivation in voltage clamped node of
    Ranvier. Nature,Lond. 265:177-179.
20. CONTI,F. NEUMCKE,B., NONNER,W. & STÄMPFLI,R. (1979). Low
    frequency fluctuations of Na current in myelinated nerve.
    Pflügers Arch. 370, R40.
21. ARMSTRONG,C.M. & BEZANILLA,F. (1977). Inactivation of the sodium
    channel.II. Gating currents experiments. J.Gen.Physiol.70-567-
    590.
22. NEHER,E. & SAKMANN,B. (1976). Single-channel currents recorded
    from membrane of denervated frog muscle fibers. Nature,Lond.,
    260:779-802.
23. NEHER,E. & CONTI,F. (1979). Discrete current fluctuations
    produced by single K^+-channels in the squid axon membrane.
    Annual meeting of the German Biophysical Society B61 (Adam,G.
    & Stark,G. ed.), Springer Verlag, Heidelberg.

INTERNAL pH AND K$^+$ CHANNEL RATE CONSTANTS

P.L. Testa, E. Carbone and E. Wanke

Laboratorio di Cibernetica e Biofisica
Consiglio Nazionale delle Ricerche
16032 Camogli, Italy

INTRODUCTION

The effects of changing the proton content of solutions bathing excitable membranes has been the subject of many reports in the past. (1-7). Two main effects were commonly observed although they differed quantitatively upon changing the nerve preparation: a) a shift on the voltage axis of the parameters characterizing nerve excitability (similar to the action of divalent ions (8)), b) a variation of the steady-state maximum conductance of the open channels selective for sodium and potassium ions. A third type of effects concerns phenomena such as the voltage-dependent block of Na$^+$-channels observed by Woodhull (9) in the node of Ranvier.

In the present work we will limit our study to the analysis of the pH effects on the kinetics parameters of the potassium channels of the squid giant axon membrane. The results are shown to have several common aspects which can be interpreted by assuming that the proton concentration influences the rate of opening and closing of the channels in three different ways: a) changing the intramembrane electric field, b) altering the lipid matrix fluidity and, c) affecting the chemical nature of the channel voltage-sensor.

MATERIALS AND METHODS

Loligo Vulgaris giant axons of 400-600 μm in diameter were voltage clamped in the presence of 10^{-6} M TTX, upon internal perfusion at different pH, at a temperature of 3°C. To avoid potassium accumulation, membrane conductance were measured as described elsewhere (10) and the following buffers were used: K-glutamate (pH range: (9-10.8), K-glutamine (8.5-10.2), K-glycylglycine (7.3-9.2),

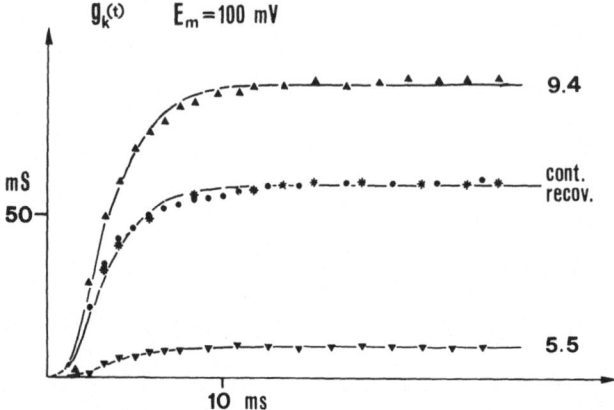

Figure 1.   Potassium conductance records taken from one axon de-
            polarized to a membrane potential E=+100 mV, at the pH
            indicated (filled triangles). Filled circles and stars
            refer to the control and recovery experiment, respectively
            Solid lines are the results of a curve-fitting based on
            the Hodgkin-Huxley equations. Holding potential, -90 mV.
            Temperature 2°C.

K-phosphate (6-7.8), K-citrate (2.5-6) and K-succinate (3.5-6).

RESULTS

    Fig. 1 illustrates the effects of acid and basic internal so-
lutions on the membrane conductance of Loligo axons depolarized to
a potential of E=+100 mV (E, absolute membrane potential). As previously reported (10), the very large reversible change in membrane
conductance is accompanied by a slight variation in the rate of
activation when the pH is changed from 5.5 to 9.4. This observation
is further confirmed by the results of a simple curve fitting based
on the Hodgkin and Huxley type of analysis as illustrated in fig. 2a.
At very high pH values the time constant for the opening of channels
is shifted by only few millivolts to the right while at pH 5.5 a
slight deformation of the $\tau_n(E)$ curve is observed. With respect to
the control experiment, $\tau_n(E)$ is lower at negative potentials and
higher at larger positive voltages with a crossing over point at
arount 0 mV. The voltage dependence of the potassium conductance,
$g_K(E)$, is shown in fig. 2b. The $g_K(E)$ curves show: 1) a vertical

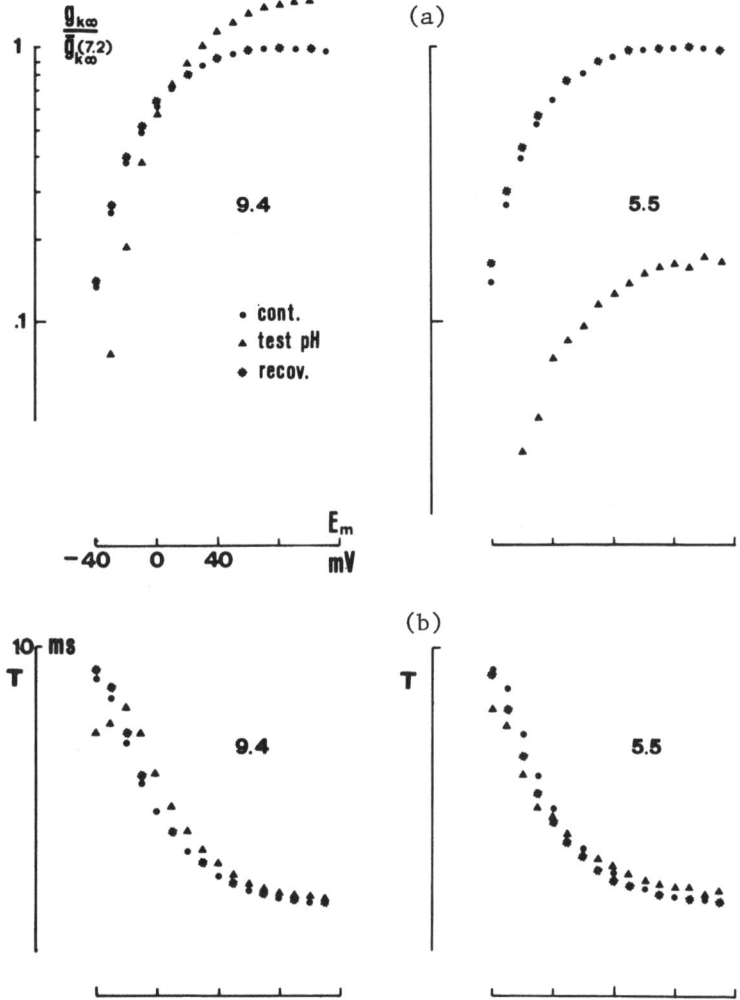

Fig. 2 – Voltage-dependence of potassium conductance $g_K(E)$ and time
constant, $\tau_n(E)$, for the same axon of fig.1 as obtained from
the curve fitting. The steady-state conductance values are
normalized with respect to the $\bar{g}_K$ at the reference pH, 7.2

shift on both directions due to the action of a titratable group
which influences, independently of the membrane potential, the availa
ble number of fully open channels (10), 2) a horizontal shift to
the right which is more evident at basic pH (20 mV).

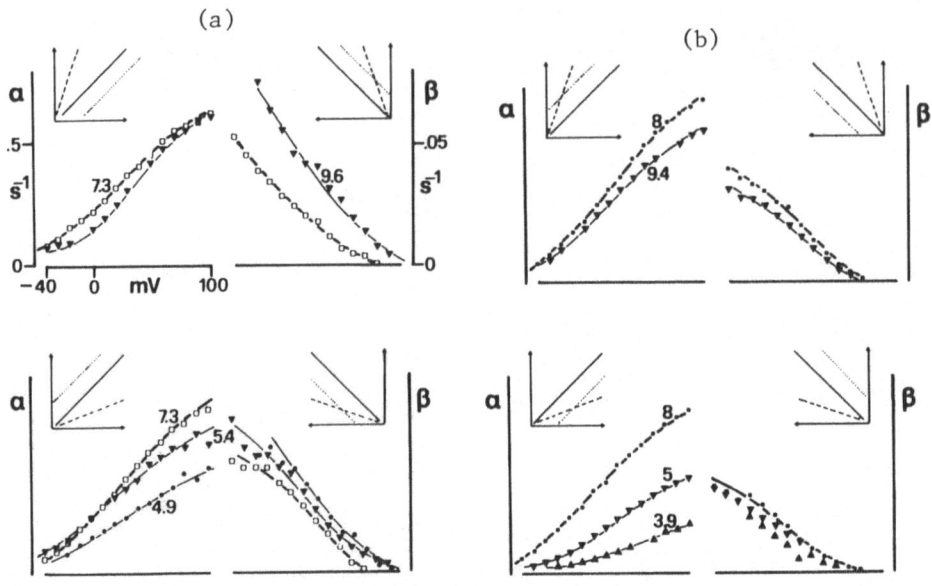

Fig. 3 - Voltage dependence of the potassium rate constants α and β
obtained from experiments in which the pH was changed as
indicated: a) internally and b) externally.
Insets: see Discussion.

    In order to study into more details the mechanisms involved in
the process of opening and closing of the potassium channel a complete
analysis of the Hodgkin and Huxley rate constants as a function of
potential, was done. The results are shown in fig. 3a. At pH 9.6
both the backward, $\beta_n(E)$, and the forward rate constant, $\alpha_n(E)$, are
shifted to the right and slightly tilted in the upward direction.
At low pH the results are different. The effects are more pronounced
for $\alpha_n(E)$ which does not shift appreciably but changes its slope proportionally to the pH. $\beta_n(E)$ shows only a small shift to the right.

    Although data on the effects of extracellular pH (5) on α_n and
β_n are now available for intact squid axons we found more profitable
for the sake of clearness to repeat those measurements under the
present experimental conditions (perfused axons, different g_K-measur-
ing system and off-line data analysis). The results from three axons
are summarized in fig. 3b. At basic pH α_n and β_n are barely affected
while at low pH only α_n is seen to be tremendously depressed. At
pH 9.4 both rates have slightly different slopes and at low pH the
slope of $\alpha_n(E)$ decreases proportionally to the pH. As previously
noticed at low pH β_n is weakly influenced.

    A rough comparison of the results of fig. 3 suggests that the effects of the internal and external pH cannot be simply explained as due to a neutralization charge effect resulting from the binding of hydrogen ions to fixed negative charges located on both sides of the membrane. Other causes besides electrostatic interactions have to be considered in order to account for the slope variations of $\alpha_n(E)$ observed at low pH. In general (11), the rate of opening and closing of the channels are thought to depend, through a Boltzman factor, on a free energy term comprehensive of both an electrostatic and a not-well specified factor representative for all the non-electrostatic energies influencing α_n and β_n. Under these hypothesis, a slope variation and/or a vertical shift of both rates is expected to occur by either changing the temperature coefficient kT or the non-electrostatic energy term. The first possibility was verified under the present experimental conditions in a series of measurements in which the temperature of the bath was changed from 0° to 13°C. The results are summarized in fig. 4. α_n and β_n are seen to change their voltage-dependency merely by a slope increase.

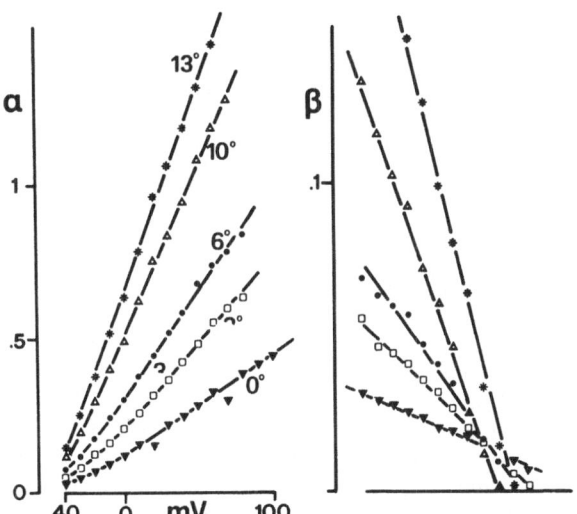

Fig. 4 – Effects of temperature on the potassium rate constant α_n and β_n.

DISCUSSION

The presence of negative fixed charges on both sides of biological membranes is a widely accepted concept (12). The evidences for
this come from the results of either altering: a) the intracellular
ionic strength (13), b) the concentration of external divalent
cations (8), or c) the extracellular pH (1-7,9,12). Although the
interpretations of the effects of divalent cations are still controversial (14,15,12), pH data are commonly interpreted as mainly due
to a neutralization rather than a screening effect. In other words,
it is postulated that the intramembrane electric field sensed by the
ionic channels can be varied as a consequence of a change in the
number of neutralized charged, resulting from a one-to-one binding
reaction between H^+ ions and fixed negative charges. The analysis,
however, has never been extended to an accurate evaluation of the
effects of pH on the H.H.'s rate constants α_n and β_n. In doing this
(see fig. 3) one has to face a rather provocative result, i.e., at
different pH α_n and β_n do not simply shift along the voltage axis
as one would expect on the basis of a simple surface charge model.

As mentioned in the results, according to the Eyring's rate
theory (11) a change in the intramembrane potential difference is
expected to produce a voltage shift of both α_n and β_n while a change
in the non-electrostatic energy term would produce a slope variation
of α_n and β_n in the same direction. In a different formalism, the
former change would cause an identical shift of conductance, $g_K(E)$,
and time constant, $\tau_n(E)$, versus E while the latter would only generate large variations of τ_n. Evidences for the latter case were first-
ly reported by Shrager (3) for crayfish axons and successively by
Carbone et al. (5) for the squid giant axon. In both papers $\tau_n(E)$
was reported to shift more consistently than $g_K(E)$.

Following the above arguments an explanation for the results
of fig. 3 should be found among the possible non-electrostatic energy
sources which might be influenced by the pH. Two possibilities (not
necessarily mutually exclusive) seem to exist. Either the pH affects
specifically the proteic structure of the ionic channels or the pH
modifies the phospholipid arrangement around the channel. The experiments of low internal and external pH tend to exclude the first possibility unless it is assumed the existence of two or more independent
groups one on each side of the membrane that will change the probability of opening of the channels depending on its degree of protonation (3). The second possibility is certainly more attractive,
especially on the view of recent works on artificial membranes (16-
18). Thus, for phosphatidyl-glycerol bilayers there are evidences
that above the ordered-fluid phase transition temperature the fluidity
of the membrane is greater when the phospholipid head groups are
mostly charged, due to the increased intermolecular separation caused
by electrostatic repulsion. On the contrary when polar heads are
mainly discharged the bilayer will appear smooth and more compact

as if a cooling effect had taken place. Although not yet supported
by further experimental data, it seems reasonable to suppose that
the same holds true for the squid axon membrane, i.e., decreasing
the negative surface charge density by increasing the hydrogen ions
concentration produces effects similar to lowering the temperature.

Under these conditions, a correct analysis of the results of
fig. 3 has to take into consideration the effects of pH on both the
fluidity of the lipid-matrix and intramembrane electric field. The
insets in fig. 3 should help to clarify this point. Three lines are
drawn representing schematically: a) the experimental results (solid-
line), b) the slope variation expected from the fluidity change
(dashed-line) and c) the shift relative to changes of the membrane
surface potential (dotted-line). In drawing them we assumed that
low pH values produce: 1) an increase in the slope of both α_n and
β_n either for intra- or extracellular pH changes and 2) a shift to
the right of both α_n and β_n for internal or to the left for external
pH variations. Along with this, high pH are thought to generate
opposite effects in either cases. Thus, at high intracellular pH,
$\alpha_n(E)$ is expected to shift right and bend upward as found experimentally. The same is true for $\beta_n(E)$ which shows large variations
at all potentials. Consistent with the interpretation are also the
results of low internal and external pH on $\alpha_n(E)$. In fig. 3a, $\alpha_n(E)$
is seen to shift left with a decreased slope. In fig. 3b, the slope
variation and the shift are concomitant, resulting into a dramatic
depression of $\alpha_n(E)$ over the entire range of potentials. In good
agreement are also the results of $\beta_n(E)$ at high and low external pH
(fig.3b). No reasonable explanations can be found for the results
of $\beta_n(E)$ at low internal pH and $\alpha_n(E)$ at high external pH. Most
likely, specific interactions with the channel might occur.

CONCLUSIONS

The most relevant part of the present work is the finding that
the pH effects on the squid axon membrane cannot be simply interpreted
in terms of voltage shifts of the α_n and β_n parameters. The qualitative explanation given for the slope variations observed at low
and high pH can account for most of the present results but leaves
unanswered important questions at the moment, such as: "How much
is our interpretation depending on the type of theoretical model
used to fit the experimental data?" or: "Are the same effects noticed
for α_n and β_n valid for the opening and closing rate constants of
the Na⁺-channels?"

Concerning the former not much can be said at the moment.
Several attempts will be done to fit the $g_K(t)$ records with models
having different reaction-schemes from the classical Hodgkin-Huxley
one. To answer the second question we can anticipate that (although
the data on our hands are still widely incomplete) similar effects
are observed for the opening and closing rates of the sodium channel.

This would give further support to the idea that the origin of the
slope variations of α_n and β_n concerns more the properties of the
lipid matrix than the proteic structures of the ionic channels.

REFERENCES

1. HILLE,B. 1968. Changes and the nerve surface. Divalent ions
    and pH J. Gen. Physiol. 51:221.
2. MOZHAYEVA,G.N., and A.P. NAUMOV 1970. Effect of surface charge
    on the steady-state potassium conductance of nodal membrane.
    Nature (Lond.) 228:164.
3. SHRAGER,P. 1974. Ionic conductance changes in voltage clamped
    crayfish axons at low pH. J. Gen. Physiol. 64:666.
4. SCHAUF,C.L. and F.A. DAVIS 1976. Sensitivity of the sodium and
    potassium channels of Myxicola giant axons to changes in ex-
    ternal pH; J. Gen. Physiol. 67: 185.
5. CARBONE,E., R.FIORAVANTI, G.PRESTIPINO, and E.WANKE 1978. Action
    of extracellular pH on Na^+ and K^+ membrane currents in the giant
    axon of Loligo Vulgaris. J. Membr. Biol. 43:295.
6. EHRENSTEIN,G., and H.M.FISHMAN 1971. Evidence against hydrogen-
    calcium competition model for activation of electrically excita-
    ble membranes. Nature (Lond.) 233:16.
7. BRODWICK,M.S. and D.C.EATON 1978. Sodium channel inactivation
    in squid axon is removed by high internal pH or tyrosine-specific
    reagents. Science (Wash.D.C.) 200:1494.
8. FRANKENHAEUSER,B., A.L.HODGKIN 1957. The action of calcium on
    the electrical properties of squid axons. J. Physiol. London
    137:217-43.
9. WOODHULL,A.M. 1973. Ionic blockage of sodium channels in nerve.
    J. Gen. Physiol. 61:687-708.
10. WANKE,E., E.CARBONE and P.L.TESTA 1979. K^+ conductance modified
    by a titratable group accessible to protons from the intracellu-
    lar side of the squid axon membrane. Biophys.J. 26:319.
11. TSIEN,R.W., and D.NOBLE 1969. A transition state theory approach
    to the kinetics of conductance changes in excitable membranes.
    J. Membr. Biol. 1:248.
12. HILLE,B., A.M.WOODHULL, B.I.SHAPIRO 1975. Negative surface charg
    near sodium channels of nerve: Divalent ions, monovalent ions,
    and pH. Philos. Trans. Soc. 270:301.
13. CHANDLER,W.K., and H.MEVES 1965. Voltage clamp experiments on
    internally perfused giant axons. J. Physiol. London 180:788-820
14. MCLAUGHLIN,S.G.A., G.SZABO and G.EISENMAN 1971. Divalent ions
    and the surface potential of charged phospholipid membranes.
    J. Gen. Physiol. 58:667.
15. SCHAUF,C.L. 1975. The interactions of calcium with Myxicola
    giant axons and a description in terms of a simple surface
    charge model. J. Physiol. 248:613.
16. TRÄUBLE,H., and E.HAUSJÖRG 1974. Electrostatic effects on lipid
    phase transitions: membrane structure and ionic environment.
    Proc.Nat.Acad.Sci. USA, 71:214.

17. WATTS,A., K.HARLOS, W.MASCHKE and D.MARCH 1978. Control of the structure and fluidity of phospho-tidyl-glycerol bilayers by pH titration. Biochim. Biophys. Acta 510:63.
18. VERKLEIJ,A.J., B.DE KRUYFF, P.H.J.T. VERVERGAERT, J.F.TOCANNE and L.L.M. VAN DEENEN 1974. The influence of pH, Ca^{2+} and protein on the thermotropic behaviour of the negatively charged phospholipid, phosphotidylglycerol. Bioch.Bioph.Acta, 339:432.
19. BEZANILLA,F. and C.M.ARMSTRONG 1977. Inactivation of sodium channel. I. Sodium current experiments. J. Gen. Physiol. 70: 549.

THE EFFECT OF HYDROSTATIC PRESSURE ON THE VOLTAGE-CLAMP CURRENTS

OF THE SQUID GIANT AXON

F.Conti, R.Fioravanti, J.R.Segal[xo], W.Stühmer[‡]

Laboratorio di Cibernetica e Biofisica, CNR

16032 Camogli, Italy

[x]Biophysics Laboratory, Veterans Administration,
Medical Center, New York, N.Y. 10010, USA

[‡]Dept. of Physics, Technical University of München,
8046 Garching, W. Germany

[o]Supported by the Veterans Administration Medical
Research Division.

INTRODUCTION

   The effect of hydrostatic pressure upon the kinetics of a chemical reaction is directly related to the volume change, $\Delta V^{\ddagger}$, associated with the formation of the intermediate activated complex (1)
and can provide useful information about the molecular mechanism
of the reaction. Substantial activation volumes associated with
changes in enzyme conformation accompanying catalysis (1) are easily
measured, and it is possible to separate different contributions
to $\Delta V^{\ddagger}$ arising from either the movement of water-sensity-modifying
groups or changes in the intrinsic volume (packing efficiency) of
the protein itself (2).

   It has been known since 1935 that hydrostatic pressure strongly
affects the physiology of whole nerve and muscle (3,4). These effects were later attributed to changes in the behavior of single
nerve fibers (5,6), as characterized by a number of parameters, such
as rheobase, threshold membrane potential, conduction velocity and
action potential duration – quantities not easily related to the
molecular events underlying excitation.

More recently Henderson & Gilbert (7) studied the voltage-clamp currents of squid giant axons subjected to helium pressures of up to 204 atm. The results were analyzed in terms of the membrane parameters describing electrical activity setforth in the Hodgkin-Huxley (HH) equations (8). According to contemporary views these parameters correspond to the rates of isomerizations of specific membrane proteins whose conformational changes are responsible for the opening and closing of cation selective pathways (channels) across the nerve membrane (9). This notion is supported by the experimental results of binding studies (10), noise analysis (11-15) and gating current measurements (16,17). Based on the effects of helium pressure, the estimated values of the activation volumes associated with channel protein isomerization are 30 to 75 cm^3/mole.

In this report we describe the results of voltage-clamp measurements in squid giant axons exposed to hydrostatic pressures of up to 612 atm. Using oil as the pressure transmitting medium, as in early works (2-6), we were able to explore a wide pressure range while avoiding possible collateral complications of gas pressurization such as changes in concentration of dissolved gas, and mechanical effects due, e.g., to bubble formation on decompression. Although the effects we observed are in qualitative agreement with the previous study (7) ours are systematically smaller, corresponding to activation volumes in the range of 25 to 30 cm^3/mole. Furthermore, in contrast to Ref.(7) our results indicate that the effect of pressure on the kinetics of sodium inactivation is at least as great as that on sodium activation.

METHODS

The experiments were performed on giant axons, 400 to 700 µm in diameter, from the hindmost stellar nerve of the squid, Loligo Vulgaris, available in Camogli. The axons were mounted within a small perspex chamber which could be separated into a number of parts during preliminary operations. Standard voltage-clamp techniques (18), including the use of lateral guard electrodes and series resistance compensation, were employed. After being inserted into the axon, the internal electrode assembly was fixed with screws to the chamber and disconnected from the manipulator. The chamber was sealed with a perspex cover, great care being taken to exclude air bubbles. The chamber-electrodes assembly was disconnected from the preparatory set-up and gently immersed in mineral oil within the pressure bomb. The hydraulic pressurizing system was very similar to that used by Segal (19) for measurements on frog-skin. Temperature was detected with a thermistor placed near the axon, and was regulated by circulating thermostated water around the pressure bomb. Pressure within the bomb was raised or lowered in 10-20 sec, but electrical measurements were made only after 2 to 3 minutes had elapsed in order to allow for temperature re-equilibration. The experiments were performed either in artificial sea water (ASW) or

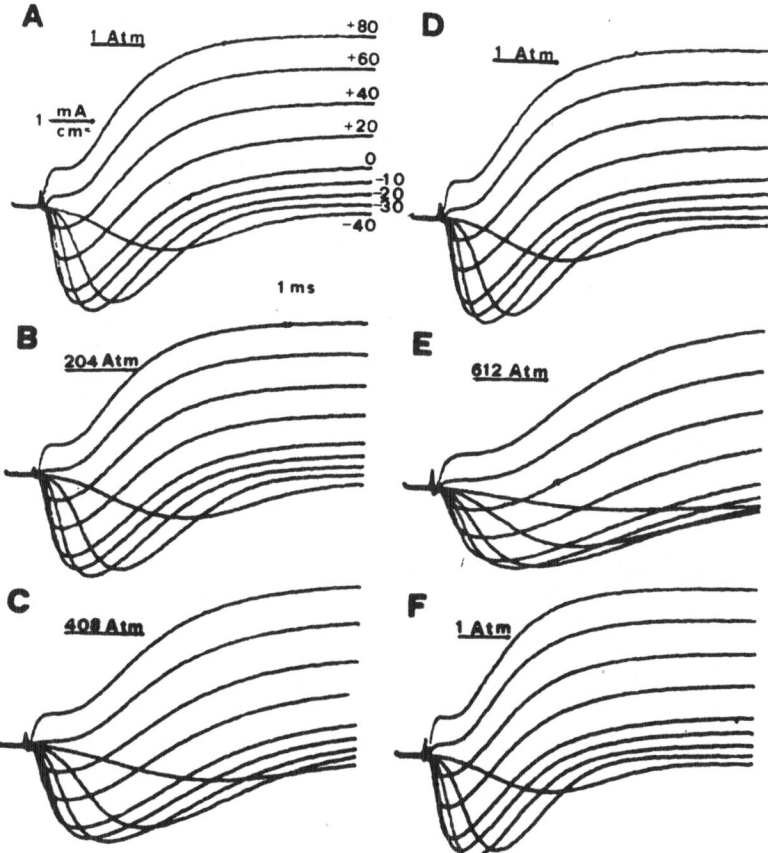

Fig. 1 – Overall effect of pressure on voltage-clamp currents. Voltage steps from E=-80 mV to the potential (in mV) indicated in the figure. The records were obtained from the same axon in the sequence A to E. T=10°C.

in ASW containing 300 mM-tetrodotoxin (TTX). ASW had the following composition: 450 mM-NaCl; 10 mM-KCl; 50 mM-CaCl$_2$; 1 mM-TrisCl; pH 7.8. Most of the linear passive components were subtracted from the total voltage-clamp currents with an analog compensating circuit. The non-compensated currents are treated as potassium currents in the TTX experiments and as the sum of sodium and potassium currents in the experiments without TTX.

RESULTS

General effects

Figure 1 shows records of voltage-clamp currents of an axon immersed in ASW and exposed to different hydrostatic pressures. The following qualitative conclusions can be drawn from the data of Fig.1. 1. Pressure slows down the overall time course of both early and late currents. 2. Pressure has very little influence on the amplitudes of late outward (potassium) currents. 3. The maximum peak inward (sodium) current is not appreciably affected by raising the pressure to 204 Atm and it is reduced by only 25% at 612 Atm. This small decrease is associated with a clear increase of the potential for which the maximum occurs, although no significant difference in reversal potential for early currents, E_{Na}, are seen in records D(1 atm), E(612 atm) and F(1 atm).

K$^+$ currents

In TTX experiments the steady-state potassium currents for large depolarizations were often slightly increased by pressure. However, as judged by the current jump following the end of the voltage clamp pulses, this effect appears to arise from a smaller potassium accumulation in the outer Schwann space (20) due to the slower development of the potassium currents. Thus, increasing pressure from 1 atm to 612 atm has a very small, if any, effect on the maximum potassium conductance, $\bar{g}_k$. A possible small shift of the potassium activation curve, n (V), (less than 10 mV, in the depolarizing direction, at 612 atm) was observed but could not be analyzed in detail because our current records were too brief for a correct estimation of steady-state conductances at small depolarizations.

The half time, $t_{1/2}$, of the increase in potassium conductance following steps to membrane potential, E, of +20 mV to +80 mV, increased with pressure, P, by a factor, $\theta_n(P) = t_{1/2}(P)/t_{1/2}(1atm)$, fairly independent of E. θ_n had a mean value of about 2 at 612 atm. Similar values of θ_n were also obtained in TTX-free ASW at depolarizations very close to the sodium reversal potential. The mean values of all the determinations of $\theta_n(P)$ are plotted semilogarithmically as a function of P in Fig. 2A.

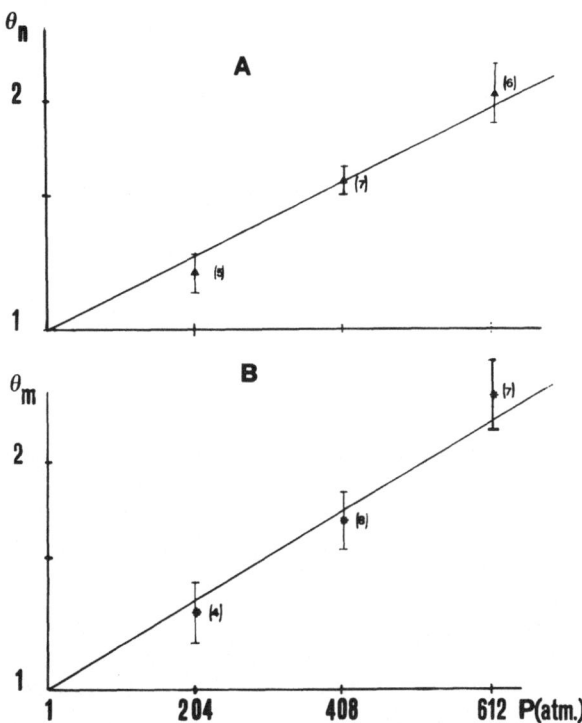

Fig. 2 – Effect on potassium and sodium activation kinetics.
The data represent means and standard errors from the number
of measurements given in parenthesis, obtained in seven
experiments at 10°C and five experiments at 18°C.
θ_n and θ_m are the fractional increase of the half time of
potassium conductance and of the time to peak sodium current,
respectively.

Na currents

The most pronounced effect of pressure on the sodium current is the slowing of its time course. This was characterized by three different parameters: the time to peak, t_p; the time to the inflection point of the rising phase, t_f; the ratio, I_p/I_f, of the peak current, I_p, to the maximum time derivative of the current, I_f. For small depolarizations (E < -10 mV) all the above parameters increased with pressure by about the same factor, $\theta_m(P)$, which had a mean value of about 2.4 at 612 atm. According to the HH equations (8), t_p, t_f and I_p/I_f are all proportional to the time constant, τ_m, of sodium activation, with different factors depending on the ratio of τ_m to the time constant, τ_h, of sodium inactivation. Thus, the above results indicate that pressure has a comparatively little effect on the ratio τ_h/τ_m for E < -10mV. At larger depolarizations our measurements of t_f and I_p/I_f were affected by large errors. However, systematic differences in the pressure dependence of t_p, t_f and I_p/I_f could be observed, in the direction expected if τ_h/τ_m increases with P. At E=10 mV our data indicate a 50% increase of τ_h/τ_m for P=612 atm. The logarithm of the mean values of $\theta_m(P)$ for small depolarizations are plotted versus P in Fig. 2B.

The conclusion that τ_h is affected by pressure at least as much as τ_m is also consistent with the observed effects of pressure on peak sodium currents. The voltage dependence of the peak sodium conductance, $g_p=I_p/(E-E_{Na})$, was affected by pressure in a way that is incompatible with the hypothesis that τ_h is pressure independent, as suggested by Henderson & Gilbert (7). The maximum value of g_p was practically independent of pressure, while the voltage for which g_p reached half of its maximum was about 7 mV more positive at 612 atm than at 1 atm. Although such a simple shift does not describe completely the effect of pressure on $g_p(E)$ (a decrease in the steepness of the E dependence was also observed at high pressures), it implies that the main cause of the decrease in the maximum I_p value is the occurrence of a similar shift in the sodium activation curve, $m_\infty(E)$.

A small, reproducible effect of pressure on the sodium inactivation curve, $h_\infty(E)$, was observed. For P=612 atm the voltage yielding 50% inactivation was shifted by about 2 mV in the hyperpolarizing direction and the steepness of the $h_\infty(E)$ curve at the midpoint was decreased by about 26%.

DISCUSSION

The main purpose of this preliminary report is to demonstrate that reproducible measurements of voltage-clamp currents in single nerve fibers at high hydrostatic pressures are possible and can provide useful information about the molecular mechanisms responsible for nerve excitation. All the effects that we have observed were fully reversible up to 612 atm, the highest pressure that we could apply with our present set-up. Irreversible phenomena, reported in Ref. (6) to occur above 4oo atm, were not seen.

A full quantitative analysis of our measurements, and of their possible implications regarding the molecular mechanism of channel gating cannot be presented here. This discussion will be limited to major qualitative conclusions.

The largest effects of pressure were on the time course of sodium and potassium currents. These effects, summarized in Fig. 2, can be interpreted to a first approximation as evidence that the isomerization or gating steps of the opening or closing of ionic pathways have activation volumes of about 30 cm^3/mole for sodium channels and of about 26 cm^3/mole for those of potassium. When the α's and β's of the HH equations are viewed as rate constants of gating isomerizations we have (1):

$$\frac{\partial \ln\alpha}{\partial P} = - \frac{\Delta V_o^{\ddagger}}{RT} \tag{1}$$

$$\frac{\partial \ln\beta}{\partial P} = - \frac{\Delta V_c^{\ddagger}}{RT} \tag{2}$$

where RT is equal to 2.324 cm^3 atm/mole at 10°C, $\Delta V_o^{\ddagger}$ is the volume change associated with the transition of a gating group from the closed to the activated (intermediate) configuration, and $\Delta V_c^{\ddagger}$ refers to the open to activated transition. From eqs. (1) and (2) the gating constant, $\tau = (\alpha+\beta)^{-1}$, is an exponential function of pressure when $\Delta V_o^{\ddagger} = \Delta V_c^{\ddagger}$, independent of pressure. Fig. 2 shows that our data is well-fitted by a linear dependence of $\ln\theta_n$ and $\ln\theta_m$ on pressure; the slopes yield the above estimates of activation volumes.

The data of Fig. 2 are not accurate enough to reveal possible deviations from a linear $\ln\theta$ versus P relation arising from small differences between $\Delta V_o^{\ddagger}$ and $\Delta V_c^{\ddagger}$. However, the observed shift in the $m_\infty(E)$ curve indicates that for the sodium channel $\Delta V_o^{\ddagger} > \Delta V_c^{\ddagger}$, so that at high pressure α_m is reduced more than β_m. Assuming an effective valence of 1.36 for the charged groups which control the m gating process (22) a shift of 7 mV in the $m_\infty(E)$ curve for P=612 atm implies $(\Delta V_o^{\ddagger} - \Delta V_c^{\ddagger}) \simeq 5$ cm^3/mole.

Another important conclusion gleaned from our data bears on the question of the effect of pressure on the kinetics of sodium inactivation. Henderson & Gilbert (7) reported no significant modification of τ_h by pressure, whereas they estimated activation volumes of 30 to 75 cm^3/mole from changes of τ_m. This finding would provide very strong support to the idea that sodium activation and inactivation are uncoupled, independent processes. In contrast, recent measurements of gating currents (16) and noise (15) imply that the two processes are coupled. Our results, indicating that the inactivation process is slowed down by high pressures by at least as much as is the activation process, are more consistent with the latter notion. Qualitatively, our pressure effects are to be expected if the inactivation step occurs only after partial (or total) activation. Such coupling would also account for the closer link between the two processes found at small depolarizations, where the activation steps are expected to be rate limiting for the whole process. At larger depolarizations the finding that τ_h has a greater pressure dependence than τ_m suggests that the final steps leading to inactivation become rate limiting and have larger activation volumes.

It seems unavoidable to conclude from our present data that previous results obtained using helium gas pressures (7) may have been adversely affected to a large extent by the high helium concentration per se and/or secondary phenomena consequent to the prolonged exposure (>20 minutes) to high pressures.

REFERENCES

1. Johnson, F.A., Eyring, A. & Polissar, M.H.. The kinetics basis of Molecular Biology. New York: Wiley, 1954.
2. Low, P.S. & Somero, G.N. Activation volumes in enzymes catalysis: their sources and modification by low-molecular-weight solutes. Proc.Nat.Acad.Sci. U.S.A. 72, 3014-3018 (1975).
3. Ebbecke, V. & Shaefer, H. Über den Einfluss hoher Druke auf den Aktionsstrom von Muskeln und Nerven. Pflügers Arch.Ges. Physiol. 236, 678-682 (1935).
4. Grundfest, H. Effects of hydrostatic pressure upon the excitability, the recovery and the potential sequence of frog nerve. Cold Spring Harbor Symp. Quant. Biol. 5, 179-187 (1936).
5. Spyropoulos, C.S. Responses of single nerve fibers at different hydrostatic pressures. Am.J.Physiol. 189, 214-218 (1957).
6. Spyropoulos, C.S. The effects of hydrostatic pressure upon the normal and narcotizated nerve fiber. J.Gen.Physiol. 40, 849-857 (1957).
7. Henderson, J.V.Jr. & Gilbert, D.L. Slowing of ionic currents in the voltage-clamped squid axon by helium pressure. Nature 258, 351-352 (1975).
8. Hodgkin, A.L. & Huxley, A.F. A quantitative description of membrane current and its application to conduction and excitation in nerve. J.Physiol. 117, 500-544 (1952).
9. Hille, B. Ionic selectivity of Na and K channels of nerve membranes. In: "Membranes: Lipid Bilayers and Biological Membranes: Dynamic Properties" (G. Eisenman ed.), vol. 3, p.255-323, New York, Dekker, 1957.
10. Ritchie, J.M. & Rogart, R.B. The binding of saxitoxin and tetrodotoxin to excitable tissue. Rev.Physiol.Biochem.Pharmacol. 79, 1-50 (1977).
11. Conti, F., De Felice, L.J. & Wanke, E. Potassium and sodium ion current noise in the membrane of the squid giant axon. J.Physiol.

248, 45-82 (1975).

12. Conti, F., Hille, B., Neumcke, B., Nonner, W. & Stämpfli, R. Measurement of the conductance of the sodium channels from current fluctuations at the node of Ranvier. J. Physiol. 262, 699-727 (1976).

13. Begenisich, T. & Stevens, C.F. How many conductance states do potassium channels have? Biophys.J. 15, 843-846 (1975).

14. Sigworth, F.J. Sodium channels in nerve apparently have two conductance states. Nature 270, 265-267 (1977).

15. Conti, F., Neumcke, B., Nonner, W. & Stämpfli, R. Conductance fluctuations from the inactivation process of sodium channels in myelinated nerve fibers. J.Physiol. (1980)

16. Armstrong, C.M. & Bezanilla, F. Inactivation of the sodium channels. II. Gating current experiments. J.Gen.Physiol. 70, 567-590 (1977).

17. Neumcke, B., Nonner,W. & Stämpfli, R. Gating currents in excitable membranes. Int.Rev.Biochem. 19, 129-155 (1978).

18. Moore, J.W. & Cole, K.S. Voltage clamp techniques. In "Physical Techniques in Biol. Research", vol. VI, p. 263-321, New York, Academic Press, 1963.

19. Segal, J.K. Pressure jump relaxation kinetics of frog skin open circuit voltage and short circuit current. Bioch.Biophys. Acta 471, 453-465 (1977).

20. Frankenhaeser, B. & Hodgkin, A.L. The after effects of impulses in the giant nerve fibers of Loligo. J. Physiol. 131, 341-376 (1956).

21. Conti, F., Fioravanti, R. & Wanke, R. Caratteristiche elettriche dello stato attivo dei siti del sodio nella membrana nervosa. In: "Atti della I Riunione Scient.Plenaria della Società Italiana di Biofisica Pura ed Applicata", p. 413-421, Parma; tipo-lito tecnografica (1973).

22. Keynes, R.D., Rojas, E. Kinetics and steady-state properties of the charged system controlling sodium conductance in the squid giant axon. J. Physiol. 239, 393-434 (1974).

SINGLE CHANNEL CONDUCTANCE MEASUREMENTS AT THE NEUROMUSCULAR JUNCTION

Erwin Neher

Max-Planck-Institut für biophysikalische Chemie

34 Goettingen, W. Germany

The neuromuscular junction is the link between nerve and muscle. An action potential arriving at the nerve ending liberates a chemical transmitter substance, which diffuses across the synaptic cleft to interact with specific receptors in the postsynaptic membrane. This leads to the opening of ionic channels in this membrane. The resultant conductance increase depolarizes the muscle fiber, which in turn triggers an action potential and thereby initiates contraction (For a review article see Steinbach & Stevens (1976)).

This report is concerned exclusively with one step out of this complicated sequence of events: the transmitter-induced conductance increase of the muscle membrane. Since the amphibian neuromuscular junction is the best studied synapse, the molecular mechanisms underlying the conductance increase are known in great detail. The findings are condensed in the simplified scheme of action of fig. 1, which is based on the work of Katz & Miledi (1972), and Magleby & Stevens (1972). In fig. 1 the membrane is shown to contain an integral protein, the acetylcholine-receptor channel, which has a specific binding site for the transmitter substance acetylcholine (ACh). Upon binding of ACh, the protein undergoes a conformational change, which opens a pore for the passage of ions. This mode of action can be written down as follows:

$$A + R \underset{k_2}{\overset{k_1}{\rightleftharpoons}} AR \underset{k_{-2}}{\overset{k_2}{\rightleftharpoons}} AR^* \tag{1}$$

where A is the agonist molecule (ACh), R is the receptor, AR is the

Fig. 1 - Simplified schematic representation of a membrane with
         Acetylcholine-receptor channel incorporated. At the left
         of the receptor has a vacant binding site for Acetylcholine
         (ACh), in the center an Acetylcholine is bound, however,
         the channel is still in its closed conformation, on the
         right the channel is open. This scheme neglects the fact
         that there are more than one ACh-molecule required to open
         up a channel.

receptor-transmitter complex in the closed conformation and AR^* the
open channel. This scheme is certainly over-simplified. It dis-
regards the well known fact, that more than one molecule of ACh is
required to open a channel (Dreyer, Peper, & Sterz 1978; Dionne,
Steinbach & Stevens 1978) and it does not include the phenomenon
of desensitization (Adams 1975).

The technique of noise analysis (see e.g. the article by Conti,
this volume) has been applied to this kind of membrane conductance
(Katz & Miledi 1972; Anderson & Stevens 1973) and has provided estimates for the conductance value γ of a single channel and also of
the mean time $\bar{t}_o$, such a channel stays open, once it has been
activated by ACh. Anderson and Stevens found values of γ = 25 pS
and $\bar{t}_o$ = 8 msec.

These values are well within the range of what can be resolved
with present day electronics. Single channel responses of comparable
magnitude have been measured in artificial membranes. It should be
a challenge to any biophysicist to develope methods for direct observation of these elementary events in order to get a description independent of all the assumptions that are implied in noise analysis.

In the following, I will outline some of the basic considerations
that are required for single channel resolution. Based on that I
will describe the principles of the patch clamp method and then illustrate some of the properties of Acetylcholine receptor channels by
means of single channel measurements.

CONSIDERATIONS ON HIGH RESOLUTION MEASUREMENTS

Apart from considerations of amplifier noise, good shielding and protection from other external interferences, there is one basic limitation of resolution to any electrical measurement, which is the discrete nature of charge carriers. The best known example of its consequences is the so-called resistor noise or thermal (Johnson) noise which is given by:

$$u_N = \sqrt{4kT\Delta f \cdot R} \tag{2}$$

where u_N is the rms-noise voltage across the ends of a resistor of value R, measured with a bandwidth of Δf. This readily transforms into an expression for current noise in a current measuring configuration

$$i_N = u_N/R = \sqrt{4kT\Delta f/R} \tag{3}$$

This is the minimum background noise that any signal source with equivalent internal resistance R can have. Equ.(3) says that the internal resistance of the signal source should be as high as possible for low background noise. Putting the required limits of resolution (e.g. $i_N \leq 0.5pA$; $\Delta f \simeq 500$ Hz) into equation (3) yields a condition for the internal resistance of the membrane, which is

$$R_M > 50 \ M\Omega \tag{4}$$

Condition (4) would be sufficient, if the membrane were a pure resistor. However, membranes also have a capacitance associated with them. This, together with series resistance of electrodes and amplifier noise, makes up for additional noise contribution. Again, inserting the requirements for both amplitude and time resolution into an appropriate equation yields an upper limit for the capacitance of the membrane, which is approximately:

$$C < 10 \ pF \tag{5}$$

Since biological membranes quite generally have a specific capacitance of 1 $\mu F/cm^2$, this number transforms into a limit for the size of the membrane and this is 1000 μm, which is, for instance, a circular patch of 35 μm diameter or else a spherical cell with 18 μm diameter. This small size cannot be successfully impaled with electrodes of low enough internal resistance. Therefore measurements of single channel responses seem pretty hopeless with conventional techniques of recording.

We therefore developed a technique adapted to the above requirements. (Neher & Sakmann 1976, Neher, Sakmann & Steinbach 1978) Thereby a fire-polished micropipette of about 1-2 μm tip diameter is

placed onto the surface of a muscle fiber, isolating electrically a
small patch of membrane. Membrane currents, originating from the
isolated patch of membrane can be recorded in the circuit connected
to the pipette (fig. 2). The requirements (4) and (5) are now well
met concerning the membrane patch ($R_{M,PATCH} \simeq 8000$ MΩ; $C_{M,PATCH} \simeq$
0.1 pF). However, the contact between the glass pipette and the
membrane is never complete, which leaves a shunt pathway between
pipette interior and Ringer bath (fig.2B). This shunt pathway is,
electrically speaking, in parallel to the patch membrane and is the
major source of background noise and current drift problems in patch
recordings. The main difficulty in obtaining good recordings lies
in reducing this shunt pathway to values higher than 50 MΩ. Various
measures had to be taken in order to obtain this goal. These included extensive enzymatic cleaning of the membrane surface and
proper shaping of the pipette tip (Neher, Sakmann & Steinbach 1978).

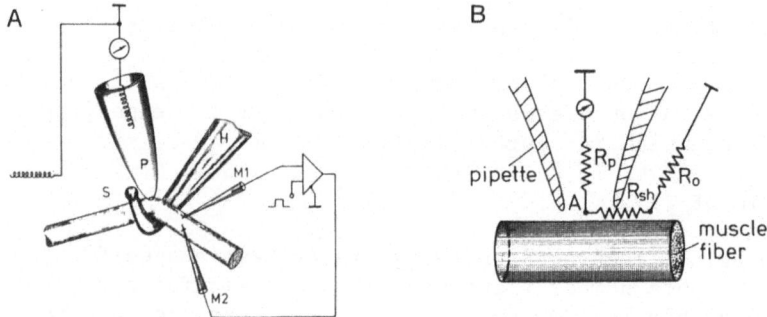

Fig. 2 - Part A. Schematic representation of the patch clamp method.
The fiber is lifted up by a glass Hook (H). Microelectrodes
(M1 and M2) and pipette P are placed within 50-100 um of
the supporting hook.
Part B. Equivalent circuit for the extracellular space in
the vicinity of the membrane-pipette-contact (reproduced from
Neher, Sakmann & Steinbach 1978, by permission).

RECORDINGS OF ACETHYLCHOLINE-RECEPTOR-CHANNELS

The recordings, described following, were done on cutaneous pectoris muscles of frog or on rat muscle, in collaboration with B. Sakmann and J. Patlak. The agonist for activation of the channels was contained in the pipette at very dilute concentrations (see below). Thus, placing the pipette onto the surface both applied agonist to the patch of membrane under study and initiated the recording. In order to obtain long well-resolved channel openings, we used Suberyl-dicholine mostly instead of Acetylcholine. This agonist was known to display slow kinetics (Katz & Miledi 1973). For technical reasons we worked with chronically denervated muscles, where due to "dener-vation hypersensitivity" (Axelsson & Thesleff 1969) the Acetylcholine sensitivity extends over a large fraction of the muscle fiber. Thus we did not have to lacate endplates for our measurements and could select sites of moderately low channel density, appropriate for single channel recording. Fig. 3 shows traces of pipette recording with Suberyldicholine at a concentration of 500 nM in the pipette. At this low concentration the probability for any channel to be in the open state is very low. The figure has three records, taken at different membrane potentials. One can see that the size of the elementary contributions is more or less constant at a given potential.

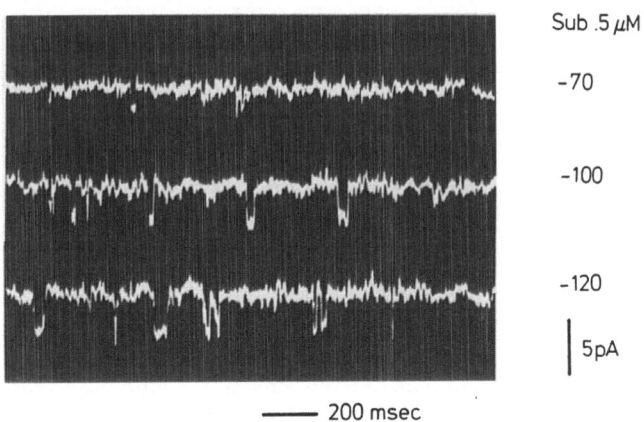

Sub .5 μM

-70

-100

-120

5pA

——— 200 msec

Fig. 3 - Recording of patch current at three different hyperpolarising membrane potentials, 12°C, rat muscle. The baseline in each trace corresponds to the "resting" state where no channel is open. Occasionally, a steplike downward deflection occurs, representing the opening of a single channel, followed by a steplike return to the baseline after some variable time interval. Downward deflection represents an increase in membrane current (reproduced from Neher, Sakmann & Stein-bach, 1978, by permission).

However, it increases with increasing hyperpolarization in an approximately linear fashion. This means that membrane pore has ohmic properties. The reversal potential of currents through the pore can be estimated from records like these to be around zero mV. This is expected, since it is known from macroscopic measurements that the ACh channel is about equally permeable to K^+- and Na^+-ions.

The duration of the channel openings, on the other hand, is a stochastic quantity. Channels appear at random and close again at random, as expected for a molecular process. Individual open times (lifetimes of the open state) can be measured in order to calculate their mean. The mean is the reciprocal of the backward rate in scheme (1):

$$\bar{t}_o^{-1} = k_{-2} \tag{6}$$

Equ.(6) is a special case of the more general statement, that the lifetime distribution of a given state ν is (in simple cases) an exponential with a mean   :

$$\bar{t}_\nu^{-1} = \Sigma_i \, k_{\nu i} \tag{7}$$

where the right hand side is the sum over all the reaction rates leading away from state ν (Colquhoun & Hawkes 1977; Neher & Steinbach 1978). It turns out that the mean channel open time is identical to the one inferred from noise analysis (Neher & Sakmann 1976). It depends on membrane voltage and on the chemical nature of the agonist applied (Suberyldicholine, Acetylcholine or Carbachol) in exactly the same way as the noise estimate does. Furthermore, Anderson & Stevens (1973), have shown that, under physiological conditions, the latter one agrees with the decay time constant of miniature end plate currents.

Thus, closing of channels is the rate limiting step during the physiological response (Anderson & Stevens 1973). Speaking in terms of single channels one can therefore describe the generation of a miniature end plate current (mepc) in the following way:
After liberation of a transmitter package at the nerve ending, ACh diffuses to the muscle membrane and is also degraded rapidly by Acetylcholinesterase. This all happens in about 0.1 to 0.3 msec, a time span which is short compared to the channel open time. However, it is long enough for about 1000 receptors to interact with ACh, to bind it (and thereby protect it from hydrolysis) and to open up the ionic channel. After that time ACh has disappeared from the cleft and 1000 channels are open, constituting the peak response of the mepc In the following 2-5 msec, channels close again according to their life time distribution, resulting in the well known exponential decay of the mepc.

The above paragraphs have been concerned with time intervals of channel openings t_o, which gave a measure of the channel closing rate k_{-2} of scheme (1). Likewise, the measurement of intervals of channel closing can give information on the channel opening rate k_2 according to equ.(7). However, here, considering scheme (1), we are faced with the problem that as long as the channel is closed we cannot distinguish whether it is in state $[A + R]$ or in state $[AR]$. We also do, generally, not know how many channels are active. Therefore equ.(7) has to be supplemented:

$$\bar{t}_c^{-1} = Np([AR] \mid [A+R],[AR]) \, k_2 \tag{8}$$

where N is the number of channels being able to open and $p([AR] \mid [A+R],[AR])$ is the conditional probability of a channel being in state $[AR]$ under the condition that it is either in state $[A + R]$ or in state $[AR]$. Assuming that the states $[AR]$ and $[A + R]$ are in fast equilibrium, one can replace $p([AR] \mid [A+R], [AR])$ by the equilibrium thermodynamic expression. Furthermore, we have found experimental situations, where we can be sure that there is only one functioning channel under the patch (see legend to fig. 4). Then equ.(8) simplifies to

$$\bar{t}_c^{-1} = \frac{C_A \, k_1/k_{-1}}{1 + C_A k_1/k_{-1}} \, k_2 \tag{9}$$

where C_A stands for concentration of Acetylcholine. Comparing equs. (6) and (8), one can see that, on variation of the concentration of ACh, the channel open time, $\bar{t}_o$ should be constant whereas the channel closed time should decrease with increasing concentration. In other words: the frequency of channel openings should increase at higher Acetylcholine concentration.

In fig. 3 the concentration of Suberyldicholine (replacing ACh) is low enough such that $\bar{t}_c$ is much larger than $\bar{t}_o$. Individual channel openings appear as discrete square pulses. In fig.(4), however, concentration is high enough for $\bar{t}_c$ to be comparable to $\bar{t}_o$. Furthermore, there is only one functioning channel present during the recording. In part A of fig. (4), taken at 5 μM Acetylcholine the time between openings is appreciably longer than the open time. In part B, however, taken at 30 μM Acetylcholine the opposite is true: the channel spends most of its time in the open state, intervals between successive openings are short.

Quantitative analysis of records like those in fig. (4) indicate that equ.(6) holds over the concentration range up to 50 μM (Sakmann, Patlak & Neher, 1980). Equ.(9), however, predicts too small a concentration dependence of $\bar{t}_c$. In first approximation $\bar{t}_c^{-1}$ varies with the 1.4th power of C_A, indicating that more than one ACh molecule

is required for opening of a channel. Such a conclusion had been
reached before on the basis of dose-response studies (Dreyer et al.
1979, Dionne et al. 1978).

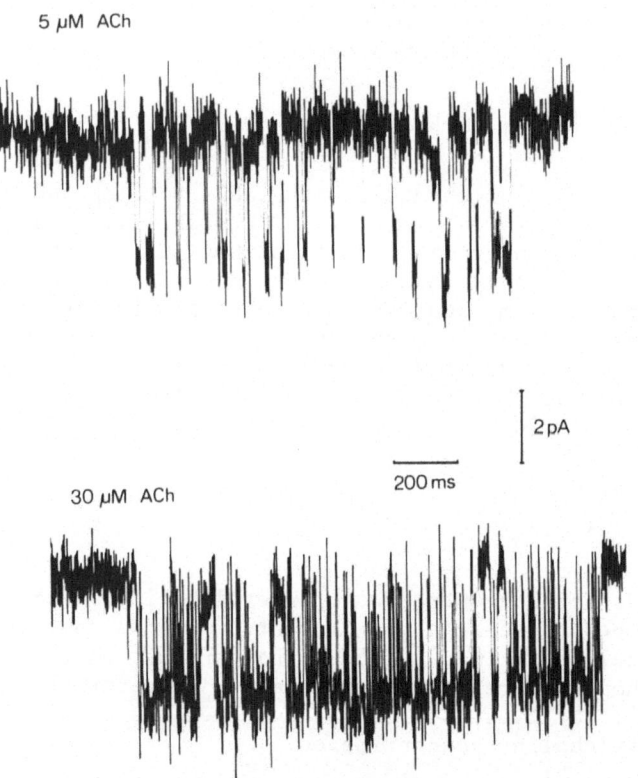

Fig. 4 - Recording from an individual channel at concentrations of
5 μm ACh (part A) and 30 μm ACh (part B).
The recordings are taken at a time when due to desensitization (Axelsson & Thesleff, 1959) all channels had been
inactivated. Then, occasionally, a single channel "reacti-
vates", functions in its normal way for periods up to seconds
(for the major part of both traces above), and desensitizes
again. The record is taken at low time resolution, such
that individual openings appear just as single downwards
spikes. At the lower concentration (part A) the spikes are
well separated in time, whereas they merge at the higher
concentration (part B). Temperature 10°C, membrane potential
-130 mV (Rana temporaria).

REFERENCES

1. ADAMS, P.R. (1975) A study of desensitization using voltage clamp. Pfluegers Arch. ges. Physiol. 360, pp. 135-144.
2. ANDERSON, C.R. & STEVENS, C.F. (1973) Voltage clamp analysis of Acetylcholine produced end-plate current fluctuations at frog neuromuscular junction. J. Physiol. 235, pp. 655-691.
3. AXELSSON, J. & THESLEFF, S. (1959) A study of supersensitivity in denervated mammalian skeleton muscle. J. Physiol. 147, pp. 178-193.
4. COLQUHOUN, D. & HAWKES, A.G. (1977) Relaxation and fluctuations of membrane currents that flow through drug-operated channels. Proc. R. Soc. Lond. B 199, pp. 231-262.
5. DIONNE, V.E., STEINBACH, J.H., & STEVENS, C.F. (1978) An analysis of the dose-response relationship at voltage-clamped forg neuromuscular junctions. J. Physiol. 281, pp. 421-444.
6. DREYER, F., PEPER, J.H. & STERZ, R. (1978) Determination of dose-response curves by quantitative ionophoresis at the frog neuromuscular junction. J. Physiol. 281, pp. 395-419.
7. KATZ, B. & MILEDI, R. (1972) The statistical nature of the acetylcholine potential and its molecular components. J. Physiol. 224, pp. 665-669.
8. KATZ, B. & MILEDI, R. (1973) The characteristics of "end-plate noise" produced by different depolarizing drugs. J. Physiol. 230, pp. 707-717.
9. MAGLEBY, K.L. & STEVENS, C.F. (1972) A quantitative description of end-plate currents. J. Physiol. 223, pp. 705-729.
10. NEHER, E. & SAKMANN, B. (1976) Noise analysis of drug induced voltage clamp currents in denervated frog muscle fibers. J. Physiol. 258, pp. 705-729.
11. NEHER, E., SAKMANN, B. & STEINBACH, J.H. (1975) The extracellular patch clamp: A method for resolving currents through individual open channels in biological membranes. Pfluegers. Arch. 375, pp. 219-228.
12. SAKMANN, B., PATLAK, J. & NEHER, E. (1980) Single acetylcholine-activated channels show burst-kinetics in the presence of desensitizing agonist concentrations. Submitted.
13. STEINBACH, J.H. & STEVENS, C.F. (1976) Neuromuscular Transmission. In: "Frog Neurobiology, a handbook" (Llinas, R. and Precht, W. eds.) Springer-Verlag-Berlin-Heidelberg-New York, 1976, pp. 33-92.

THE TIME COURSE OF MEMBRANE CONDUCTANCE CHANGE EVOKED BY IONTOPHORE-

TICALLY APPLIED ACETYLCHOLINE

Maria Scuka

Istituto di Fisiologia umana

Università degli Studi di Trieste - Italia

Resent theoretical analysis (Purves, 1977; Fradisek, Kordas and
Svetina, 1978) clearly showed that in the end-plate the receptor
response is dependent on many parameters, e.g. duration of the pulse,
receptor distribution, position of the microelectrode tip, kinetics
of receptor-transmitter interactions and cooperativity. In order
to compare these mathematical analysis with the experimental data,
the current evoked by iontophoretically applied transmitter was
studied.

The experiments were carried out on the frog sartorius muscle,
perfused with Ringer solution. The end-plate region was convention-
ally voltage clamped with two intracellular microelectrodes (3M KCl,
resistance 8-15 MΩ). Acetylcholine (ACh) was applied by a third
microelectrode (2M acetylcholine chloride, resistance 15-20 MΩ) to
the chemosensitive area of the muscle fibre. The iontophoretic
micropipette tip was moved towards the point of maximum response as
to achieve the maximum receptor density in the epicenter of the
micropipette.

The time course of receptor response, i.e. the time course of
conductance change was studied by varying the duration of transmitter
release and the distance of the microelectrode tip from the receptive
surface.

The log of the rise time (T max) upon application of ACh versus
log of the distance of drug release (r) is shown in Fig. 1. It can
be observed that, at higher values of r, the log T max depends
linearly on log r, as it was suggested by Peper, Dreyer and Müller
(1975). However, this relation is no more linear at small values
of r, when the release of ACh is not instantaneous. It can be

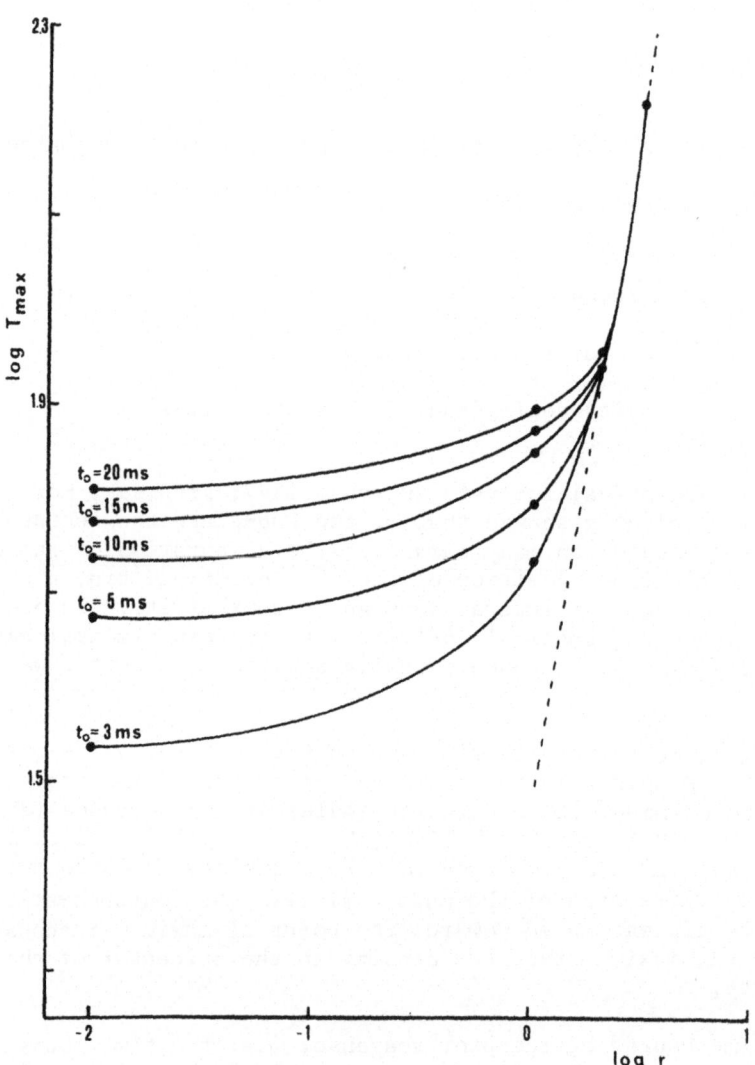

Fig. 1 - Rise time (T max) of ACh-current evoked by varying the
duration of transmitter release pulse (while the amplitude
remained constant) versus distance of drug release.

noticed that the non-linear part of this relation highly depends
on the release time of ACh.

REFERENCES

1) PURVES R.D., The time course of cellular responses to iontophoretically applied drugs, J. theor. Biol. 65, 327, 1977.
2) GRADISEK A., KORDAS M. and SVETINA S., A mathematical analysis
   of factors determining the time-course of membrane conductance
   change evoked by electrophoretic drug application, J. theor.
   Biol. 71, 311-322, 1978.
3) PEPER K., DREYER F. and MÜLLER K., Analysis of cooperativity
   of drug-receptor interaction by quantitative iontophoresis at
   frog motor end plate, Cold Spring Harbor Symp. Quant. Biol. 40,
   187-192, 1975.

THE CAPACITY OF CELL MEMBRANES: GATING CURRENT MEASUREMENTS

R.H. ADRIAN

University of Cambridge – Physiological Laboratory

Downing Street – Cambridge (Great Britain)

The electrical capacity of biological membranes was actively investigated by several authors in the nineteen-twenties and their results contributed to our knowledge of membrane structure (Fricke, 1925; Fricke & Morse, 1925; Cole, 1928; see also Cole, 1968). In particular these studies supported the idea that the cell membrane was a very thin lipid layer (Gorter & Grendel, 1925; Davson & Danielli, 1943). A lipid layer some 5 mm thick would be expected to have a capacitance in the region of 0.9 $\mu F/cm^2$ (assuming a dielectric constant of 5). Experimentally a figure of about 1 $\mu F/cm^2$ came to be regarded as more or less universally applicable. In 1968 Cole wrote "the work by Fricke and many others has ..... produced so wide a variety of experiments as to make a membrane capacity (within a factor of 2) of 1 $\mu F/cm^2$ a physical constant of most living cells".

The demonstration by Cole & Curtis (1939) that the membrane capacitance did not change by more than a few per cent in the course of an action potential in the squid giant axon seemed to confirm the idea that the capacity reflected mainly the structure of the lipids in the membrane. Cole & Curtis (1939) also showed that the membrane conductance increases dramatically during an action potential. The ionic currents which underlie this conductance change were analysed by Hodgkin & Huxley (1952). They showed how the conductances to sodium and potassium ions increase on depolarization and their analysis has become the basis of our understanding of excitability and propagation in nerve and muscle. The great success of the Hodgkin-Huxley analysis in explaining the behaviour of many excitable cells in terms of ionic conductance diverted interest from the dielectric properties of the cell membrane. For most purposes it was quite adequate to assume that the membrane dielectric was both linear and ideal, even though previous studies had shown that

the phase angle was less than 90°, implying a wide dispersion of
dielectric relaxation times (Cole, 1968).

These assumptions about the dielectric behaviour of the cell
membrane became part of both the methods of measuring membrane capacity and the analysis of those measurements. It is the purpose
of this article to suggest that since the assumptions turn out to
be falsifiable some of the conclusions will need changing. One
should probably add that the required changes will be in the details
of our picture of excitability and certainly not in the picture itself.

If a constant current is passed, via an intracellular microelectrode, across the membrane of a spherical cell, the potential
across the membrane will change from its resting value to a new
value. If the membrane behaves as a linear resistance and capacity
in parallel, the charging will be exponential and the membrane capacity can be obtained from the time constant of either charging
or discharging. If current is passed into an infinitely long cylindrical cell at the mid-point one can also derive the membrane
capacity from the magnitude of the potential change and its time
course. Measurements of this kind have been very widely made in
nerve and muscle fibres from many species following the very careful
study of the resting membrane resistance (R_m) and capacity (C_m) in
lobster axons by Hodgkin & Rushton (1946). They showed that, within
their experimental accuracy and at the normal resting potential this
membrane did behave as if it had an equivalent circuit of R_m in
parallel with C_m, and that C_m was about 1 μF/cm^2.

When the analysis, using a microelectrode to pass the current,
was extended to striated muscle by Katz (1948), it became clear that
muscle was anomalous in that the membrane capacity was substantially
larger than in nerve fibres. In frog sartorius muscle fibres C_m
appeared to be about 5 μF/cm^2 of fibre surface. In time it became
clear that this was due to the extensive membranes of the transverse
tubular system in muscle fibres. Several studies using A.C. impedance methods were undertaken to define the equivalent circuit of the
morphologically complex membranes in a muscle fibre (Falk & Fatt,
1964; Schneider, 1970; Valdiosera, Clausen & Eisenberg, 1974).
Since the transverse tubules have a very small diameter their luminal
resistance will be in series with a substantial fraction of the
membrane capacity, and one would expect an equivalent circuit for
the membrane more complex than a resistance and capacity in parallel.
This is indeed found to be the case. However conclusions of such
studies depend heavily on the assumption of ideal behaviour of the
membrane dielectric since it may be difficult to distinguish between
an ideal capacity in series with the resistance of the transverse
tubular lumen and a lossy membrane dielectric. Undoubtedly the
capacity of the transverse tubular membrane contributes to the
measured capacity of muscle fibres but it is probably rash to draw

precise parallels between the observable morphology and elements of
an equivalent circuit which depends so heavily on the assumption of
ideal behaviour.

      To avoid these uncertainties we have tried (Adrian & Almers,
1974) to use methods which make no assumptions about equivalent
circuits and in addition look specifically for non-linear behaviour
instead of assuming linearity. We can define an effective capacity
of unit length of fibre as the ratio of the charge required to the
potential displacement produced, given that the potential displacement is longitudinally uniform. In terms of the membrane current
i(t) in Amp/cm fibre across a membrane which is uniformly polarized
by a voltage displacement V_o at t = 0

$$C_{eff} = \frac{1}{V_o} \int_o^\infty \{ i(t) - i(\infty) \} dt$$

where $i(\infty)$ is the steady current across the membrane.
At the break of the voltage step

$$C_{eff} = \frac{1}{V_o} \int_o^\infty i(t) dt$$

assuming that the steady current at the undisplaced potential is
zero.

      In a three electrode clamp for muscle fibres (Adrian, Chandler
& Hodgkin, 1970) a measure of the capacity is obtained directly from
integrating the transient part of the current (ΔV) record, but it is
sometimes convenient to use electrodes in the middle of a fibre and
to define C_{eff} in terms of the current $I_o(t)$ delivered by the microelectrode and the step voltage displacement V_o at the middle of a
fibre which extends in both directions far in comparison with the
length constant of the fibre (λ). Then for the make of a step change
of potential

$$C_{eff} = \frac{1}{V_o \lambda} \int_o^\infty \{ I_o(t) - I_o(\infty) \} dt$$

and for the break of a voltage step

$$C_{eff} = \frac{1}{V_o \lambda} \int_o^\infty I_o(t) dt.$$

These simple definitions can be extended to non-infinite cables and
to constant currents. For a constant current delivered to the middle

of an infinite cable and beginning at t = 0

$$C_{eff} = \frac{I_o}{V_o \lambda} \int_o^\infty \{ 1 - \frac{V_o(t)}{V_o} \} dt.$$

In this case V_o is the steady potential displacement for $t \to \infty$.

Provided the integration is over a sufficiently long interval C_{eff} gives a measure of the zero frequency capacity and therefore of the total polarizability of the membrane dielectric. In striated muscle which is morphologically complex there is an important additional proviso. The ionic current across the tubular membrane must not be large enough to cause significant steady state radial potential gradients in the tubular lumen.

Armed with this measure of polarizability we can ask whether polarizability is a function of membrane field. Essentially we seek experimental conditions which reduce as far as possible ionic current across the membrane. Large ionic currents give rise to trouble for two reasons: they may change with membrane potential and with time. In the former case if the current becomes large one cannot guarantee radial uniformity of potential in the transverse tubular system. In the latter case a time dependent ionic current may alter measured transient current. Both sources of error are reduced if the ionic currents are small.

If the dielectric behaviour of striated muscle were capable of representation by an equivalent circuit containing only linear resistances and capacitors, then no matter what the complexity of the equivalent circuit the current response to a small potential step imposed across the membrane should be independent of the existence or magnitude of a steady potential across the membrane. In terms of a muscle fibre, if the membrane dielectric were linear, the measured C_{eff} should be independent of the membrane potential. The value of C_{eff} measured with a 10 mV step from -180 mV should exactly equal that measured with a 10 mV step from any other membrane potential. That this is not, in fact, the case, was first shown by Schneider & Chandler (1976). Adrian & Almers (1976) and Adrian & Peres (1979) have confirmed and extended their findings. If the membrane potential range from -200 mV to 0 mV is explored by using a 10 mV step to measure the capacity at each membrane potential, the capacity rises by about 10% between -200 mV and -80 mV; between -80 mV and 0 mV the capacity goes through a peak at -40 mV, some 1.7 times the capacity at the resting potential (-90 mV). At 0 mV the capacity has about its normal value but it falls for positive values of the internal potential. In the physiologically interesting range of membrane potential where several processes are regulated by the membrane potential the membrane capacity is not a constant but varies by as much as 70%. Calculations of action potentials

and propagation velocity are likely to be affected by such large
changes (Adrian, 1975).

When one considers the multiplicity of proteins in the membrane
postulated for excitable membranes, or indeed for cell membranes in
general, it is perhaps hardly surprising that the membrane should
show a complex dielectric behaviour. Each ionic channel is believed
to be a large and complex lipoprotein structure, part of which probably permanently spans the membrane and so must be subject in some
way to the electric field within the membrane. These proteins cannot
be non-polar because their potential sensitivity must depend on some
alteration of their structure or position. Membranes have a multiplicity of functions of which regulating ionic currents are only
among the most obvious and most obviously potential regulated.
There are thus likely to be many membrane-associated proteins which
extend if not across the membrane at least part way into the lipid
layers of the membrane. Intra-membrane proteins, whether held across
the membrane or free to rotate within it are likely to provide extra
polarizability in that their structure or orientation could be affected by the membrane field (Almers, 1978). Since the fields can
be made large ($\pm$ 10^5 V cm^{-1}) one might expect dielectric saturation
when the distortion or reorientation is no longer increased by increasing the field. The greater part of the peak in capacity at
-40 mV appears to represent a component in the membrane dielectric
which goes from one saturated state to the other when the potential
is changed from -80 to 0 mV. This component has been tentatively
associated with the regulation of contraction (Schneider & Chandler,
1973; Chandler, Rakowski & Schneider, 1976).

In striated muscle there is clearly no range of membrane potentials in which the polarization is a linear function of the imposed voltage. Over the whole of the attainable range of membrane
potentials there is evidence of dielectric saturation, and it is
clear that the complex dependence of polarizability on potential
implies more than one, and probably several, polarization processes.
This need not surprise us since there are several voltage regulated
processes, but it requires careful experimental analysis to separate
and identify the component polarization currents, often called
membrane charge movements or gating currents.

We can examine the total polarization current at any particular
membrane potential by looking at the transient current produced by
a 10 mV step, always assuming that transient ionic current has some-
how been eliminated. Such a record includes linear and non-linear
polarization current and these currents may last for some time if
the dielectric is lossy or if the potential changes with a delay in
any part of the system (e.g. the transverse tubular membrane). If
we subtract transient currents for two 10 mV steps from different
membrane potentials we eliminate linear polarization currents and
what is left is related to saturable, and therefore non-linear

processes. But it is important to remember that for any given subtraction we are left with the difference in non-linear events. It
is not safe to assume that any one membrane potential gives us a
control record in the sense that the polarization currents in it
are present at all other voltages. Adrian & Peres (1977 and 1979)
have used this approach to examine the apparently complex kinetic
behaviour of polarization currents in a small range of voltages
between -50 mV and -40 mV. In this range polarization current for
a 10 mV potential step, in the sense that the inside of the fibre
becomes more positive, takes place in two fairly distinct components.
When the potential changes in the reverse direction by 10 mV the
polarization current resembles more or less a single exponential
component as would be predicted for a simple dielectric relaxation.

The interpretation of this finding is uncertain and one needs
more experimental work to examine the dielectric behaviour of cell
membranes with specifically different electrical properties. But
it is already clear that the membrane capacity is neither linear
nor ideal. It shows both saturation and loss.

REFERENCES

Adrian, R.H. (1975). Conduction velocity and gating current in the
    squid giant axon. Proc. Roy. Soc. B, 189, 81-86.
Adrian, R.H. & Almers, W. (1974). Membrane capacity measurements
    on frog skeletal muscle in media of low ion content. J. Physiol
    237, 573-605.
Adrian, R.H. & Almers, W. (1976). The voltage dependence of membrane
    capacity. J. Physiol. 254, 317-338.
Adrian, R.H. & Peres, A.R. (1977). A gating signal for the potassium
    channel? Nature, Lond. 267, 800-804.
Adrian, R.H. & Peres, A. (1979). Charge movement and membrane ca-
    pacity. J. Physiol. 289, 83-97.
Adrian, R.H., Chandler, W.K. & Hodgkin, A.L. (1970). Voltage clamp
    experiments in striated muscle fibres. J. Physiol. 208, 607-644
Almers, W. (1978). Gating currents and charge movements in excitable
    membranes. Review of Physiology, Biochemistry and Pharmacology
    82, 96-190.
Chandler, W.K., Rakowski, R.F. & Schneider, M.F. (1976). A non-
    linear voltage dependent charge movement in frog skeletal muscle
    J. Physiol. 254, 245-283.
Cole, K.S. (1928). Electric impedance of suspensions of spheres.
    J. gen. Physiol. 12, 29-36.
Cole, K.S. (1968). Membranes, Ions and Impulses. University of
    California Press, Berkeley & Los Angeles.
Cole, K.S. & Curtis, H.J. (1939). Electric impedance of the squid
    axon during activity. J. gen. Physiol. 22, 649-670.
Davson, H. & Danielli, J.F. (1943). The Permeability of Natural
    Membranes. Cambridge University Press.

Falk, G. & Fatt, P. (1964). Linear electrical properties of striated muscle fibres observed with intracellular electrodes. Proc. Roy. Soc. B 160, 69-123.

Fricke, H. (1925). The electric capacity of suspensions with special reference to blood. J. gen. Physiol. 9, 137-152.

Fricke, H. & Morse, S. (1925). The electric resistance and capacity of blood for frequencies between 800 and 4½ million cycles. J. gen. Physiol. 9, 153-167.

Gorter, E. & Grendel, F. (1925). On bimolecular layers of lipoids on the chromocytes of the blood. J. Exp. Med. 41, 439-443.

Hodgkin, A.L. & Huxley, A.F. (1952). A quantitative description of membrane current and its application to conduction and excitation in nerve. J. Physiol. 117, 500-544.

Hodgkin, A.L. & Rushton, W.A.H. (1946). The electrical constants of a custacean nerve fibre. Proc. Roy. Soc. B 133, 444-479.

Katz, B. (1948). The electrical properties of the muscle fibre membrane. Proc. Roy. Soc. B 135, 506-

Schneider, M.F. (1970). Linear electrical properties of the transverse tubules and surface membranes of skeletal muscle fibres. J. gen. Physiol. 56, 640-671.

Schneider, M.F. & Chandler, W.K. (1973). Voltage dependent charge movement in skeletal muscle: a possible step in excitation-contraction coupling. Nature, Lond. 242, 224-246.

Schneider, M.F. & Chandler, W.K. (1976). Effects of membrane potential on the capacitance of skeletal muscle fibres. J. gen. Physiol. 67, 125-163.

Valdiosera, R., Clausen, C. & Eisenberg, R.S. (1974). Impedance of frog skeletal muscle in various solutions. J. gen. Physiol. 63, 460-491.

CHARGE MOVEMENT AND CONTRACTION THRESHOLD IN SKELETAL MUSCLE FIBRES

A. Peres and A. Ferroni

Istituto di Fisiologia Generale e Chimica Biologica

Università di Milano, Via Mangiagalli, 32 Milan, Italy

INTRODUCTION

Currents due to the movement of charged particles in the membrane thickness have been shown to exist in skeletal muscle fibers by many authors (Schneider & Chandler, 1973; Chandler, Rakowski & Schneider, 1976a; Adrian & Almers, 1976a). The voltage range in which this charge movement appears suggests that it may have important physiological role. In fact in excitable cells the membrane potential controls many physiological functions. In particular, in skeletal muscle fibres not only the conductances of K^+ and Na^+ ions but also the activation of mechanical contraction is under the control of the electric field existing across the cell membrane (Hodgkin & Horowicz, 1960).

The total charge movement appears to be the sum of at least two (Adrian & Almers 1976b) or possibly three (Adrian & Peres, 1977, 1979) different components, presumably due to the movement of different species of particles.

Adrian & Peres (1979) on the basis of the resemblance of the kinetics of a slow charge movement (which they called Q_γ) with $d(n^4)/dt$, tentatively suggested that it might be associated with the opening of the potassium channel. The remaining charge movement (Q_β) which has a faster kinetics and a different voltage distribution remained as a possible candidate as a regulator of activation of contraction (see also Adrian et al., 1976).

All the cited experiments have been performed with the technique of the three microelectrodes voltage clamp. In order to avoid membrane damage, contraction was blocked with hypertonic solution

or prevented by stimulating the fibres only with contraction sub-
threshold stimuli. In the present work experiments are reported
in which, making use of a technique recently introduced by Kovàcs
& Schneider (1978), it is possible to measure charge movement in
contracting fibres.

METHODS

The methods used in the present experiments are essentially
the same as in Kovàcs & Schneider (1978). Semitendinous muscles
were dissected from frogs of the species Rana esculenta. Single
fibres were isolated for a length of about 1 cm from the proximal
tendon of the ventral branch. The Ringer solution was changed to
a solution containing 120 mM K-glutammate, 2 mM $MgCl_2$, .01 mM EGTA,
5 mM Tris maleate buffer at a pH of 7 (relaxing solution). Contraction of the fibre after this solution change indicated that the
fibre was undamaged. The fibre was then cut and transferred to the
experimental chamber. The chamber had two compartments separated
by a wall. The fibre was mounted in the chamber across the wall
and then the two compartments were electrically isolated with vaseline. The left compartment contained the cut end of the fibre, while
the right compartment contained the intact end. The fibre was held
in position with a tiny spring pinched to the tendon. The solution
in the intact end compartment was then changed to one containing
90 mM TEA_2SO_4, 10 mM RbCl, 8 mM $CaCl_2$, 5 mM Tris maleate buffer at
a pH of 7 and 2×10^{-7} g/ml TTX. This solution minimizes ionic
currents through the membrane. When it was desired to block contraction also the solution in the cut end compartment was changed
with one identical to the relaxing solution but containing 20 mM
EGTA (intracellular solution).

Two pairs of Ag-AgCl electrodes connected to the solutions
through Agar bridges were used to measure the potential difference
between the compartments and to inject current. An electronic
circuit allowed to take into account and to compensate the effects
of imperfect sealing of the vaseline and of the resistance in series
with the membrane (mainly intracellular resistance). A microelectrode was inserted in the intact end of the fibre while injecting
square current pulses and the compensation circuit was set to give
an output identical to the microelectrode recording.

After this preliminary stage the voltage clamp was turned on
using the output of the compensating circuit as a measure of the
membrane potential. The holding potential was then set to - 100 mV
and small voltage steps were given with the microelectrode still in
the fibre in order to check that the membrane was clamped properly.
The microelectrode was then removed and the measures initiated.
All experiments have been done at temperatures ranging between 4.8
and 5.6 °C.

The experiments were under the control of a Zilog MCZ 1/05 computer. Analog signals of membrane currents were fed to the computer through an interface based on a AD570JD Analog Devices analog to digital converter at a sampling frequency of 4kHz. Digitally converted signals were stored on disk for both on line elaboration and subsequent off line analysis.

RESULTS

The experimental protocol was as usual for charge movement measurements. Five control pulses + 20 mV from the holding potential were given to the fibre; then two test pulses to the same test voltage were given. The sequence was repeated for different test voltages. The membrane current signals were elaborated by the computer which made a scaled difference between test and control pulses and produced a signal due only to non-linear membrane electrical components. All currents measurements were normalized to membrane

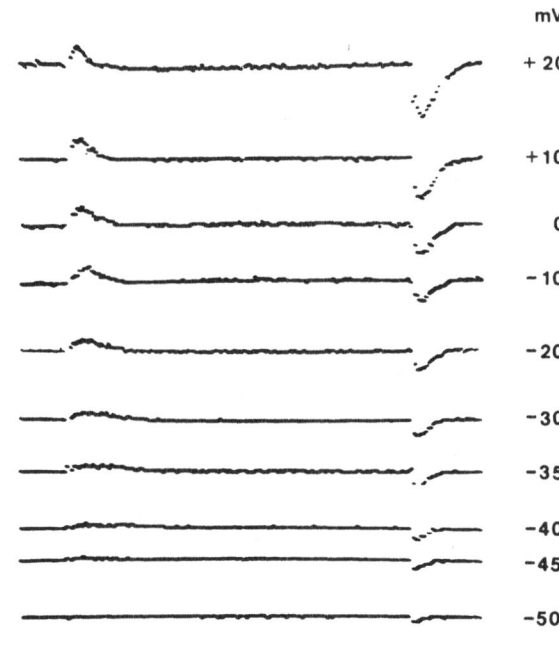

Figure 1. Charge movement records in non-contracting fibres. Values of test voltages are shown on the right.

capacity which was measured integrating the capacitative transient
of the control pulses after leakage subtraction.

Fig. 1 shows the results from a typical charge movement experiment. Duration of the pulses was 150 msec. It is clear that this
technique allows measurements of charge movement which shows the
same main features of the one determined with the three microelectrodes technique.
In these experiments the intracellular solution was used at the
cut end in order to block contraction.

Some consideration is needed on the reliability of current
kinetics. It is apparent in Fig. 1 that the off transients are not
only single decreasing exponentials and also that the on transients
increase slowly and never start abruptly from a positive value.
This is due to a non perfect regulation of the compensating circuit;

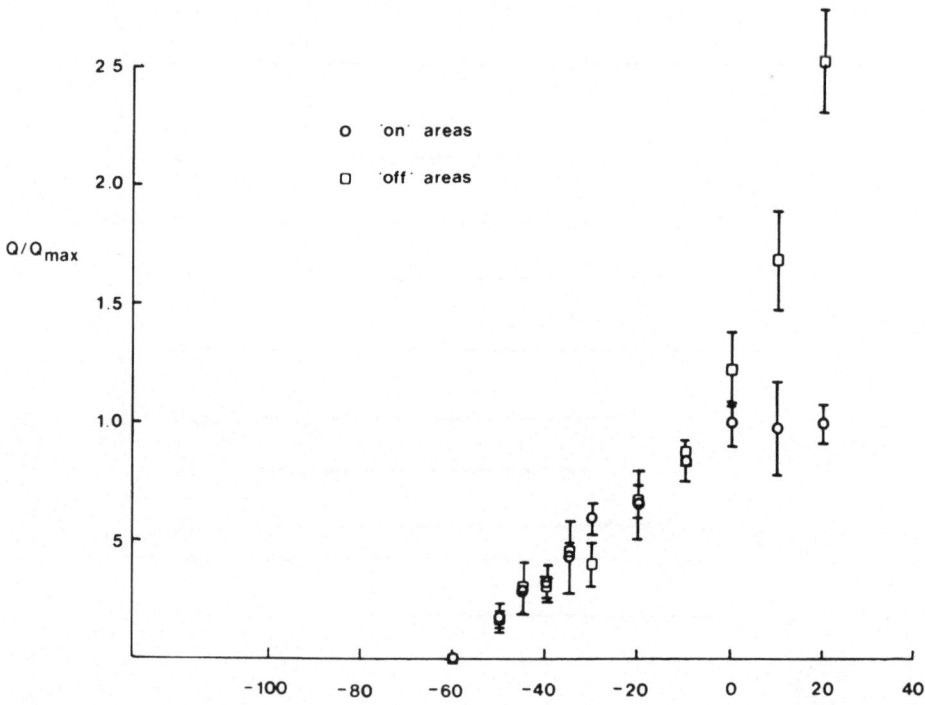

Figure 2.   Voltage distribution of charge movement from experiments
            of the kind of Fig. 1. Results pooled from four fibres
            have been normalized to maximum Q_{on} value.
            Bars are S.E. of the mean.

in fact perfect compensation caused damped oscillations during fast
voltage changes. The system was not fully compensated to avoid these
oscillations sacrificing the steepness of the voltage changes. This
led to a definite alteration of the kinetics during the first milli-
seconds after a voltage change. Insofar total charge movement is
concerned the shape of the voltage change is not relevant, provided
it reaches the desired final value.

Integration of the on and off transients to obtain the amount
of charge which has been moved by the test pulses was performed
using Simpson's rule after leakage subtraction. Results pooled
from four fibres are shown in Fig. 2 where on and off areas are
plotted separately to show that they are quite similar in the range
from − 60 to 0 mV. For more positive voltages on areas saturate
while off areas increase greatly. This behaviour has been already
described and the most likely explanation seems to be that TEA block
of K^+ conductance is not complete especially for large depolarizations (Stanfield, 1970; Adrian & Almers, 1976b). The mean value
of the Q_{max} relative to on areas from the four fibres is 21.15 ±
6.53 nC/µF.

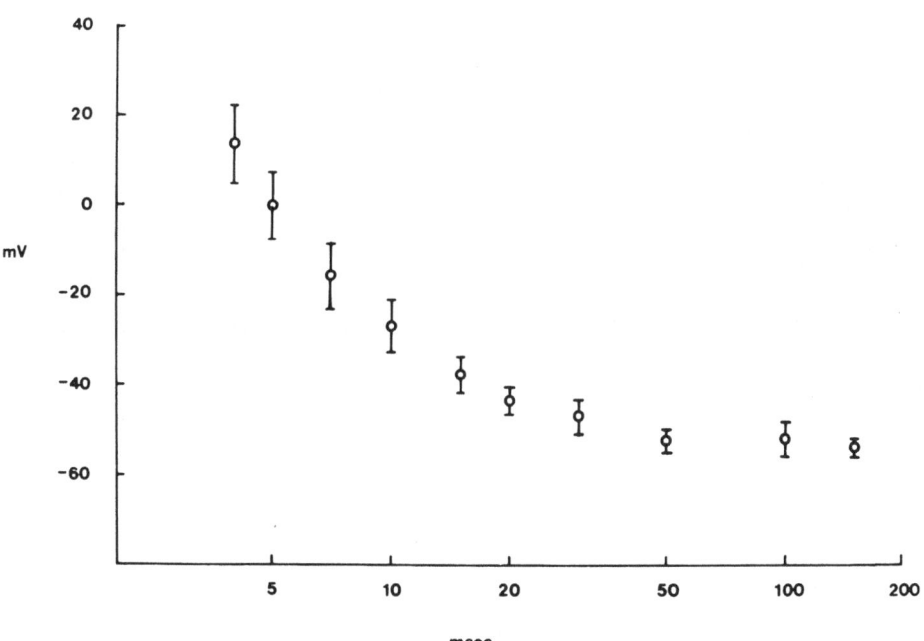

Figure 3. Strength-duration curve for a just visible contraction.
           Results pooled from three fibres.
           Bars are S.E. of the mean.

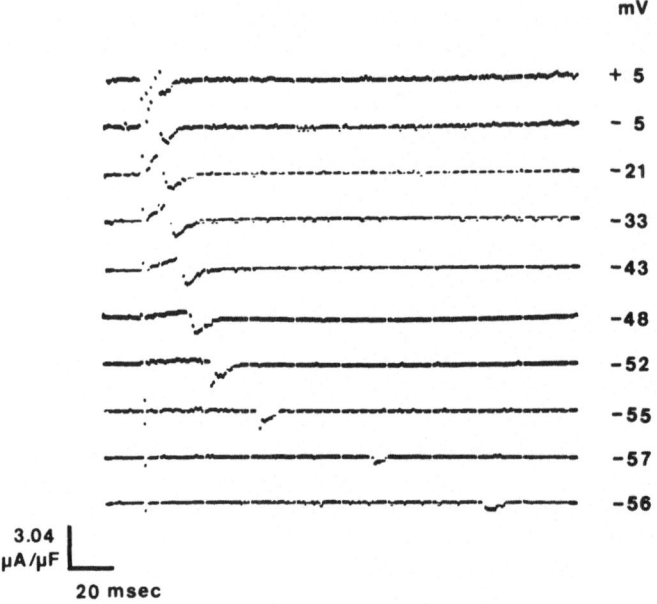

Figure 4.   Charge movement associated with pulses of different
            amplitude and duration, all giving just visible con-
            tractions.

     In another set of experiments the cut end of the fibres was
immersed in the relaxing solution. In this solution fibres con-
tracted in response to appropriate stimuli. In these experiments
the charge movement produced by voltage pulses causing just detectable contraction was measured. Starting from 150 msec the duration
of the pulses was decreased and for each duration the amplitude was
changed millivolt by millivolt until a just eye-detectable contraction occurred. A sequence of control and test pulses to this
potential and with this duration was then given via the computer
and charge movement recorded.

     Fig. 3 shows a strength-duration curve (data from three experi-
ments) on a semilog plot. For long pulses our contraction threshold
voltage seems to be more negative than the one reported by Kovàcs &
Schneider (1978) but quite similar to the one reported by Constantin
(1974) in intact fibres.

     The current difference traces from the computer are shown in
Fig. 4 and results of the integration of the on and off transients
are shown in Fig. 5 (circles).

For the two most positive voltages the saturation of the analog to

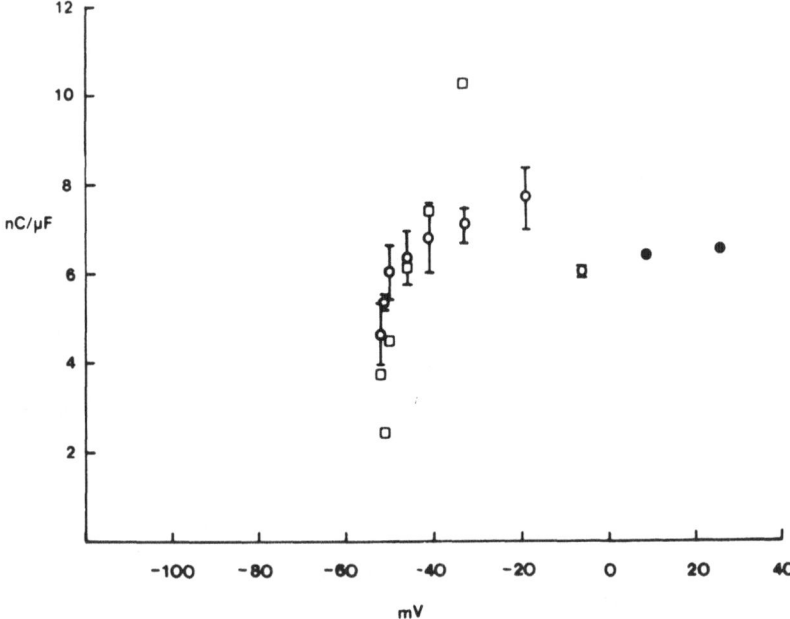

Figure 5.   Open circles: mean between on and off areas from an experiment of the kind of Fig. 4. Filled circles: only
off areas from same experiment. Squares: charge movement measured in the same fibre but with 150 msec pulses.

digital converter combined with digital filtering made impossible
to properly integrate the on transient, so that only off areas were
computed. For all the other voltages the on and off areas showed
appreciable identity as can be noted by the standard error bars.
Squares in Fig. 5 show the charge moving in the same fibre but with
a fixed pulse duration of 150 msec. It was not possible to cover
the complete range of potentials with this duration because fibre
movement produced big artifacts on the current records.

It appears from Fig. 5 that the charge moved by pulses which
cause a just detectable contraction is increasing in the range of
potentials between $-$ 55 and $-$ 45 mV, but is more or less constant
for more positive voltages. Fig. 6 compares results from two
different fibres. Circles are from a fibre in which the total
charge (i.e. the one moving with 150 msec pulses) was measured.
This fibre was chosen because it had values of total charge similar
to the one of the fibre of Fig. 5 at the voltages where it was
possible to measure them. Triangles are the same points of Fig. 5.
It is seen that the charge which is moved by pulses which give a
just detectable contraction is about 27% of the total charge.

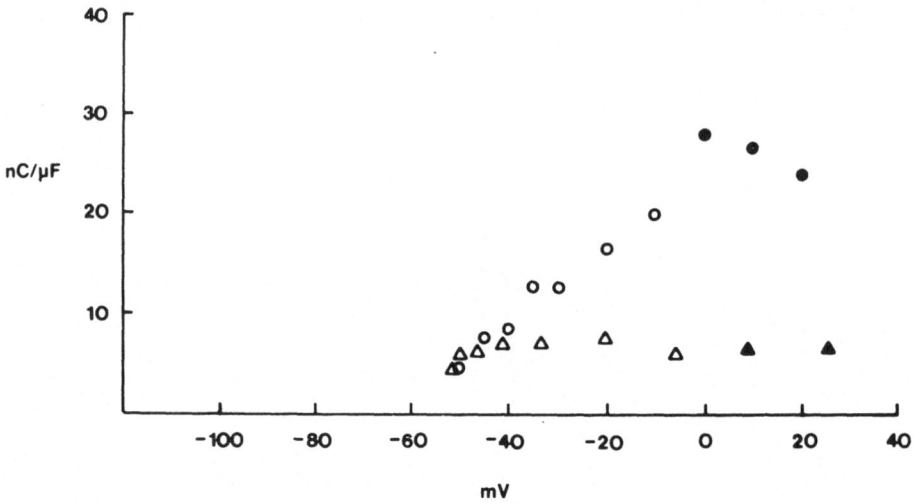

Figure 6.   Comparison between total charge and charge moving during
            pulses which give a just detectable contraction. Results
            from two different fibres. Filled circles are only on
            transients, filled triangles are only off transients.

DISCUSSION .

     The observation that the amount of charge moved by pulses which
give a just detectable contraction is not constant over the entire
voltage range seems to exclude the most obvious hypothesis that a
constant fraction of the total charge movement is needed to trigger
the events that lead to contraction. On the other hand if one
accepts the idea that the total observed charge movement is made
up of different components with different physiological roles, then
the present results are not inconsistent with the view that a
constant fraction of a particular component of charge must move in
order to activate contraction.

     From Fig. 5 it appears that the minimum amount of charge associ-
ated with contraction is (at − 52 mV) 4.6 nC/μF, while the constant
level is at about 7 nC/μF. Furthermore in the voltage range where
the charge increases it coincides with the total charge. This
voltage range is precisely the region where Q_γ (Adrian & Peres,
1979) begins to appear in a very abrupt way. The other kind of
charge described by Adrian & Peres (1979), Q_β, appears already at
more negative potentials and has a less steep voltage distribution.
It might be that a threshold amount of this latter kind of charge
is needed to trigger contraction, while the increase in the charge
associated with just detectable contraction pulses might be due to

an increase in charge Q_γ. This increase would stop at more positive potentials because the pulses are too short. In this view the constancy of the charge shown in Fig. 5 would be completely incidental.

REFERENCES

1. Adrian, R.H. & Almers, W., 1976a. The voltage dependence of membrane capacity. J. Physiol. 254:317-338.
2. Adrian, R.H. & Almers, W., 1976b. Charge movement in the membrane of striated muscle. J. Physiol. 254:339-360.
3. Adrian, R.H., Chandler, W.K. & Rakowski, R.F., 1976. Charge movement and mechanical repriming in striated muscle. J. Physiol. 254:361-388.
4. Adrian, R.H. & Peres, A., 1977. A gating signal for the potassium channel?. Nature Lond. 267:800-804.
5. Adrian, R.H. & Peres, A., 1979. Charge movement and membrane capacity in frog muscle. J. Physiol. 289:83-97.
6. Chandler, W.K., Rakowski, R.F. & Schneider, M.F., 1976. A non-linear voltage dependent charge movement in frog skeletal muscle. J. Physiol. 254:245-283.
7. Constantin, L.L., 1974. Contractile activation in frog skeletal muscle. J. gen. Physiol. 63:657-674.
8. Hodgkin, A.L. & Horowicz, P., 1960. Potassium contractures in single muscle fibres. J. Physiol. 153:386-403.
9. Kovács, L. & Schneider, M.F., 1978. Contractile activation by voltage clamp depolarization of cut skeletal muscle fibres. J. Physiol. 277:483-506.
10. Schneider, M.F. & Chandler, W.K., 1973. Voltage dependent charge movement in skeletal muscle: a possible step in excitation-contraction coupling. Nature Lond. 242:224-226.
11. Stanfield, P.R., 1970. The effect of tetraethylammonium ion on the delayed currents of frog skeletal muscle. J. Physiol. 209:209-229.

PROPERTIES OF THE CURRENT (i_f) WHICH UNDERLIES ADRENALINE ACTION IN THE $\underline{SA}$ NODE. COMPARISON WITH i_{K2} IN THE PURKINJE FIBRE

Dario DiFrancesco and Carlos Ojeda

Istituto di Fisiologia Generale e Chimica Biologica

Università di Milano

SUMMARY

The properties of the current i_f in the cardiac SA node cells are investigated in the light of its resemblance to the current i_{K2} of cardiac Purkinje fibres in their quality of "pacemaker" currents and in their role in mediating the adrenaline-induced acceleration of pacemaker activity. We find that:

- Na removal reduces i_f, as observed for i_{K2}; furthermore i_f, like i_{K2}, is blocked by cesium

- Despite these similarities, membrane conductance measurements in normal and high K show that i_f cannot be a pure K-current like i_{K2}. On the contrary, i_f behaves as an inward current activated on hyperpolarizations below -50/-60 mV from holding potentials near -40 mV.

INTRODUCTION

The current i_f has been recently described as the one which controls the pacemaker activity of cardiac SA node cells and mediates the acceleratory effect of adrenaline (Brown, DiFrancesco & Noble, 1979a). In this respect, i.e. in the voltage range of activation and in the response to adrenaline, i_f is similar to another current already described in the heart, the current i_{K2} in the Purkinje fibre (Noble & Tsien, 1968; Hauswirth, Noble & Tsien, 1968). Here we show that besides having the same voltage range of activation and the same response to adrenaline, the two currents are similar in their dependence on cesium and sodium removal. However, while i_{K2} is known to be a K-current, we have evidence that i_f depends on external K in a way opposite to that expected for a K-current, and indeed

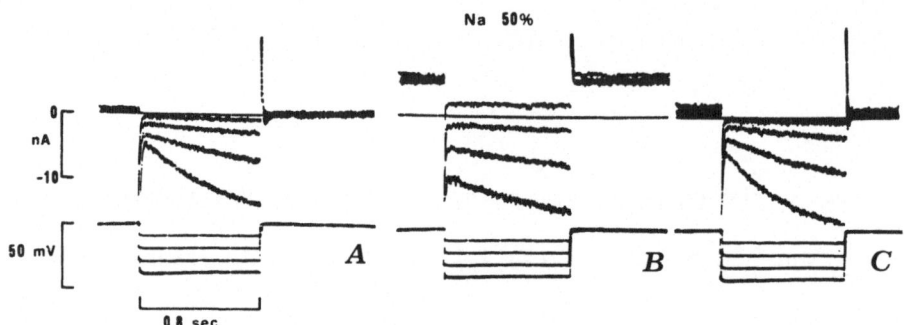

Fig. 1 - Effect of lowering Na concentration to 50% on i_f. A series
of hyperpolarizations are given from a holding potential of
-47 mV before (A), during (B) and after (C) perfusion with
50% Na-Tyrode. The decrease in i_f occurred within 5 min.

membrane conductance measurements during onset of i_f show that i_f
behaves as an inward current activated by hyperpolarizations.

METHODS

Strips of tissue from the SA nodal region of rabbit heart are
dissected and immersed in Tyrode solution at 35°C. By subsequent
cuts and ligatures a strip is reduced to a 0.25 x 0.25 mm^2 beating
preparation, which is impaled with two standard 3M KCl microelectrodes
(about 40 Mohms). Voltage clamp is performed with traditional methods
Voltage and current traces are recorded on a two-trace storage oscilloscope and on a 4 FM channel tape recorder for subsequent analysis

RESULTS

Effects of Na removal on i_f

Fig. 1 shows an experiment where 50% Na was replaced by Tris.
The current i_f, appearing in Fig. 1A as a large, slow activating
inward component on hyperpolarizing below -50/-60 mV (from a holding
potential of -47 mV in this case), undergoes a strong reduction in
50% Na (B). This phenomenon is known to occur also for the current
i_{K2} in Purkinje fibres (McAllister & Noble, 1966; DiFrancesco & Mc
Naughton, 1979), even if a satisfactory explanation for this has
not yet been found.

Effects of Cs

Cs is a blocker of the K-channel in skeletal muscle (Beaugè et
al., 1973), axon (Adelman et al., 1978), eel electroplaques (Ruiz-
Manresa et al., 1970) and starfish egg cell (Hagiwara & Takahashi,

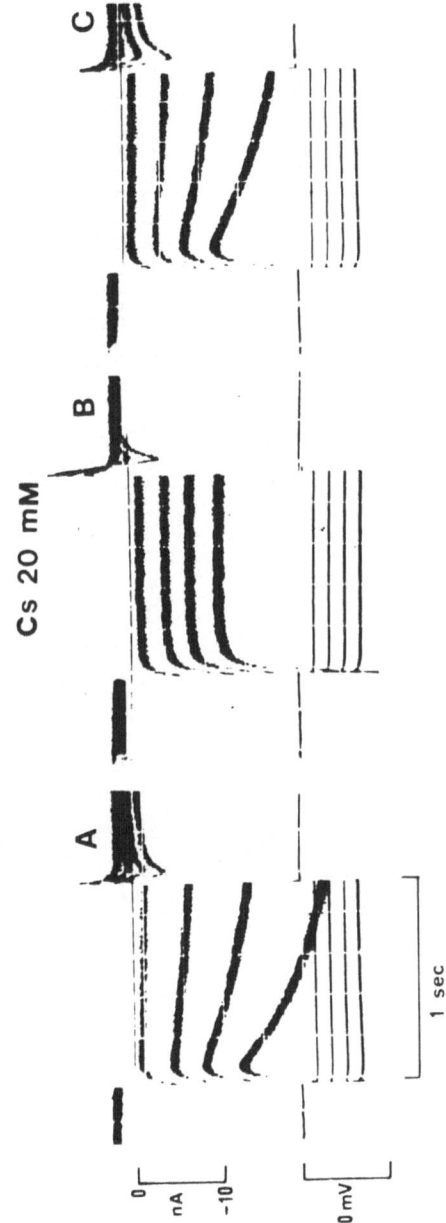

Fig. 2 – Effect of 20 mM Cs on if. From a –38 mV holding potential a series of 10 mV negative steps 1 sec long are applied in the control solution (A), 2 min after perfusion with Cs 20 mM (B) and 20 min after return to control (C). The time dependence of current completely disappears in Cs.

D. DiFRANCESCO AND C. OJEDA

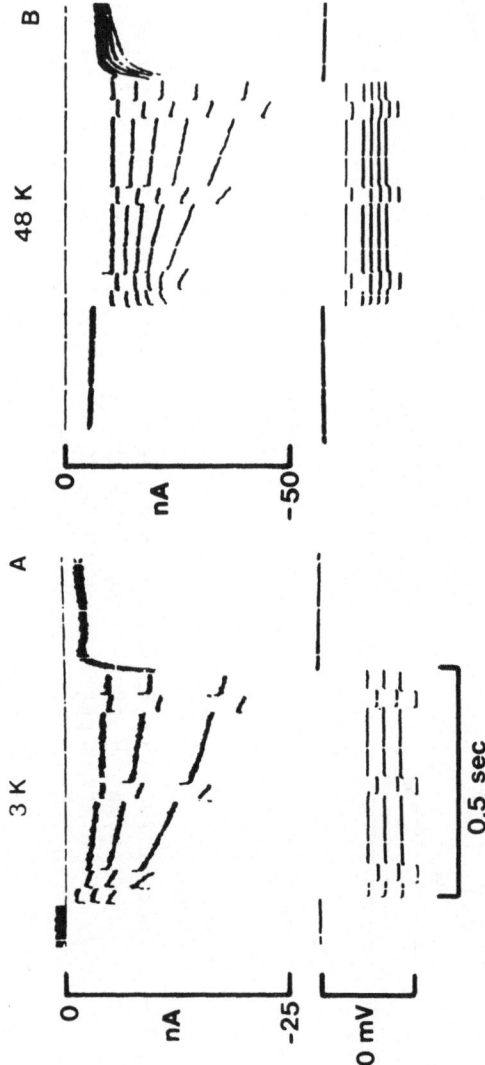

Fig. 3 — Conductance measurements in 3 mM K (A) and 48 mM K (B) solutions. Note that while changing from 3 to 48 mM K the K-current i_K is seen to reverse (first 100–200 msec in the traces in B, especially those corresponding to the largest negative pulses), the current i_f still increases in the inward direction. The membrane conductance during onset of i_f is apparently increasing in both K concentrations. Further explanation in the text.

1974), and in the Purkinje fibre it completely abolishes the i_{K1} and i_{K2} currents (Isenberg, 1976). In Fig. 2 the effects of 20 mM Cs are shown on a set of records where i_f is activated in the range -48 to -78 mV, from a holding potential of -38 mV. The complete block of i_f intervened after about 2 min. perfusion with Cs (B), and was nearly completely reversible (C), in all the voltage range investigated.

Membrane conductance measurements in normal and high potassium

The results presented above, together with the similarities between i_f and i_{K2} in their voltage dependence and in their response to adrenaline (Brown et al., 1979a), strongly suggest that we are dealing with two identical current systems in two different tissues. However, the dependence of i_f on external K shows that this is not the case, because i_f does not behave as a K-current (like i_{K2}). Fig. 3 shows two sets of traces recorded in the same preparation in 3 (A) and 48 mM K (B), where on the main hyperpolarizations small and short pulses are superimposed in order to measure the membrane conductance during the onset of i_f. First, this experiment shows that i_f cannot be carried by K, because in 48 mM K i_f is still inward increasing at potentials more negative than the expected K-reversal potential (see highest pulse at -75 mV in Fig. 3B). The fact that E_K is more positive than -75 mV is also evident from the outward decaying time course visible in the early part of the current traces in Fig. 3B (particularly the one corresponding to the highest pulse), reflecting the decay of the K-current already studied in the SA node, i_K (DiFrancesco et al., 1979), which is activated above -50/-60 mV. This feature can be better observed by giving a predepolarization before the test pulse, in order to activate more i_K (not shown). Secondly, having excluded that i_f is carried by potassium, the experiment of Fig. 3 also shows that the membrane conductance increase with time during i_f onset, both in 3 and in 48 mM K, as evident with the increase with time of the current displacement elicited by the small pulses (see current traces corresponding to highest pulses in Fig. 3A and B in particular). Thus, the behaviour of i_f during a hyperpolarizing V-clamp pulse is consistent with a channel opening process.

DISCUSSION

If one applies the pulse experiment of Fig. 3 on a HH-type current described by the general equation

$$i(E,t) = y(E,t) \ \overline{i}(E) \tag{1}$$

where y is a 1-st order kinetic variable and $\overline{i}$ the fully-activated current-voltage relation, during a main pulse to E the current displacement elicited by small pulses of amplitude δE is

$$\delta i(E,t) = \delta i_b(E) + y(E,t) \; (\overline{i}(E+E) - \overline{i}(E)) = \delta i_b(E) + y(E,t)$$
$$(d\overline{i}/dE)_E \; \delta E \tag{2}$$

δi_b being the contribution of time-independent components. Thus, the experiment of Fig. 3 permits the measurement of the slope conductance of the relation $\overline{i}(E)$, once the time-dependence of y is known. Given that in 48 mM K the IV relation of a K-current cannot possibly have a negative slope conductance at about -75 mV (most negative pulse in Fig. 3B), we conclude that the behaviour of i_f is consistent with the description of equation (1) only if i_f is an inward current activated on hyperpolarizations. This is in striking contrast with the similarity of i_f with the current i_{K2} of Purkinje fibres in their response to adrenaline, low Na, Cs and in their common characteristics of "pacemaker" currents (Brown et al., 1979b; Noble & Tsien, 1968; Hauswirth, Noble & Tsien, 1968), given that it is well established that i_{K2} is a pure K-current (but see DiFrancesco, Ohba and Ojeda, 1979, for a discussion on the interference of K-depletion phenomena in the determination of the properties of i_{K2}). More experiments are necessary to elucidate the ionic mechanism underlying the current i_f in the SA node.

REFERENCES

1. Adelman,W.J. & French, R.J.   Blocking of the squid axon potassium channel by external caesium ions. J. Physiol. 276:13-25 (1978)
2. Beaugè,L.A., Medici,A. & Sjodin, R.A.   The influence of external caesium ions on potassium efflux in frog skeletal muscle. J. Physiol. 228:1-11 (1973).
3. Brown, H.F., DiFrancesco, D. & Noble, S.J.   How does adrenaline accelerate the heart? Nature 280:235-236 (1979).
4. Brown, H.F., DiFrancesco,D. & Noble, S.J.   Frequency modulation of the cardiac pacemaker oscillation. Symposium on "Biological Oscillators", Ed. M.J. Berridge and P.E. Rapp J. Exp. Biol. 81.
5. DiFrancesco,D. & McNaughton,P.A.   The effects of calcium on the outward membrane currents in the cardiac Purkinje fibres. J. Physiol. 289:347-373 (1979).
6. DiFrancesco,D., Noma,A. & Trautwein,A.   Kinetics and magnitude of the time dependent potassium current in the rabbit sino-atrial node: effect of external potassium. Pflügers Arch.(in the press)
7. DiFrancesco,D. Ohba,M. & Ojeda,C.   Measurement and significance of the reversal potential for the pacemaker current i_{K2} in Purkinje fibres. J. Physiol. (in the press).
8. Hagiwara,S., Takahashi,K.   The anomalous rectification and cation selectivity of the membrane of a Starfish egg cell. J. Membrane Biol. 18:61-80 (1974).
9. Hauswirth,O., Noble,D. & Tsien,R.W.   Adrenaline: mechanism of action on the pacemaker potential in cardiac Purkinje fibres. Science, N.Y. 162:916-917

10. Isenberg,G. Cardiac Purkinje fibres: Cesium as a tool to block
    inward rectifying potassium currents. Pflügers Arch. 365:99-
    106 (1976).
11. McAllister, R.E. & Noble,D. The time and voltage dependence of
    the slow outward current in cardiac Purkinje fibres. J. Physiol.
    186:632-662 (1966).
12. Noble,D. & Tsien,R.W. The kinetics and rectifying properties
    of the slow potassium current in cardiac Purkinje fibres.
    J. Physiol. 195:185-214 (1968).
13. Ruiz-Manresa,F., Ruarte,A.C., Schwartz,T.L. & Grundfest, H.
    Potassium inactivation and impedence changes during spike
    electrogenesis in eel electroplaques. J. Gen. Physiol. 55:
    33-47 (1970).

A MODEL TO ACCOUNT FOR FIRING BEHAVIOUR OF NEURONS STIMULATED BY RECTANGULAR AND SINUSOIDAL CURRENTS

G. Monticelli

Istituto di Fisiologia Generale e di Chimica Biologica dell'Università,
via Mangiagalli 32.   20133 Milano

In biological systems the information is coded in terms of the interval between successive action potentials.

Nerve and sensory cells are able to transform an external stimulus into a train of action potentials. The sequence of produced action potentials is, in some way, related to the action, but little is known about the mechanisms responsible for this coding process.

A way to study these mechanisms is to apply different, known stimuli to the nerve cell and to observe the resulting electrical activity of the cell. Much emphasis, in particular, has been placed on the use of cyclic stimuli (1) which are a natural mean of exciting a system whose parameters vary periodically (2,3).

This paper is concerned with the responses of Helix pomatia neurons stimulated with rectangular and sinusoidal currents with the aim to verify a mathematical model able to describe adaptation in discharge frequency that has been demonstrated in Helix neurons (4,5,6,7,8).

THEORETICAL

The basic assumptions here adopted were those for the crayfish slowly adapting stretch receptor neurons (9).

The model was generalized to account for the frequency f_s of spontaneously firing cells.

If $G(t)$ is an excitatory current and $J(t)$ an inhibitory current, the instantaneous firing frequency $f(t)$ is then given by:

$$f(t) = K \quad G(t) - J(t) \quad + f_s \qquad\qquad -1-$$

where K is a proportionality constant (Hz/Na). The inhibitory
current is assumed to be incremented by each spike (with incremental
constant b'(nA)) and it decays exponentially (with decay time
constant τ). Thus

$$dJ(t)/dt = b'f(t) - J(t)/\tau \qquad\qquad -2-$$

The f(t) derivative can be obtained from equations -1- and -2-
as a function of the excitatory current only. We have:

$$\frac{df(t)}{dt} + \frac{Kb'\tau + 1}{\tau} \quad f(t) = K \frac{dG(t)}{dt} - \frac{KG(t) + f_s}{\tau} \qquad\qquad -3-$$

Equation -3- can be integrated for different shaped stimulation
currents.

For stimulation with constant current steps $G(t) = G, dG/dt = 0$
and the initial condition is $f(0) = KG+f_s$ if we assume that $J(0)=0$
(considering J(t) an inhibitory current in response to the applied
stimulus). In this case the instantaneous firing frequency is
given as:

$$f(t) = \quad f(o)-f(+\infty) \quad \exp(-t/\tau_f) + f(+\infty) \qquad\qquad -4-$$

where $f(+\infty)$ is the frequency steady-state value calculated as
$t \rightarrow +\infty$ and $\tau_f = \tau/1+Kb'\tau$ is the decay time constant.
The frequency decays exponentially from the initial value f(0) to
an asymptotic value $f(+\infty)$.
For a sinusoidal stimulating current $G(t)=G_0 \sin \omega(t+t_1)$ ($\omega= 2\pi\nu$,
ν is the stimulus frequency, G_0 the amplitude of sinusoidal current
and t_1 the time between the frequency response of the cell and the
applied stimulus) eq.-3- integrates to

$$f(t)=\tau_f f_s/\tau+C \exp(-t/\tau_f)+KG_0/((1/\tau_f)^2+\omega^2) \quad \omega(1/\tau_f-1/\tau)$$
$$\cos \omega(t+t_1)+(\omega^2+1/\tau\tau_f)\sin \omega(t+t_1) \qquad\qquad -5-$$

The exponential term in this equation accounts for the neuron different responses to successive stimulation cycles. As $t\rightarrow+\infty$ the cycle
is repeating but the term accounting for the "adaptation" to alter-
nating current disappears; the cell is at steady-state.
t_1 can be determined assuming constant the electrical charge Q at
which an action potential can be generated:

$$t_1 = 1/\omega \text{ Arcos } (1- \omega Q/G_0) \qquad\qquad -6-$$

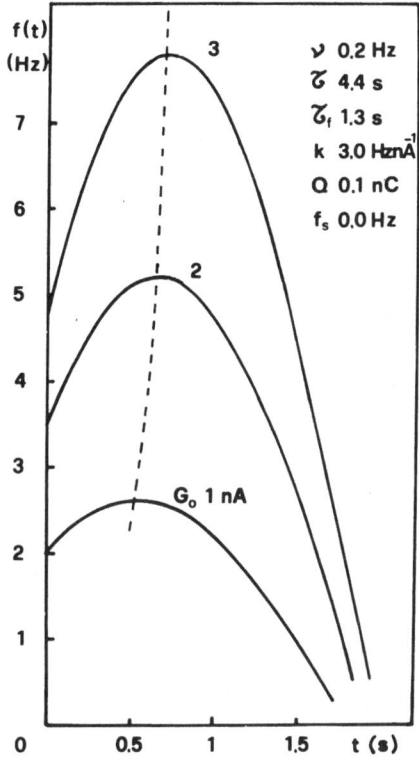

Fig. 1 - Computed curves for a sinusoidally stimulated neuron.

Fig. 1 shows the steady state behaviours of f(t) during a stimulation cycle for different G_o values.
This pattern of frequency change is similar to those of parabolic bursters found in Aplysia and Helix.

EXPERIMENTAL

To test the mathematical model suboesophageal ganglion neurons of Helix pomatia were stimulated by long lasting (9 sec) currents at different strengths and by sinusoidal currents of different amplitude and frequency. The studied cells were either beating pacemakers at low frequency or were silent at rest.

These neurons are able to exhibit repetitive activity during a sustained stimulus, but firing frequency decreases with time. This decrease in firing frequency is called adaptation.

At weak currents the initial firing rate is linearly related to the excitatory current. Fig. 2 shows current frequency relations

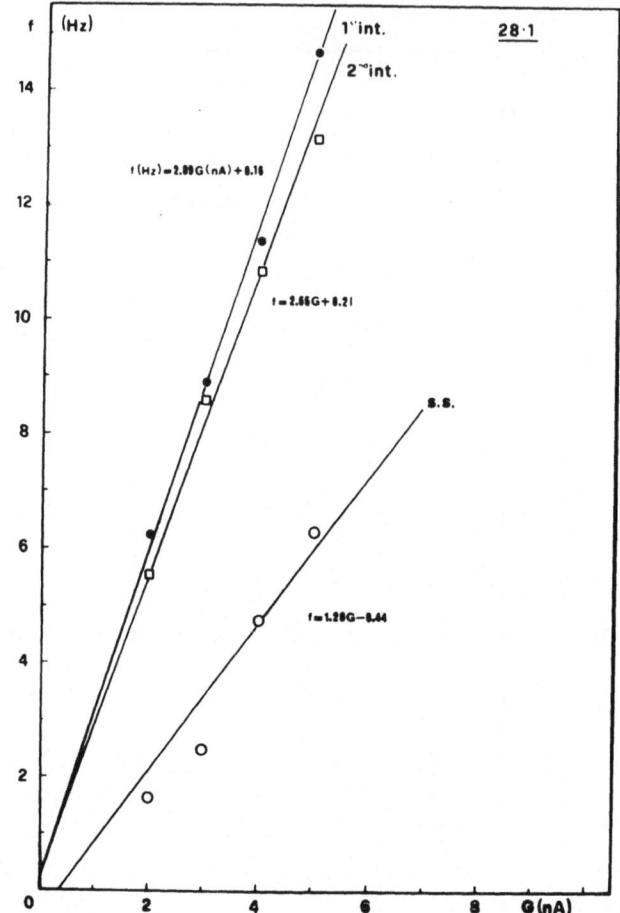

Fig. 2 - Current-frequency behaviours for first, second and steady-
         state interspike time interval.

for first, second and steady-state interspike intervals in a parietal
ganglion neuron. This linear relationship is in close agreement with
the assumption in eq.-1- and, in this case, a plot of f(o) vs. G
yielded a line with slope K=2.89 Hz/nA.

    In fig. 3 are shown the frequency responses of the same neuron
of fig. 1 to various intensities of a suddenly applied depolarizing
current. ⸳ The firing frequency jumped up to an initial value and then
declined exponentially to a lower steady-state value as obtained in
eq.-4-. K, b', τ_f and τ values can be estimated and utilized by
calculus to simulate the neuron responses. The computed curves
agreed with the experimental ones (Fig. 3).

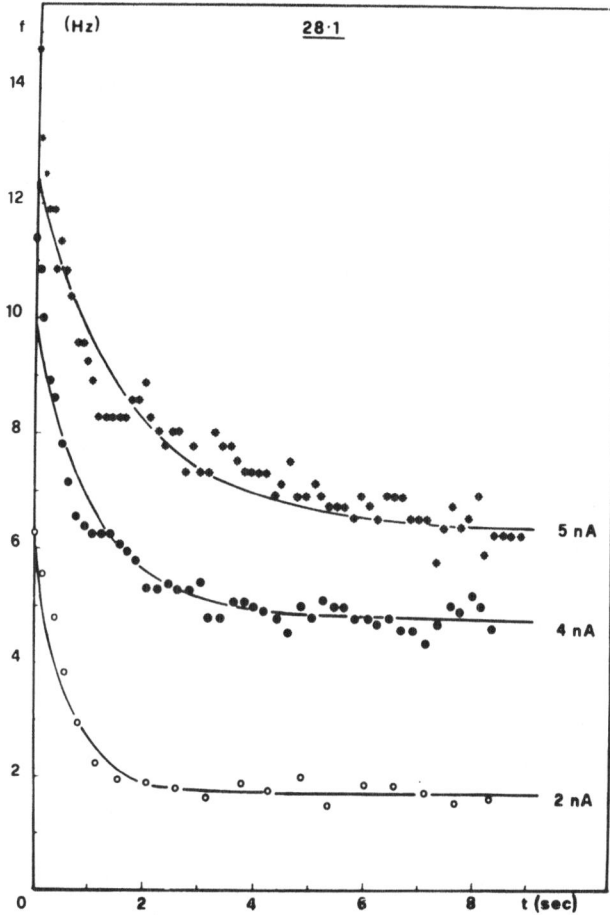

Fig. 3 – Time courses of changes in impulse frequency during intra-
          cellularly applied constant current stimulation. Instantane-
          ous frequency was obtained as the inverse of interspike time
          interval. Solid lines represent theoretical behaviours.

     Neurons stimulated by sinusoidal currents exhibit a synchronized
response to the driving cycle, consisting of a train of repetitive
action potentials as the bursting cells which rhythmically show self-
sustained membrane oscillations accompanied by burst of repetitive
impulses. The typical pattern of frequency change in response to
a sinusoidal stimulus is graphically illustrated in Fig. 4.

CONCLUDING COMMENTS

     It has been shown that the proposed model mimics the behaviour
of Helix pomatia neurons stimulated by rectangular and sinusoidal
currents. In particular both the model and the biological system

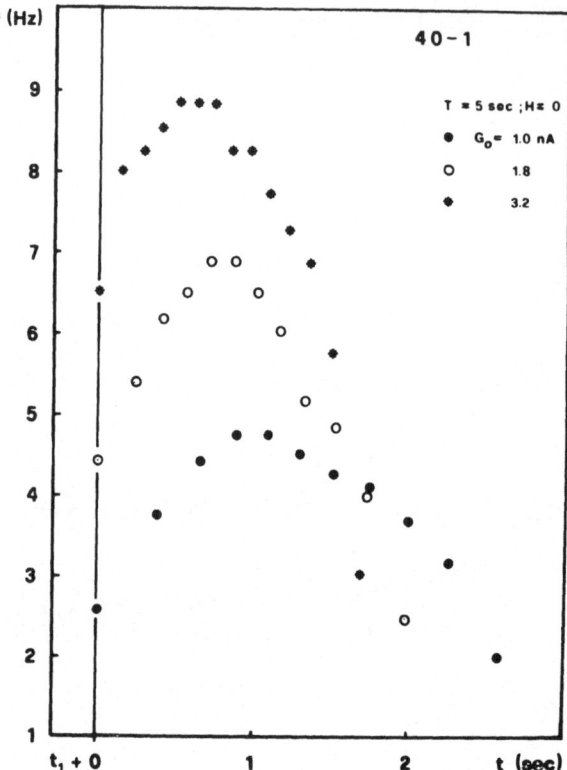

Fig. 4 - Frequency responses to sinusoidal current stimulations.
Neuron spontaneously discharging at rest; f_s=2.3 Hz.

adapt similarly in response to a constant depolarizing current if
G is linearly related to f(O). The relation between initial frequency
of firing and stimulus strength is therefore a very important one
and in the linearity range it's possible to estimate the values of
the parameters K, b', τ and $τ_f$.

In response to injected sinusoidal currents the Helix pomatia
neurons demonstrate a phase-locking behaviour consisting of a train
of repetitive action potentials whose pattern is frequency dependent.
At low values of current frequency these neurons exhibit patterns
qualitatively similar to those of normally bursting neurons.
Utilizing the parameters determined by constant current steps the
model describes these responses if G_o is in the linearity range of
(G, f(O)).

The author is indebted to Miss C. Canegallo and to Mr. F.Provenghi
for their help in numerical calculations.
This work was supported by a C.N.R. Grant.

REFERENCES

1. KIM, Y.I. and KIM M. (1977) "Modulation of repetitive action
    potentials in molluscan neurons stimulated with alternating
    currents". Brain Res. 122:361-366.
2. JUNGE, D. and STEPHENS, C.L. (1973) "Cyclic variation of
    potassium conductance in a burst-generating neurone in Aplysia.
    J. Physiol. (Lond.) 235:155-181.
3. GOLA, M. (1974) "Neurones à ondes-salves des mollusques -
    variation cyclique lentes des conductances ioniques". Pflügers
    Arch. 352:17-36.
4. ARVANITAKI, A., COSTA, H. and TAKEUCHI, H. (1965) "Fréquences
    des potentiels oscillatoires et pointes de la membrane somatique,
    par courants transmembranaires constants et linéairement
    croissants (neurone d'Helix pomatia)" Comptes Rend.Soc.Biol.
    (Paris) 159:697:703.
5. MAGURA, I.S and KRYSHTAL, O.A. (1970) "Effect of conditioning
    polarisation of soma of giant neurones of mollusks on the
    mechanism of action potential generation" Neurophysiology,
    2:91-99.
6. CONNOR, J.A. and STEVENS, C.F. (1971) "Prediction of repetitive
    firing behaviour from voltage clamp data on an isolated neurone
    soma". J. Physiol. (Lond.) 213:31-53.
7. MONTICELLI, G. (1977) "Quantitative aspects of repetitive firing
    of Helix aspersa neurons, caused by injected currents (long
    lasting)". Proc. XXVII Cong. IUPS 13:1549.
8. COLDING-JØRGENSEN, M. (1977) "Impulse dependent adaptation in
    Helix pomatia neurones: effect of the impulse on the firing
    pattern" Acta Physiol. Scand. 101:369-381.
9. SOKOLOVE, P.G. (1972) "Computer simulation of after-inhibition
    in crayfish slowly adapting stretch receptor neuron".
    Biophys. J. 12:1429-1451.

VISUAL ADAPTATION AND CHANGES IN THE TIME COURSE OF THE RECEPTOR POTENTIAL IN THE RETINULA CELLS OF THE HONEYBEE DRONE

C. Busso, M. Ferraro, and D. Lovisolo

Institute of General Physiology-University of Turin
C.so Raffaello 30, Turin, Italy
Institute of Animal Physiology -University of Turin
V.le Mattioli, 25 Turin, Italy

Studies of light adaptation in the visual cells of invertebrates have shown the occurrence of two different effects of the adaptating light on the sensitivity of the receptor cell: in some cases an increase in sensitivity called "facilitation" or "sensitivity facilitation" has been reported (Dahl, 1978), while in others the more usual decrease in sensitivity was observed, along with a shortening in the time scale of the response (Fuortes and Hodgkin, 1964; Baumann, 1968). Light adaptation in the honeybee drone photoreceptors has been investigated by Baumann (1968) and Bader et al. (1976) and has been shown that simple relation between sensitivity and time scale changes observed in Limulus by Fuortes and Hodgkin (1964) does not hold for the drone.

The aim of the experiments presented here was to analyze in more detail the changes in time scale and aplitude between the dark--adapted and light-adapted responses in the visual cells of the drone, and to try to describe quantitatively their dependence upon different conditions of adaptation.

METHODS

Responses from the retinula cells of the honeybee drone (Apis mellifera) were recorded intracellularly with glass microelectrodes filled with 3 M CK1 (DC resistance between 20 and 40 MΩ). The eye was prepared as described by Baumann (1968) and perfused with a solution of the following composition (mM): Na$^+$, 280; K$^+$, 3.2; Ca^{++}, 1.80; Cl$^-$, 287; Tris HCl, 9.0; glucose, 10 (Bader et al., 1976). The solution was equilibrated with a mixture of 95% O_2 and 5% CO_2. The pH was 7.3 ± 0.1. Light stimulation consisted of flashes

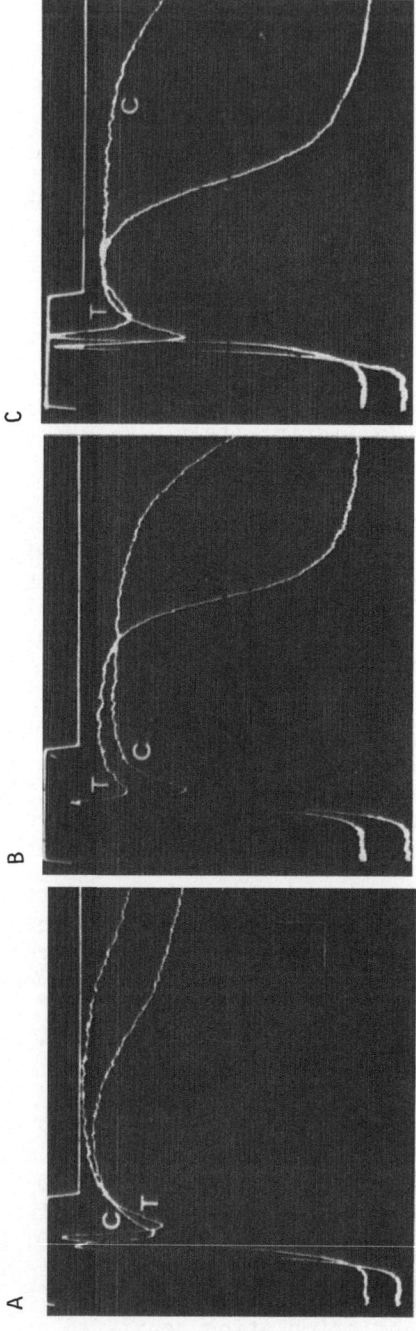

Fig. 1 – Transient phase of the response to a 20 msec control flash, C, and to a 20 msec test flash, T, presented respectively 0.5 (A), 3 (B), and 20 (C) sec after the end of a 20 sec adapting light. Calibration: 10 mV, 10 msec.

obtained from a Spindler & Hoyer halogen lamp; the light beam was interrupted by a shutter mounted on a Siemesn electromagnetic relay driven by a Romagnoli DIGIT 3T stimulator. The illuminance near the eye was 6000 lux.

RESULTS

The dark-adapted eye was stimulated with a train of flashes consisting of a first 20 msec "control" flash, followed by adapting lights of various durations, and then, after different dark intervals, by a "test" flash identical to the control one. Light intensity was the same for the three flashes. After each train, the eye was kept in the dark for about 10 min.

Fig. 1 shows the response to the control flash, C, compared with the response to the test one, T, presented to the eye respectively 0.5, 3, 20 sec after the end of a 20 sec adapting light. It can be seen in Fig. 1A that the slow transient component of the response to the control flash starts from a depolarized level, relative to the dark potential: this is due to the presence, at the end of the adapting light, of a depolarizing afterpotential (DAP, Baumann and Hadjilazaro, 1971). The level of depolarization reached by the peak of test response is slightly lower than that of the control one, and the rate of decay is slightly faster. In Fig. 1B, the response starts from an hyperpolarized level, reaches a level of depolarization higher than the control one, and appears to be much faster in the rate of decay. Fig. 1C shows that after a dark period of 20 sec the receptor potential still starts from an hyperpolarized level, the difference in the level of depolarization reached by the peak of the transient component is extinguished, but the slope of the decay phase is still faster in the test response as compared to the control one.

Experiments were performed with adapting lights of 3, 6, 10, 20 sec duration and with dark intervals of 0.5 to 60 sec. Fig. 2 shows the time course of the relative change, V_T/V_C, in the level of depolarization reached by the slow transient wave, measured as the difference between the peak of the transient and the dark potential. It can be seen that for all the values of the adapting light, for dark intervals between 1 and 5 sec there is an increase in the level of the polarization reached by the transient wave, resembling the effect called "facilitation" by some authors (Dahl, 1978), while for longer times the ratio between control and test values approaches unity. For 0.5 sec intervals the test response appears to be sometimes slightly "facilitated", some others "adapted". Although the increase in the level of the polarization could be considered quite small, it should be remembered that the responses shown are near the saturation level so that even small increases in the amplitude of the response should imply a significant increase in sensitivity.

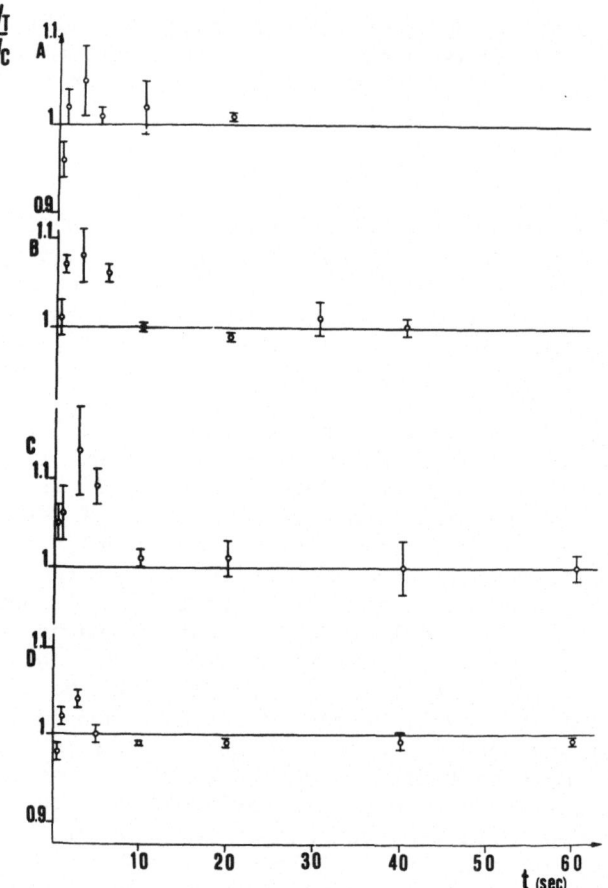

Fig. 2 - Relative change in the level of depolarization reached by
the peak of the transient component, V_T/V_C (V_T, level of
depolarization reached by the test response, V_C, level of
depolarization reached by the control one), as a function
of the dark interval. Adapting lights respectively of
3 sec (A), 6 sec (B), 10 sec (C), 20 sec (D).

Fig. 3 shows an experiment performed with flashes attenuated
by a factor of 0.06 as compared to the light intensity used throughout all other experiments; the adapting light was of 6 sec duration
and the dark interval was 1 sec. The change in the level of depolarization is more marked.

In order to describe the change in the time course of the test
response, two parameters were tested: time to peak and maximum slope
of the decay phase. Fig. 4 shows the time course of the relative
change of the time to peak, T_T/T_C, for an adapting light of 3 sec

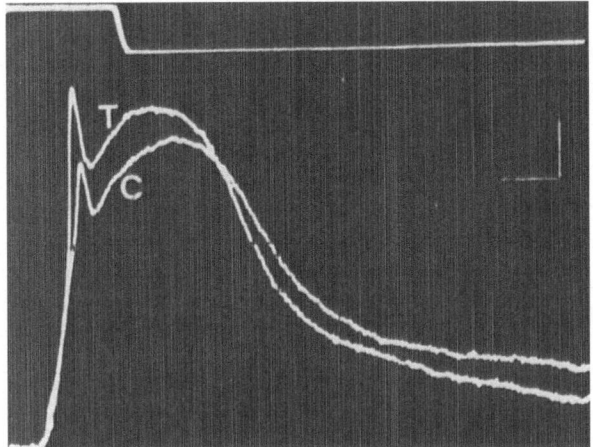

Fig. 3 - Response to a control flash, C, compared with the response to a test flash, T, presented 1 sec after the end of a 6 sec adapting light. Light intensity was attenuated by a factor of 0.06. Calibrations as in Fig. 1.

duration. (The values obtained for other adapting times were similar).

It can be seen that the time course does not resemble those in Fig. 2: while times to peak of the test response are always shorter, no straightforward time relationship could be identified. It could be observed that changes in time to peak depend not only on changes in the decay phase of the transient; but also on changes in its rate of rise. This is in contrast, as already pointed out by Baumann (1968), with the model proposed by Fuortes and Hodgkin (1964) for Limulus. So, an accurate analysis of the changes in the temporal parameters of the response should imply a study of the changes of both the rise and decay rates. The first one is not easy to describe, being masked by the spike, so we describe only the decay phase.

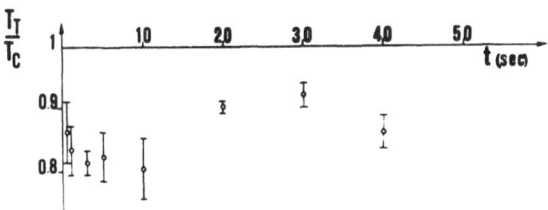

Fig. 4 - Relatively change in time to peak, T_T/T_C, as a function of the dark interval, for a 3 sec adapting light.

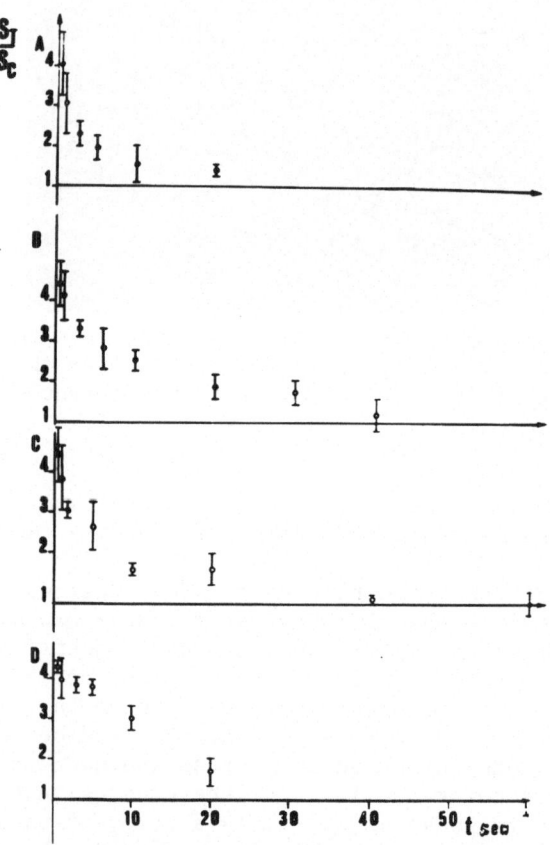

Fig. 5 — Relative change in the maximum slope of decay, S_T/S_C, as a
       function of the dark interval. Adapting lights of 3 (A),
       6 (B), 10 (C), and 20 (D) sec.

       Fig. 5 shows the time course of the relative change, S_T/S_C, in
the maximum slope of decay for the 4 adapting times. It can be
observed that the time courses of V_T/V_C and S_T/S_C differ: in Fig. 5
S_T/S_C decreases monotonically as the dark interval increases and
reaches unity in about 20-40 sec.

DISCUSSION

As pointed out by Bader et al. (1976), changes in duration of the transient phase of the receptor potential of the retinula cells of the drone are more marked and last longer than changes in the level of depolarization; in the experiments presented here, however, it is possible to observe that, while a decrease in the response amplitude is possibly present only for very short times after the end of the adapting lights, in the first few seconds of dark an increase in amplitude can be observed. This fact could be interpreted as a transient increase in sensitivity, but for this hypothesis to be confirmed more detailed experimental work is required.

BIBLIOGRAPHY

Bader,C.R., Baumann,F., and Bertrand,D., 1976, Role of Intracellular Calcium and Sodium in Light Adaptation in the Retina of the Honey Bee Drone (Apis mellifera,L.), J. Gen. Physiol., 67:475.
Baumann,F., 1968, Slow and Spike Potentials Recorded from Retinula-Cells of the Honeybee Drone in Response to Light, J. Gen. Physiol., 52:855.
Baumann,F., and Hadjilazaro,B., 1971, A Depolarizing aftereffect of Intense Light in the Drone Visual Receptor, Vision Res., 12-17.
Dahl,R.D., 1978, Facilitation in Arthropod Photoreceptors, J. Gen. Physiol., 71:221.
Fuortes,M.G.F., and Hodgkin,A.L., 1964, Changes in time scale and sensibility in the ommatidia of Limulus, J. Physiol., 172:239.

NONEXCITABLE MEMBRANES AND ARTIFICIAL SYSTEMS

PHOTOPIGMENT INDUCING PORES IN LIPID BILAYER MEMBRANES

F.Gambale, A.Gliozzi, I.M.Pepe, M.Robello and R.Rolandi

Laboratorio di Cibernetica e Biofisica, Camogli (Genova)
and
Istituto di Scienze Fisiche dell'Università di Genova,
Genova (Italy)

SUMMARY

A photopigment extracted from the honeybee compound eye was
incorporated into positively charged lipid bilayers, in the dark.
As a consequence cation selective pathways are formed, as deduced
from conductance and potential measurements. Light causes a further
increase of macroscopic conductance, associated with the formation
of ionic channels with individual conductance of about 80 pS.

INTRODUCTION

The molecular mechanism whereby a light signal is converted
into an electrical signal, in both vertebrate and invertebrate photoreceptor cells, is largely unknown. Light is absorbed by rhodopsin,
a chromophore-bearing membrane protein, and as a consequence a conformational change occurs in the protein. This is probably the
primary process leading to cellular excitation (Wald et al., 1963).

In the case of vertebrate animals a model has been proposed
according to which rhodopsin acts as an ionic channel which opens
upon illumination allowing the escape of calcium ions from the membrane of the sacs in rod outer segment (Hagins, 1972). These ions
would act as "transmitters" connecting the disk membrane, where
rhodopsin is located, to the plasma membrane. Calcium would close
the Na channels in the plasma membrane, thus producing an hyperpolarization capable to trigger the electrical signal in the optic
nerve.

Structural studies on hydrogen exchange (Englander and Englander,

1977) suggest that a large fraction of polypeptide chains of rho--
dopsin act to stabilize a channel of water penetrating into the
membrane. Such a channel should be quite large (10-15 Å in diameter)
and could be eventually formed by the cooperation of two or more
monomers. Recently Montal, Darszon and Trissl (1977) showed that
the conductance of bilayer membranes, obtained by the hydrophobic
apposition of two monolayers containing rhodopsin from bovine retinas,
increases upon illumination. The observed results are interpreted
in terms of the formation of a transmembrane channel, about 10 Å
in diameter, which closes on increasing potential. The same conclusion is reached by analyzing permeability changes to ions and
non electrolytes induced by light on rhodopsin-phospholipid vesicles
(Darszon, Montal and Zarco, 1977).

In the case of invertebrate photoreceptors the situation seems
more complex and there is not a clear evidence in favour of a particular model. However, since the photochemistry in both vertebrate
and invertebrate animals is quite similar, it seems reasonable to
look for an anlogous mechanism of phototransduction.

In the past two years we were able to incorporate into lipid
bilayers a light sensitive pigment extracted from the honeybee compound eye (Gambale et al., 1977). In this paper we report further
measurements indicating that cation selective pathways are formed
in the membrane upon incorporation of the photopigment in the dark.
The kinetics of incorporation suggests that pores should be formed
by an aggregate of two molecules. This implies a quadratic dependence of conductance on the protein concentration, which has been
experimentally observed. Light causes an increase of the macroscopic
conductance and the formation of ionic channels with conductances
of about 80 pS.

MATERIALS AND METHODS

Extraction of the photopigment

Workers honeybee heads were homogenized in 0.03% Tris (hydro-
xymethyl) aminomethane and 0.14% glycine (pH=8.4) gel buffer at 0°C
in dim red light. The homogenate was incubated for 15 minutes at
20°C with 2μCi/ml of tritiated vitamin A (all-trans retinol 15^{3}H ,
2.4μCi/mM, 0.12 mg/ml; New England Nuclear). After centrifugation
at 20,000 g for 30 mins, the supernatant was allowed to run in a
polyacrylamide gel for preparative electrophoresis. The standard
gel (7% acrylamide) was prepared following the method of Davis (1964)
and Ornstein (1964) in a tube 10 cm long and 3.5 cm of internal diameter. The radioactive protein was eluted under continuous current
(20 mA at a voltage of 400 V). In a typical extraction about 0.25 mg
of protein were obtained starting from some 400 honeybee heads.
The nature of the chromophore of the extracted protein was investigated by thin-layer-chromatography and the results indicate to be

retinaldehyde (Pepe and Cugnoli, submitted). Spectrophotometric measurements showed that the extracted pigment has an absorbance maximum at 440 nm, which bleaches in the light yielding a photo-product with λ max of about 365 nm. (Pepe, Schwemer and Paulsen, in preparation). The Schiff-base linkage between retinal and protein becomes exposed to the external medium after bleaching, following probably a conformational change of the protein. Hydroxylamine reacts with the pigment detaching the chromophore and giving rise to the formation of retinal-oxine. Hypothesis on the functional role of this photopigment, which is very similar to that isolated from the honeybee drone retina (Pepe, Perrelet and Baumann, 1976) is still an open question. It might be involved in the visual cycle as a precursor of the visual pigment of the compound eye.

Incorporation of the protein

Two different procedures were used. In the experiments of Figs. 1 and 2 the protein was added in the aqueous phase on both sides of the membrane to reach a concentration of the order of $5 \cdot 10^{-5}$ mg/ml. In all other experiments reported in this work a small volume (generally 20 μl) of a 50 μg/ml, buffered solution of the protein was added to the lipid mixture in 1:2 or 1:4 volume ratio. The resulting emulsion was stirred and then incubated in the dark at 4°C for a time varying from 0.5 to 4 hours in different experiments. During the process the lipid solution phase became radio-active while the aqueous phase decreased its radioactivity suggesting that the tritiated protein was incorporated into the organic phase. The possibility that only retinal is incorporated during this process is ruled out by control experiments described below. An independent evidence is provided by freeze-fracture studies of liposomes from the same lipid mixture which showed particulated fracture surfaces indicating that the protein is indeed embedded in the lipid matrix (I.M.Pepe and A.Perrelet, unpublished).

Membrane formation and electrical measurements

Planar lipid bilayers were obtained from mixtures, at various percentage, of decane solutions of egg lecithin (50 mg/ml, BDH) and oleylamine (17 mg/ml, Koch-Light, England), and formed on a 1.5 mm hole between two teflon chambers each of 20 ml of volume, filled with salt solutions.

Soy-bean lecithin and oxidized cholesterol membranes were also used in some experiments. The conductance was measured with two Ag-AgCl electrodes connected to an electrical circuit similar to that previously described (Ciani et al., 1975). Current fluctuations measurements were performed in voltage-clamp. The value of the current was inferred from the potential drop on a resistence R_e, in series with the membrane. In most experiments R_e was smaller than 10^7 and the time constant, τ, of the circuit was less than 10 ms.

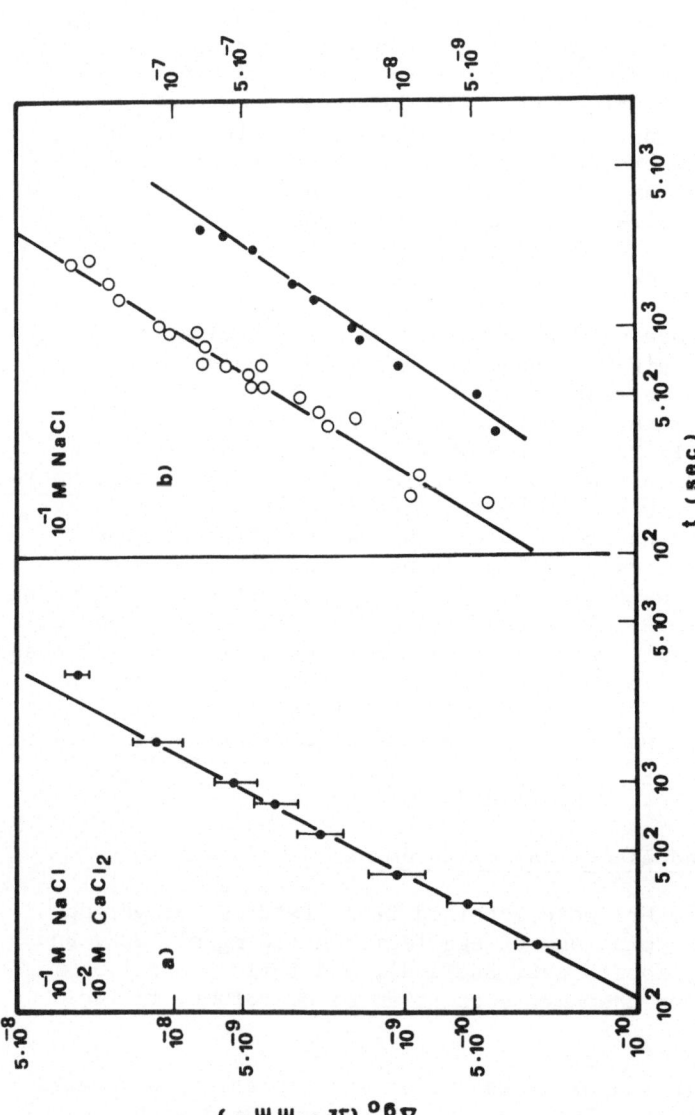

Fig. 1 – Time dependence of conductance variations induced by the photopigment added to the external solution.

a) Bilayers comprised of 20% oleylamine and 80% phosphatidyl choline. Ionic solution 0.1M NaCl, 10 mM CaCl₂ buffered at pH 7 with 10 mM Tris-HCl. Each point is the mean value of seven experiments. The bar is the standard deviation.

b) Bilayers comprised of various lipid mixtures; upper curve: 30% oleylamine + 70% phosphatidyl choline; lower curve: 10% oleylamine + 90% phosphatidyl choline. Each point is the mean value of three experiments. Ionic solution 0.1M NaCl buffered at pH 7 with Tris-HCl 10 mM. Experiments performed in the dark.

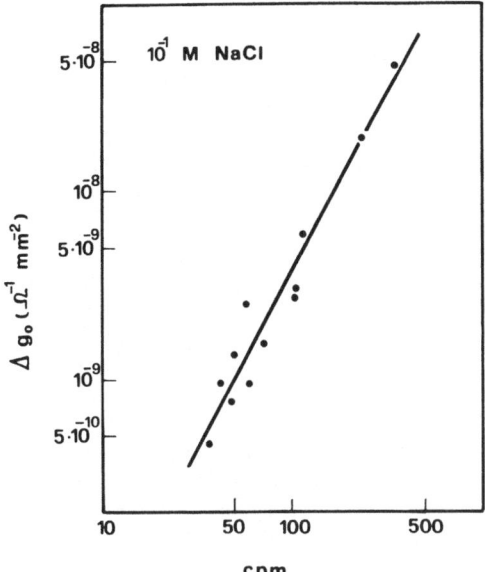

Fig. 2 - Concentration dependence of conductance in photopigment-
         lipid bilayers. The increase in steady state conductance
         induced by photopigment incorporation, Δg_o (where $g_o(o)$=
         300 pS is the mean value of the bare membrane conductance),
         is plotted as a function of the radioactivity (cpm.) of
         10 µl of the photopigment-lipid solution employed to form
         the membrane. Each point corresponds to a different mem-
         brane. Different values of photopigment concentration were
         obtained with various (from 0.5 to 3 hrs) incubation times.
         Ionic solutions 0.1 M NaCl buffered at pH 7 with 10 mM
         tris HCl. Lipid mixture 20% oleylamine and 80% phospha-
         tidyl choline. Experiments in the dark.

A voltage amplifier was used to record the signal either on a dual
beam storage oscilloscope Tektronix 5103N, or by an ink-writing
oscillographic recorder Hp 7402 or by a potentiometric recorder
Hp 680M. The amplifiers used were type 42J from Analog Devices
(Norwood, Mass.).

The teflon cell was inserted in a black metal box, with a hole for illumination and mounted on a shockproof table. The light source was a 100 mW white light pipe or a commercial xenon electronic flash (1000 W, 1 msec. duration). Light was focused on the membrane through an optical system which cut wavelengths below 400 nm. All experiments "in the dark" were performed in dim red light ($\lambda > 650$ nm). The diameter of the membrane was measured during the experiments with the light pipe covered by cut-off gelatin filter (Kodak n.70). During light stimulation the latter was replaced by an infrared absorbing filter.

RESULTS

a) Membrane conductance in the dark

When added to both sides of a pure lecithin bilayer, the photo-pigment induced little or negligible conductance changes. Conversely with lecithin-oleylamine bilayers a large increase in the electrical conductance was obtained. Taking as t=0 the time at which the prote-in was added, the specific "zero current" conductance g_o(t=90 mins) was a factor $5 \cdot 10^2$ larger than $g_o(0)$, which was of the order of $3 \cdot 10^{-8}$ S/cm^2. The time course of the specific conductance variation, $\Delta g_o = g_o(t) - g_o(0)$, induced by the addition of an equal amount of photo-pigment on both sides of the membrane is illustrated in Fig. 1a. It is interesting to observe that in the logarithmic scale, Δg_o, is a straight-line with a slope, m, of about 2 (the best fitting of the experimental points yields m=1.9). Values of m in the range 1.5-2 have been found in most experiments performed in different conditions. As discussed later such kinetics of incorporation suggests that Δg_o is associated with ionic pathways arising from the formation in the dark of dimeric photopigment aggregates.

The photopigment incorporation is greatly enhanced by the electrostatic interaction with the membrane, as suggested by the increase in Δg_o upon increasing the membrane surface charge density as illustrated in Fig. 1b. Membranes comprised of different values of positive charge density were obtained by mixing, at the indicated percentages, the positively charged lipid oleylamine with the zwitterionic lipid phosphatidyl choline. Increase of charge density induced a shift along the positive ordinates of the log-log plot of g_o vs time, whereas the slope remained unchanged.

In control experiments, the addition of retinal (dissolved in a small volume of ethanol) or bovine serum albumin to the aqueous bath did not produce significant changes in conductance.

b) Concentration dependence of conductance in the dark

In order to test the hypothesis of a dimer formation, which implies a quadratic dependence of conductance on protein concentration, we have performed the following experiment.

Small amounts of lipid solution were mixed with the protein, as described in the Methods. After incubation, the lipid solution was divided into two samples. One of them was employed to form the membrane (whose conductance was found to reach a steady value about 10 mins. after the bilayer formation); the other part was mixed with a liquid scintillation solution in order to measure the tritium activity. In this way the conductance as well as the relative protein concentration into the lipid phase could be measured. Fig. 2 shows the results. The conductance variation Δg_0 (where $g_0(0)=300$ pS.mm^{-2} is the mean conductance of the bare membrane) is plotted as a function of the number of counts per minute (corrected for the background activity) of 10 µl of the organic solution. The best fitting of the points, in a double logarithmic plot, is a straight line with a slope of about 2 (1.9) in agreement with the expected quadratic dependence of conductance on protein concentration.

To rule out the possibility that only tritiated retinal is incorporated during the incubation process, control experiments were performed dissolving retinal, in the same or higher range of concentration, in the organic phase. The conductance variations of the retinal-lipid bilayers are given in Fig. 3. Notice that a 10^3 fold

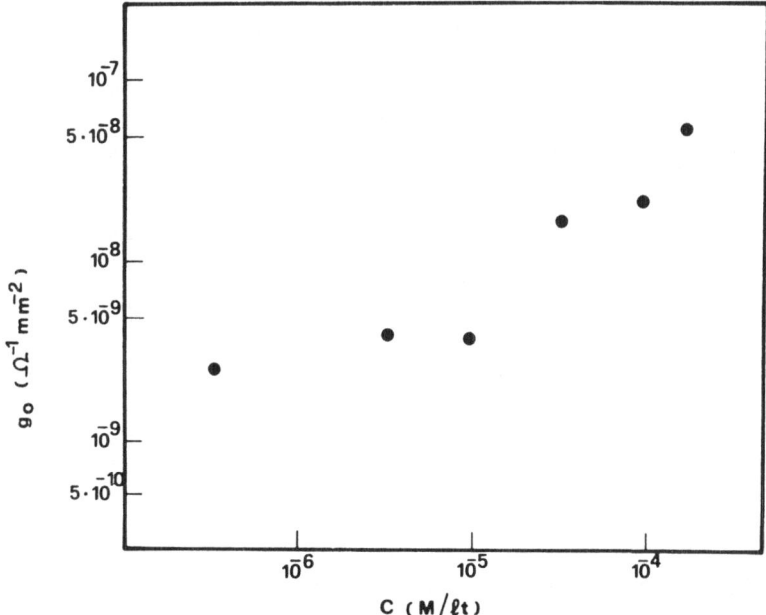

Fig. 3 - Conductance variations are plotted as a function of retinal concentration in the lipid mixture. Other conditions as in Fig. 2.

increase in retinal concentration is needed to increase the conductance by a factor of about 20, while in the case of photopigment
only a four-fold increase in concentration is required to get a
similar conductance variation.

c) <u>Membrane conductance in the light</u>

When membranes were formed in the dark with the photopigment
directly added to the lipid (see Methods), an increase in conductance
was observed for about 10 minutes. This is shown in the first part
of Fig. 4, where conductance changes are plotted as a function of
the time after the membrane formation.

Once the steady state was reached, the photopigment-lipid bilayer was illuminated with a continuous white light of approximately
150 $\mu W/mm^2$, at the time indicated by the arrows. A further increase
in conductance was then observed as shown in Fig. 4.

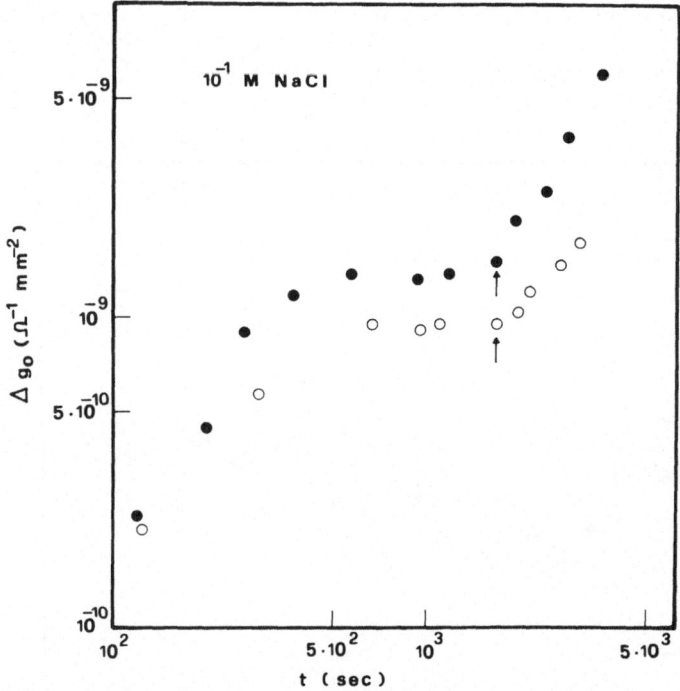

Fig. 4 – Conductance changes in the dark and in the light in photo-
          pigment-lipid bilayers after membrane formation. The arrow
          indicates the point at which light was switched on. Ionic
          solution 0.1 M NaCl buffered at pH 7 with 10 mM tris HCl.
          Lipid mixture 20% oleylamine and 80% phosphatidyl choline.

A time lag was observed between the onset of the light and the
first appreciable conductance variation; this time varied from about
10 to 150 sec. in different experiments and decreased the higher
were the values of membrane conductance in the dark.

Control experiments performed with lipid mixtures containing
free retinal or protein with the chromophore detached by hydroxylamine at about the same concentrations as those used in experiments
with the photopigment, showed no light-induced conductance variation.

d) Current fluctuations

Voltage-clamp currents were recorded from lipid bilayers in
which a small amount of photopigment was previously incorporated
(see Methods). In these measurements the membrane was formed in
red light and successively illuminated either with a continuous
white light or with 1-4 flashes of light of a Xenon lamp (1 msec
duration, 1 W/mm^2). In 80% of the measurements light elicited a
"response", after a lag time varying between some seconds and 150
seconds with a mean value of 70 seconds. Two types of responses
were observed:

i) Noisy current fluctuations, corresponding to conductance jumps
of about 200 pS, combined with an increase of the mean membrane
conductance.
ii) Square-shaped current fluctuations between two constant levels
and multiple step transitions, like those shown in Fig. 5, combined
with a very small increase of mean membrane conductance.

These fluctuations were persistent under continuous light, while
they lasted only for some minutes after a flash. This behaviour
was observed in 50% of the membranes displaying fluctuations.
The square-shaped current transitions may be interpreted as associated with the opening and closing of ionic channels. Their amplitude
varies between 50 to 500 pS, with a peak around 80 pS.

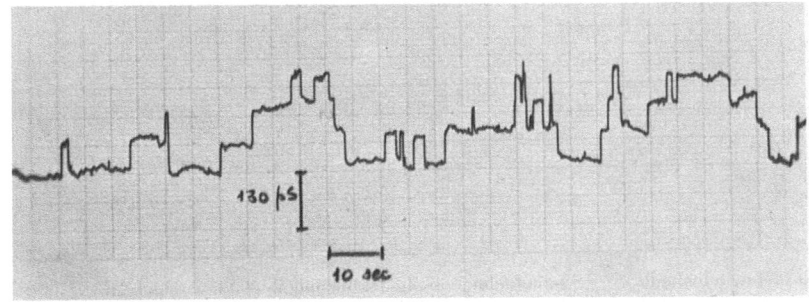

Fig. 5 - Current fluctuations in voltage clamped membranes.
         Fluctuations induced by continuous light (150 μW/mm^2) in
         photopigment-lipid bilayer clamped at 40 mV.

Control experiments were also performed using the protein with the chromophore detached by hydroxylamine, inserted into the lipid bilayer with the same incubation procedure employed for the photo-pigment. Light-independent current transitions between two constant levels, corresponding to conductance jumps of about 500 pS were exhibited by these membranes.

We have not found a correlation between the latency time, the i) or ii) response and the membrane conductance (and consequently the photopigment concentration) before illumination. Therefore it is not easy to give a simple interpretation to the above phenomena. Further experiments with monochromatic light using noise analysis are required.

DISCUSSION

Evidence of channel formation

In a previous work (Gambale et al., 1977) it was suggested that the observed conductance changes in the dark, induced by the photopigment incorporation, might be due to the formation of trans-membrane pores in the lipid bilayer. The time course of conductance variation, shown in Fig. 1a,b, provides the opportunity to propose a simple heuristic model on channel formation in the dark (see Appendix A). According to this model we assume that channels are formed by an aggregation of two photopigment molecules. The equation which describes the number of conducting dimers yields to an ex-pression of the macroscopic conductance variation, with a time dependency, similar to that obtained by the best fitting of the experimental data, which are shown to vary as t^2. Moreover, the conductance, at a given time, should depend quadratically on the aqueous concentration of the protein. In the case the protein is directly incorporated into the lipid, it should depend on the protein concentration in the lipid bulk phase. Experiments of Fig. 3 show that, in the steady state, the conductance varies indeed as the square of concentration, in agreement with the predictions of the model.

A further point in favour of such a model is provided by pre-liminary electron microscopy freeze-fracture studies, which show that lipid vesicles, comprised of the same lipid mixture and incu-bated with the photopigment, contain aggregates, embedded in the lipid matrix, whose linear dimensions are of the order of 100 Å. However, as already stated, current fluctuations in the dark did not show square-shaped current trnaistions, providing a direct measure of channel conductance. Conversely, square-shaped fluctu-ations were observed in the light. The observed transitions (cf. Fig. 5) are similar in magnitude of those induced by molecules (like e.g. alamethicin, hemocyanin) which are known to form dynamic channels in the membrane (Ehrenstein and Lecar, 1977). Such values

pertain to a rate of ionic permeation consistent with the formation
of a channel, but incompatible with a carrier mechanism, like that
of valynomicin, where the rate of ionic translocation is 10^3 times
smaller (Laüger, 1972; Gambale et al., 1973). Potential measurements
reported elsewhere (Gambale et al., 1979) indicate that these
channels are cation selective.

Light-induced conductance variations

The results shown in Fig. 4 indicate that the photopigment is
able to induce increases in membrane conductance upon illumination,
once the steady state is reached in the dark. The mechanism underlying this phenomenon seems to be related to the ability of the
system to form dynamic channels[1]. Such a behavior offers ground
for speculations on possible mechanisms of photoinduced electrical
excitation in photoreceptors. The discrepancy arises from the
comparison of the time lag between the onset of light and the ap-
pearance of the current fluctuations and the delay of the physiological response in photoreceptors.

A similar delay was also found by Montal et al. (1977) with
bovine rhodopsin incorporated in planar lipid bilayers. A more
detailed discussion on the possible occurrence of a diffusion-limited
process of molecular aggregation is provided by these authors.

One may also suggest that the latency represente not only a
diffusion-limited formation of aggregates but perhaps a rate-limited
acquisition of the protein of the "conformation" which allows for
the pores formation.

We have not treated, in this work, the wave-length dependency
of the phenomenon, which seems to be of crucial importance. The
photopigment has an absorbance maximum at 440 nm, in an aqueous
medium. In our artificial system different conductance changes may

---

[1] Although in the present experimental conditions retinal is not
detached from the protein, a minor detachment of retinal from some
protein molecules might occur. In this case free retinal could block
the amino groups of the oleylamine lipid making the membrane less
positively charged (Bonting and Bangham, 1967). This screening effect could be responsible of the observed conductance increase upon
illumination. Control experiment with membranes made with oxidized
cholesterol (containing no amino groups) showed a behavior similar
to that in Fig. 4. This result, together with the finding that light
induces opening of channels in the membrane, rules out the possibility
that the observed conductance increase may be merely due to the detaching of free retinal from the protein. Nevertheless a contribution
of this phenomenon to the overall drift of the membrane conductance
towards higher values during illumination cannot be excluded.

be obtained illuminating the membrane with various wavelengths.
A refinement of these preliminary measurements, together with the
spectral absorbance of the photopigment in the lipidic medium, will
enlighten on this important point.

Final remarks

The above findings illustrate a mechanism of ion translocation
through a lipid bilayer, mediated by a photopigment. Possibly the
protein opens cation selective channels also in the dark, as suggested by conductance variations as a function of t^2 and of c^2. After
illumination at low photopigment concentration and in voltage-clamp
conditions single channel fluctuations may be recorded. The channel
structure is still matter of study. If a progressive increase of
channel conductance, and therefore of the equivalent pore radius,
from dark, to illuminated photopigment, to protein with the chromophore detached by hydroxylamine will be established, an hypothesis
on the importance of retinaldehyde in the channel structure will
be consequent.

ACKNOWLEDGEMENTS

We thank Dr. F.Conti for a critical reading of a first draft
of the manuscript. The technical help of C.Cugnoli is also greatly
acknowledged. This work was partially supported by grant N.3.709.76
from the Swiss National Science Foundation.

APPENDIX A

Channel formation

During the incorporation process, the rate of change of photopigment concentration inside the membrane, C_i, may be written as:

$$\frac{dC_i}{dt} = K_1 C_e - K_2 C_i \tag{1}$$

where K_1 and K_2 are rate constants of the incorporation reaction,
and C_e the interfacial concentration in the external solution.
We may assume that

$$K_1 C_e \gg K_2 C_i , \tag{2}$$

which means that the number of molecules entering into the membrane
per unit time is much greater than that leaving the membrane. This
fact is certainly true in the first stage of the reaction. Moreover
we may assume $C_e \simeq$ cost, owing to the great ratio between the external volume and the membrane volume.
Therefore, under these conditions, we may write

$$C_i = K_1 \, C_e \, t. \tag{3}$$

We shall assume that the pore formation is due to the reaction of two monomers, M, to for a dimer, D:

$$M + M \rightleftharpoons D. \tag{4}$$

Since the process of pores formation is much faster than the incorporation of new molecules, we may assume that the reaction (4) is a "quasi equilibrium" state and therefore, from the mass action low, we may write:

$$C_D = K \, C_i^2 , \tag{5}$$

where K is the equilibrium constant of the reaction (4), and C_D is the dimer concentration. The number of conducting dimers, or channels, per unit area at a given time t, n(t), will be given therefore by:

$$n(t) = b \, t^2, \tag{6}$$

where b is a constant defined as:

$$b = K \, K_1^2 \, C_c^2 \, N, \tag{7}$$

and N is the Avogadro's number.
The specific conductance variation, g_0, may be written as:

$$\Delta g_0 = n(t)\lambda, \tag{8}$$

where λ is the single channel conductance. Inserting eq.(6) into eq.(8) one obtains the empirical law which fits the data of Fig. 1 i.e.:

$$\Delta g_0 = a \, t^2 \tag{9}$$

Eq.(8) allows to evaluate the mean number of open pores at a certain time t, $\bar{n}(t)$. Referring to data of Fig. 1a we have, after 30 mins: $\Delta g_0 = 17$ nS/mm^2, taking e.g. $\lambda \approx 80$ pS we have $\bar{n}(30') \approx 200$ pores/mm^2.

REFERENCES

Bonting, S.L. and Bangham, A.D.   On the Biochemical Mechanism of the
    visual process. Exptl. Eye Res. (1967) 6, 400-413.
Ciani, S., Gambale, F., Gliozzi, A. and Rolandi, R., 1975. Effects
    of unstirred layers on the steady-state zero-current conductance
    of bilayer membranes mediated by neutral carriers of ions.
    J. Membrane Biol. 24:1.
Ciani, S., Laprade, R., Eisenman, G., Szabo, G., 1973. Theory for
    carrier-mediated zero-current conductance of bilayers extended
    to allow for nonequilibrium of interfacial reactions, spatially
    dependent mobilities and barrier shape. J. Membrane Biol. 11:255.
Darszon, A., Montal, M. and Zarco, J., 1977. Light increases the
    ion and non-electrolyte permeability of rhodopsin-phospholipid
    vesicles. Biochem.Biophys.Res.Comm. 76: 820.
Davis, B.J., 1964. Disc electrophoresis II. Method and application
    to human serum proteins. Ann. N.Y. Acad. Sci. 121:404.
Ehrenstein, G. and Lecar, H., 1977. Electrically gated ionic
    channels in lipid bilayers. Quart. Rev. Biophys. 10:1.
Englander, J.J. and Englander, S.W., 1977. Comparison of bacterial
    and animal rhodopsin by hydrogen exchange studies. Nature 265:
    658.
Fettiplace, R., Andrews, D.M. and Haydon, D.A., 1971. The thickness,
    composition and structure of some lipid bilayers and natural
    membranes. J. Membrane Biol. 5:277.
Gambale, F., Gliozzi, A., Pepe, I.M., Robello M. and Rolandi, R.,
    1977. Incorporation into lipid bilayer membranes of a photo-
    sensitive pigment from the honeybee compound eye. Biochim.
    Bioph. Acta 467:103.
Gambale, F., Gliozzi, A., Robello, R., 1973. Determination of rate
    constants in carrier-mediated diffusion through lipid bilayers.
    Biochim. Biophys. Acta 330:325.
Gambale, F., Gliozzi, A., Pepe, I.M., Robello, M. and Rolandi, R.,
    1979. Transport properties induced in lipid bilayer membranes
    by an Insect Photopigment. La Gazzetta Chimica Italiana, 8.
Goldsmith, T.H., 1958. The visual system of the honeybee. Proc.
    Natl. Acad. Sci. USA 44:123.
Hagins, W.A., 1972. The visual process; excitatory mechanisms in
    the primary receptor cells. Ann. Rev. Biophys. Bioeng. 1:131.
Laüger, P., 1972. Carrier mediated ion transport. Science 178:24.
McLaughlin, S.G.A., Szabo, G., Eisenman, G., 1971. Divalent ions
    and the surface potential of charged phospholipid membranes.
    J. Gen. Physiol. 58: 667.
Montal, M., Darszon, A. and Trissl, H.W., 1977. Transmembrane
    channel formation in rhodopsin-containing bilayer membranes.
    Nature 267:221.
Ornstein, L., 1964. Disc electrophoresis I. Background and theory.
    Ann. N.Y. Acad. Sci. 121:321.
Pepe, I.M. and Cugnoli, C. Photopigment from the honeybee compound
    eye (submitted for publication)

Pepe, I.M., Perrelet, A. and Baumann, F., 1976. Isolation by poly-
    acrylamide gel electrophoresis of a light-sensitive vitamin A-
    -protein complex from the retina of the honeybee drone.
    Vision Res. 16:905.
Wald, G., Brown, P.K., Gibbon, I.R., 1963. The problem of visual
    excitation, J. Opt. Soc. Am. 53:20.
Yau, K.W., Lamb, T.D., Baylor, D.A. Light induced fluctuations in
    membrane current of single toad rod outer segments. Nature
    269:78.

IONIC TRANSPORT PROPERTIES OF THE HEMOCYANIN CHANNEL

R. Antolini and G. Menestrina

Dipartimento di Fisica, Università di Trento

38050 Povo (Trento) ITALY

INTRODUCTION

Hemocyanins are large copper proteins, with molecular weights ranging from 10^5 to 10^7, which occur freely dissolved in the hemolymph of a wide variety of invertebrate species, where they play the role of oxygen carriers.

Electron micrographs and ultracentrifugal analyses show that the shape of gastropod hemocyanins is roughly that of a hollow cylinder, whose height may vary depending on solution conditions. In a diluted salt solution, near to the neutral pH, these proteins are in an aggregation state known as 100S, which means that their molecular weight is about eight million, the external diameter of the cylinder is 30 nm and its height is 35 nm (1).

Several works (2,3) have described the ability of an hemocyanin, called Keyhole Limpet Hemocyanin (KLH), to interact with black lipid membranes (BLM) giving rise to the formation of facilitated ionic pathways through the bilayer.

In our laboratory we could observe these channels working with hemocyanins extracted from the hemolymph of four different molluscs: Megatura Crenulata, Paludina Vivipara, Busycon Canalicolatum and Helix Pomatia.

We describe here the ionic conductivity characteristics of the two hemocyanin channels that we have studied most, namely Megatura Crenulata hemocyanin (MCh) channel and Paludina Vivipara hemocyanin (PVh) channel. All experiments described were performed on BLM comprised of oxidized cholesterol that was prepared following the

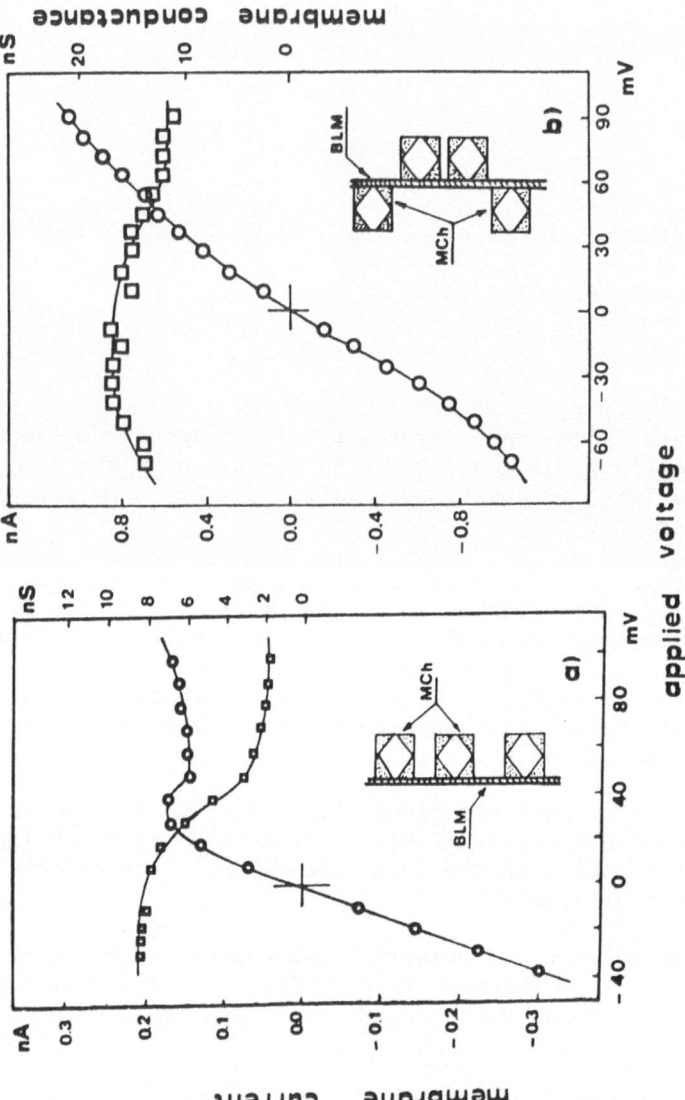

Fig. 1 – Steady state current voltage and conductance-voltage characteristics showing the difference between one-sided and two-sided action of MCh on lipid bilayer membranes:
a) the two curves are strongly non-linear in the positive part, displaying negative
   dynamic resistence and sigmoid shaped conductance typical of excitability
b) the two curves are still non-linear but now quite symmetrical indicating a random
   distribution of MCh molecules on the two sides of the membrane.

procedure of Tien (4). The BLM were formed in a teflon cup with the usual technique (5). Megatura Crenulata hemocyanin was purchased by Calbiochem, while Paludina Vivipara hemocyanin was a gift of prof. B. Salvato (University of Padova).

MCh CHANNEL.

The addition of small amounts (1 µg/ml) of MCh to one of the two electrolytic solutions, after the complete formation of the membrane, yields a step-wise increase of the current corresponding to constant increments of the membrane conductance. Each increment may be interpreted as due to a stable interaction between the molecule of the protein and the lipid bilayer, which gives rise to the formation of a channel.

After this interaction the I-V steady state characteristic of the membrane becomes asymmetrical and strongly non-linear, showing a region with negative dynamic resistance. The corresponding conductance versus voltage curve shows the sigmoid shape which is typical of excitable tissues.

The asymmetry of I-V and G-V curves (fig. 1a) has to be related to an asymmetrical distribution of the protein between the two sides of the membrane, that is MCh interacts with the bilayer opening a channel but cannot pass throughout it. In this way all the molecules maintain the same orientation: one face immersed into the hydrophobic region, where they suffer almost the whole potential drop, and one face in the hemocyanin containing water solution.

Actually experiments performed in symmetrical conditions, obtained for example adding the protein to both sides of the membrane, give as a result I-V and G-V characteristics which are still non--linear but now quite symmetrical (fig. 1b) as the effects of molecules with both possible orientations are now superimposed.

In order to study the nature of the voltage dependence of the MCh doped membrane conductance, we made experiments with few channels. In such conditions applying to the membrane potential steps of different heights and sign, one can observe that with positive potentials (referred to the protein containing compartment) all the channels tend to close, passing through some discrete conductance levels whose occupation probabilities are voltage dependent. With negative potentials, on the contrary, all the channels stay in the upper conductance level. The relaxation time for these discrete transitions is in the range of 10^2 sec (fig. 2). Besides each discrete level displays a non-linear I-V characteristic due to a continuous variation of the conductance which has a relaxation time of 10^{-4} sec (3). Both these phenomena have relaxation times which are much longer than the time of ionic redistribution in the channel (about 10^{-9} sec), and therefore they are to be related to a configurational transition

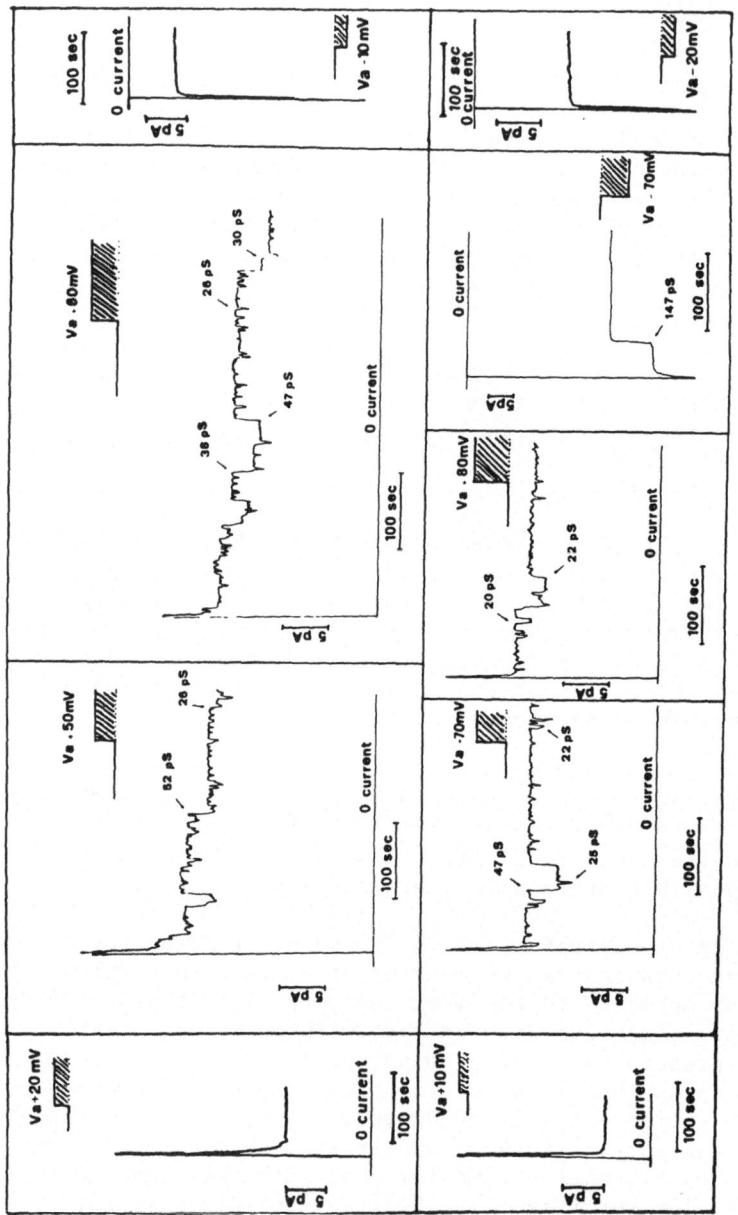

Fig. 2 – Membrane current relaxation after a potential step. One can notice that for low positive potentials (+10, +20 mV) there is a linear increase of the current; for high positive potentials (>30 mV) there is a relaxation towards lower levels of conductance which takes place by fluctuations through discrete levels (some of the conductance changes are indicated in the figure). For negative potentials there is always a linear increase of the current, without any fluctuation. At -70 mV one can see a great step-decrease of current which corresponds to the closure of one channel.

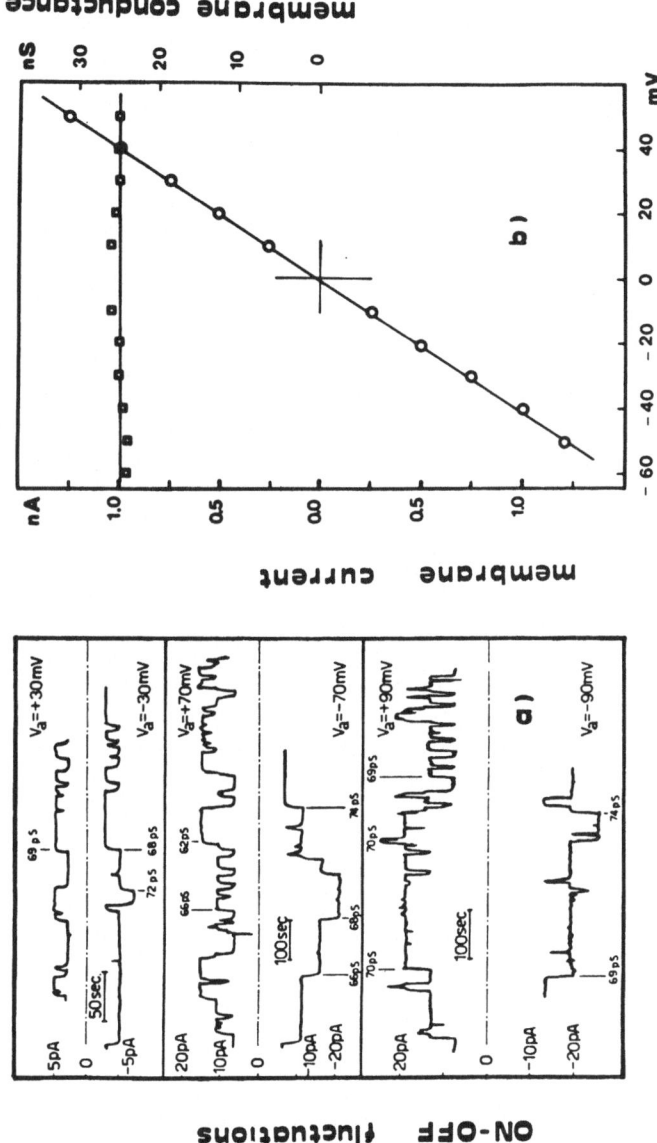

Fig. 3 — a) Current fluctuations in oxidized cholesterol membranes modified with PVh. One can observe the current jumps, followed by ON-OFF fluctuations, resulting from the formation of some channels, at different membrane potentials. Applied voltage and changes in membrane conductance are reported. (The figure is taken from (7)).
b) Steady state current voltage and conductance-voltage curves of PVh doped bilayer membranes. One can notice that I-V characteristic is linear and therefore G-V is a constant.

of the MCh channel itself, that may include different factors:
lipid protein interaction, ternary and quaternary structure changes
of the protein, position of the molecule in the membrane, etc.

I-V characteristics obtained with one single channel and with
many channels membranes, about 2.10^2 channels present at a time,
agree fairly well, implying that all channels act independently one
from the other.

In order to understand something more on the channel structure
we studied its ionic selectivity and could establish that the channel
is quite impermeable to anions while cations can pass through it with
permeabilities directly proportional to the respective mobilities
in free solution (6). This fact indicated that diffusion of cations
in the channel obeys to the same mechanism which occurs in the aqueou
solution and therefore that the channel is completely hydrated.
Taking into account the strong discrimination between anions and
cations one can think that the channel is a cylindrical pore filled
with water and with a negative charge distribution on the inner wall,
a picture that agrees with the known informations on the structure
of hemocyanin in 100 S aggregation state, and with the fact that at
a neutral pH the molecule bears a net negative charge. This charge
distribution creates at the entrance of the channel a potential
barrier which prevents the entry of anions while favours that of
cations.

The existence of such a potential barrier at the edge of the
channel is probably the main cause of a saturation effect which can
be observed studying the single channel conductance versus electrolyte concentration curve. Actually we found that instead of linear
dependence the channel conductance is proportional to the square
root of the solution conductance (fig. 4a).

PVh CHANNEL.

The interaction of PVh, with lipid bilayers is less stable than
that of MCh; the channel formation is reversible and each channel
disappears after a period of few minutes. Soon after the formation
the channel fluctuates between two conductance levels, one of high
conductance, we call it ON-state, and one with almost null conductanc
OFF-state.

Experiments performed in the presence of few channels (typicall
3 or 4) with varying applied potentials, between −100 mV and + 100 m
have shown that the membrane conductance fluctuates between the diff
ent levels following quite well a pure binomial. This implies that
all the channels act independently and that each channel has a probability 1/2 to be open. Single channel conductance is also independent from membrane potential, as one can see in fig. 3a (7).
All these facts allow us to foresee that many channel membranes

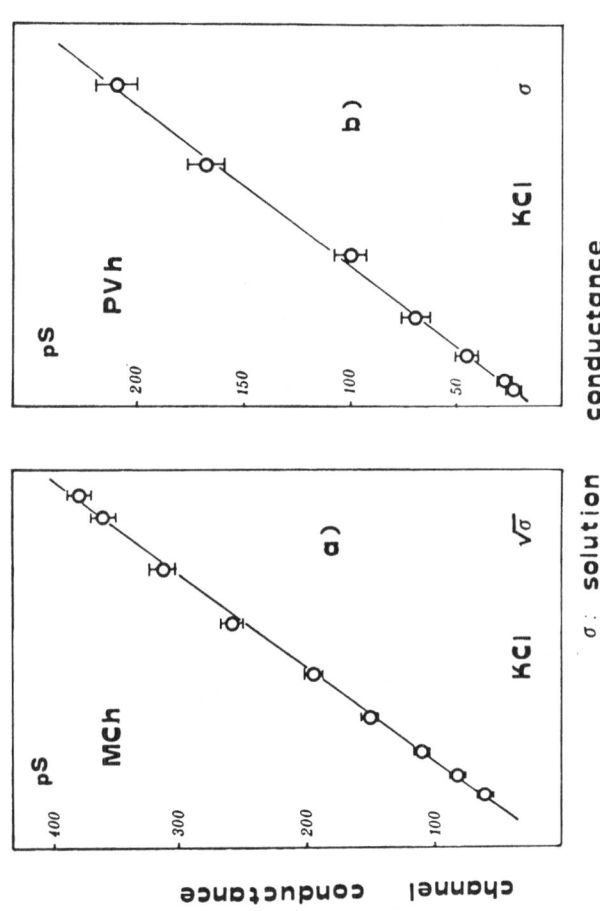

Fig. 4 – a) Linear dependence of the MCh single channel conductance from the square root of the solution conductance. This indicates a saturation effect probably due to the inter-action of ions with the channel.

b) Linear dependence of the PVh single channel conductance from the solution conductance. No saturation effect is present.

should have a linear I-V characteristic, which we could actually
observe (fig. 3b).

These voltage independent ON-OFF fluctuations seem to indicate
a less strong interaction between the bilayer and the protein, than
in the case of MCh. The molecule of PVh might, for example, move on
the lipid surface and penetrate more or less into the hydrophobic
region, giving rise in this way to the two different values of conductance.

Another difference between MCh and PVh channels regards the
channel conductance versus electrolyte conductance curve; with PVh
there is no observable saturation effect but, on the contrary, there
is a direct proportionality as one can see in fig. 4b.

The absence of phenomena such as the voltage dependent gating
and the saturation of conductance while increasing salt concentration
both presumably due to the presence of some charged groups, makes us
think to the PVh channel as a simple pore, filled with water solution
and spanning through the membrane. Further studies are planned, with
these and other hemocyanins, in order to understand the nature of
these charged groups, and of the gating mechanism that they induce
in the membrane conductance.

REFERENCES

1. Van Holde, K.E. and Van Brugge, EVF.J. (1971) in: "Biological
    Macromolecules Series" (Timasheff, S.N. and Fasman, G.D. eds)
    vol. 5, pp. 1-53, Marcel Dekker, New York.
2. Pant, M.C. and Conran, P. (1972) J. Membr. Biol. 8, 357-362.
3. Latorre, R., Alvarez, O., Ehrestein, G., Espinoza, M. and Reyes,
    J., (1975) J. Membr. Biol. 25, 163-182.
4. Tien, H.T. (1974) "Bilayer Lipid Membranes (BLM); theory and
    practice", Marcel Dekker, New York.
5. Szabo, G., Eisenman, G. and Ciani, S. (1969) J. Membr. Biol.
    1, 346-382.
6. Antolini, R. and Menestrina, G. (1979) FEBS Lett. 100, 377-381.
7. Menestrina, G. and Antolini, R. (1979) Biophys. Biochem. Res.
    Comm. 88, 433-439.

ON THE STRUCTURE OF MELITTIN IN AQUEOUS SOLUTIONS AND UPON INTERACTION WITH MEMBRANE MODEL SYSTEMS

Roberto STROM[+], Carlo CRIFO'[+], Vincenza VITI[o], Laura GUIDONI[o] and Franca PODO[o]

[+] Istituti di Chimica Biologica, Università di Roma e
L. Istituto Univ. di Medicina e Chirurgia, L'Aquila,
and Centro di Biologia Molecolare del CNR, Roma, Italy

[o] Laboratorio di Biologia Cellulare e Immunologia,
Istituto Superiore di Sanità, Roma, Italy.

INTRODUCTION

Melittin, the main constituent of bee venom, is a powerful lytic agent, able to induce a conspicuous disruption of lipid membrane systems (1,2,3). It is (Fig. 1) a cationic peptide, with an uneven distribution of hydrophobic aminoacid residues (positions 1-20) and of hydrophilic ones (positions 21-36):

Gly-Ile-Gly-Ala-Val-Leu-Lys-Val-Leu-Thr-Thr-Gly-Leu-Pro-Ala-Leu-Ile-
Ser-Trp-Ile-Lys-Arg-Lys-Arg-Gln-Gln-NH_2

Fig. 1 - Primary structure of melittin from Apis mellifera (4).

The conformation of the peptide when bound either to membrane systems or to detergent micelles has been found to differ from that existing in aqueous solutions (5,6), the latter having a lower α-helix content.

High ionic strength was however found to shift the conformation of melittin from a mainly random-coiled structure to a mainly helical one (7); simultaneously, there is a self association of the peptide to a tetramer, with a concomitant burying of the tryptophan residue in a less polar medium. It seemed therefore worth investigating the

conformation(s) of melittin in aqueous solutions under various conditions of pH, ionic strength, and nature of counterions, and the
possible relevance of these different conformational states on the
interaction with phospholipid vesicles and/or with detergents.

MATERIALS AND METHODS

Melittin grade II, phospholipase-free was purchased from Sigma
Chemical Co. St. Louis, Mo. USA, and used as such. Circular dichroism
spectra were obtained at room temperature, using 1 mm path-length
cells, on a Cary Model 60 spectropolarimeter with CD attachment.
The mean residue ellipticity, θ, is expressed as deg. cm^2 $dmol^{-1}$.

Fourier transform proton magnetic resonance (PMR) experiments
were performed at 29°C at 100 MHz, using a Varian XL-100-15FT spectrometer.

Chemical shifts (in ppm) were evaluated with respect to DSS in
D_2O. Proton spin-lattice relaxation times (T_1) were measured at
100 MHz by the inversion-recovery pulse sequence t-180°-τ-90°, being
the delay between the perturbing (180°) and the monitoring (90°)
pulses, and a long delay time between the application of successive
pulse sequences ($\gg$ 5 T_1). In order to minimize the problems arising
from the limited dynamic range of the computer, the partially relaxed
spectra were obtained under double precision conditions. The accuracy
of T_1 values was estimated to be around ± 10%.

RESULTS

Structure of melittin in aqueous solutions

Fig. 2 shows how the CD spectra of the peptide backbone of
melittin in aqueous solution changes upon addition of NaCl or of
Na_2SO_4. It can be seen that the presence of $SO_4^=$ anions induces a
high helical content. This structural effect is shared by other
divalent and multivalent anions while monovalent ones exert it only
at very high concentration (Fig. 3) a notable exception being represented by $NaClO_4$, the effectiveness of which is comparable to that
of Na_2SO_4. In the case of phosphate and of EDTA the effect depends
on the ionization state, monovalent anions behaving like Cl^-.

High helical content is also found at high pH values, irrespective of the presence of particular anions.

In 0.15 M NaCl a clearcut transition occurs with a pK of 8.8;
two titratable groups with undistinguishable pK's seem to be involved
(Fig. 4). It was also found that melittin has a larger number of
titratable groups in the alkaline region when in 0.15 M NaCl than
in Na_2SO_4 (Fig. 5).

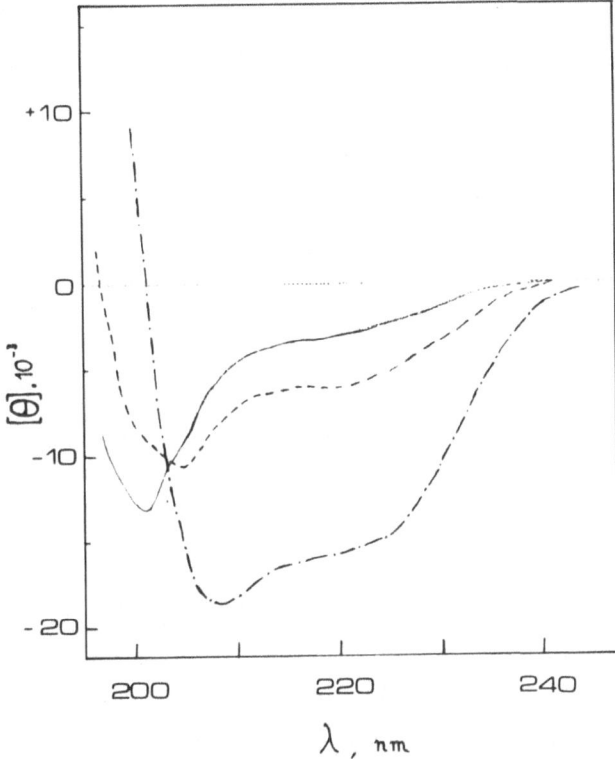

Fig. 2 - Intrinsic CD spectra of melittin (at a concentration of
0.5 mg protein. ml^{-1} in H$_2$O (——), in 0.15 M NaCl (---)
and in 0.15 M Na$_2$SO$_4$, pH ∿ 6 (—·—·—).

    Preliminary experiments indicate that treatment of melittin
with formaldehyde and subsequent reduction with sodium borohydride
so as to block the ionizable aminogroups, results in a modified
peptide having a high helical content even in pure H$_2$O. Also PMR
spectra (Fig. 6) indicate profound conformational differences according to the ionic environment of melittin in D$_2$O.

    In the absence of counterions the spectral features are typical
of a random coil structure. Upon addition of either 0.15 M Na$_2$SO$_4$
or of 2 M NaCl the spectral profile was modified by a spreading
and/or broadening of various signals in the aliphatic protons region
(1-5 ppm). Besides, some CH$_3$ signals appeared in the region between
0 and 1 ppm, indicating for these resonances un upfield shift attri-

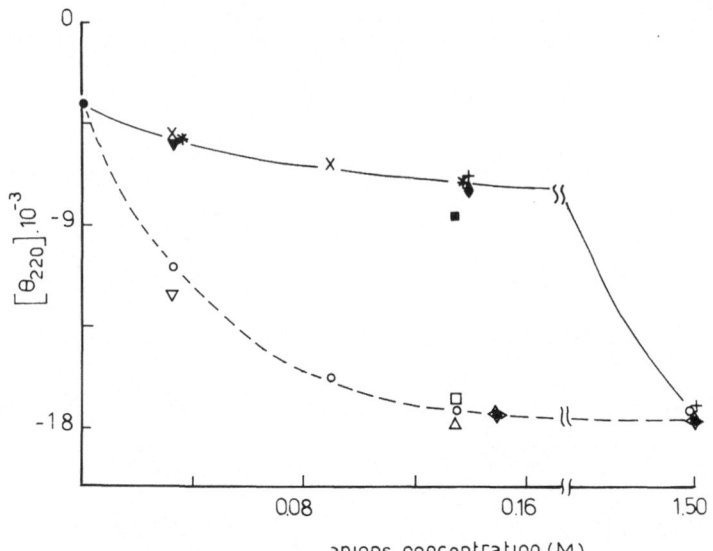

Fig. 3 - Dependence upon anions concentration of melittin ellipticity
at 220 nm: (●): H_2O, pH∿6.5; (X): NaCl, pH∿6.5; (o): phos-
phate, pH 7.4; (■): borate, pH 7; (□): borate, pH 10;
(Δ): Na_2SO_4, pH 5.5; (◆): Na isethionate, pH 6.5; (+): TES-
HCl, pH 7.5; (✱): acetate, pH 6; (◈): Na perchlorate, pH 6.5
(▼): Na-EDTA, pH 7; (∇): Ca-EDTA, pH 4. In the case of the
EDTA salts, the concentration of the carboxyl groups was
considered.

butable to ring-current effects. At the tryptophan level the main
effect was a conspicuous downfield shift of the signal assigned to
the proton in position 2 of the indole ring.

In 0.15 M NaCl the PMR profile was also markedly different from
that in pure D_2O, resembling that in Na_2SO_4 or in very high ionic
strength. It showed however only little, if any, upfield shift of
CH_3 resonances and the tryptophan signals in the region above 7.5 ppm
were definitely broader, nor did these features change with increasing
pH.

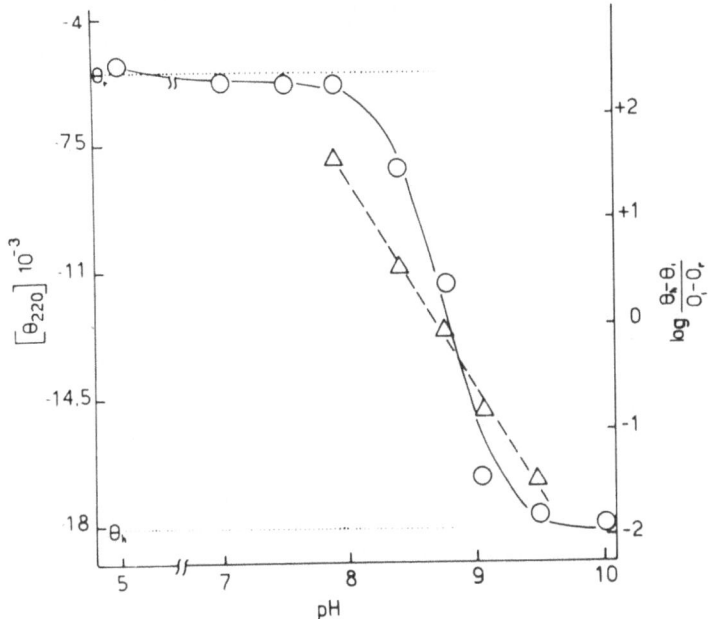

Fig. 4 - pH dependence of melittin ellipticity at 220 nm (o—o) in
0.15 M NaCl containing 5 mM TES and 5 mM piperazine. By
assuming that the equilibrium between the random coil conformation (r) and the α-helical one (h) is governed by the
ionization of n cationic groups, the ellipticities θ_i's at
pH values in the titration range are such that a plot of

$$\log \frac{\theta_h - \theta_i}{\theta_i - \theta_r} \quad \text{vs pH is a straight line } (\triangle\text{--}\triangle) \text{ with slope n=2,}$$

crossing the abscissa axis at the pK of the ionizable group(s).

Since the tryptophan signals are relatively well resolved, the
spin-lattice relaxation times of the various protons of this aminoacid residue could be determined (Table 1). Rather unexpectedly it
was found that in 0.15 M NaCl or phosphate at neutral pH the T_1 values
of all signals and particularly of that presumably arising from the
proton in position 2, were markedly lower than the values obtained
in pure D_2O. In 0.15 M sulphate, or in concentrated salt solutions,
or at high pH T_1 values increased considerably.

Interaction of melittin with phospholipids or with detergents

Upon addition of egg lecithin or of anionic detergents to
melittin solutions in H_2O or in dilute (0.15 M) NaCl the ellipticity

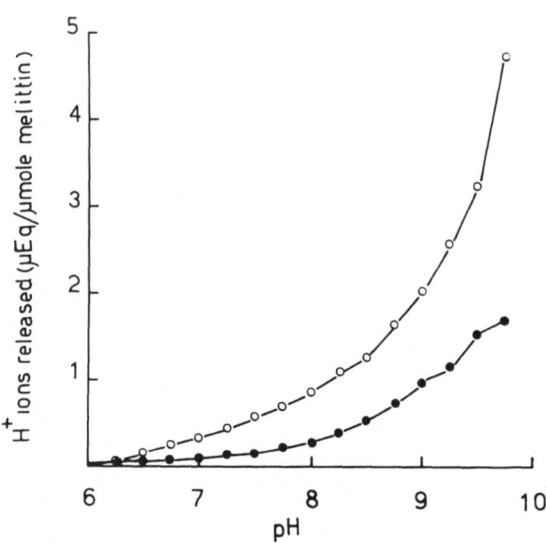

Fig. 5 - pH dependence of proton release by melittin in 0.15 M NaCl
(o—o) and in 0.15 M Na$_2$SO$_4$ (•—•).

at 220 nm became more negative, suggesting an increase of the helical
content of the peptide.

In a certain range of phospholipid (or detergent)-to-melittin
ratios the intensity of the CD signal was however considerably
quenched but reached again highly negative values when the phospholipid (or the detergent) was added in large excess. The quenching
phenomenon affected the whole spectrum in the 240-200 nm region and
was indeed more considerable at the lowest wavelengths.
A similar phenomenon took place also when divalent ions, such as
phosphate or sulphate, were present, except that the initial ellipticity values were already relatively high (Fig. 7). In fact, the
presence or absence of divalent anions did not affect the phospholipid (or detergent) to melittin ratio at which the quenching of
the CD signal took place.

The NMR spectrum of melittin in the presence of phospholipids
was essentially similar to that obtained in phosphate except for the
absence of the ring-current shifts of CH$_3$ resonances (Fig. 8).
There was also a phospholipid-dependent increase of the T$_1$ values
of the tryptophan signal assigned to the proton in position 2 and/
or 7 (Fig. 9).

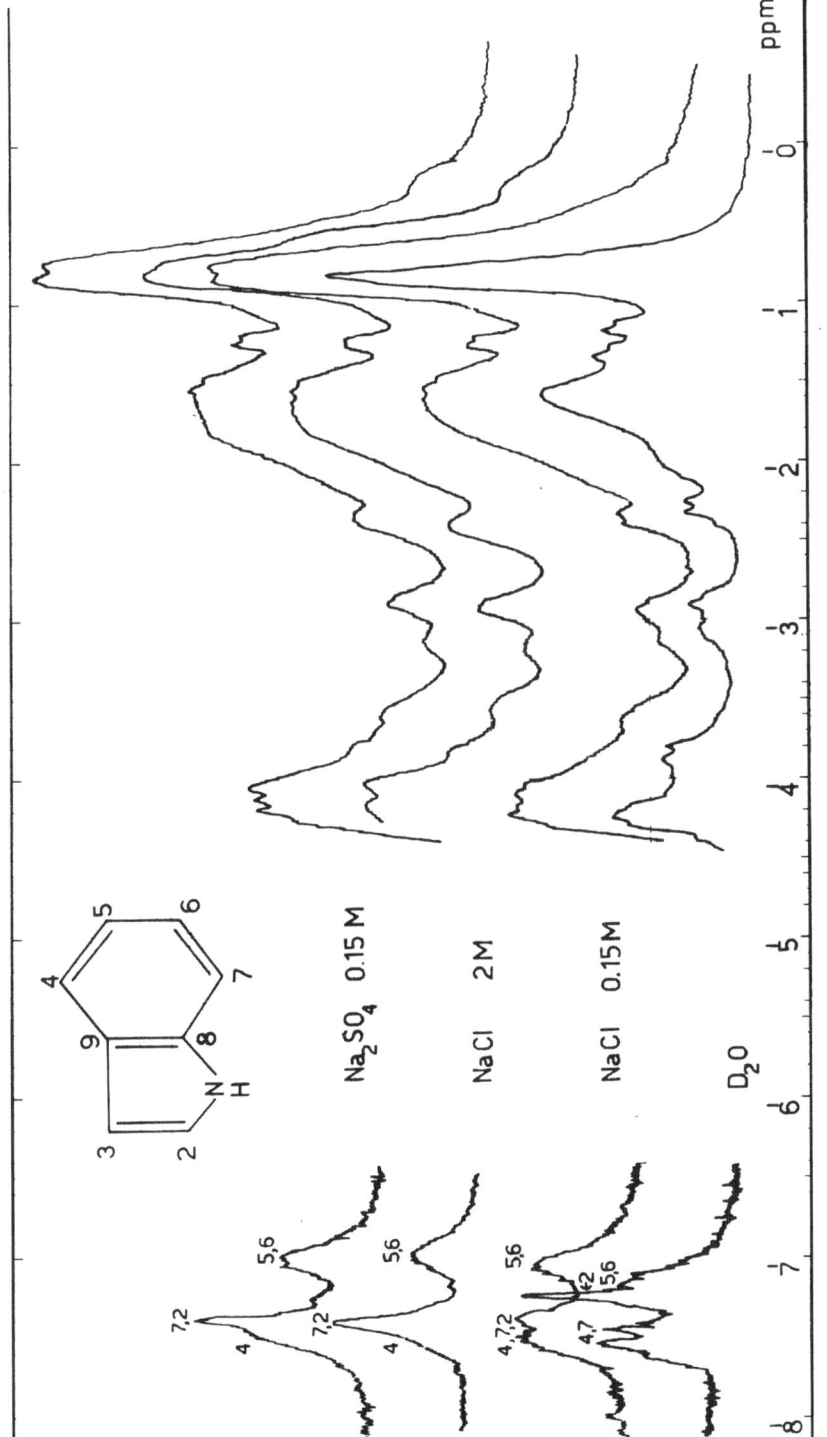

Fig. 6 – Proton magnetic resonance spectra at 100 MHz and at 29°C of melittin (10 mg/ml) in D₂O solutions of varying ionic contents and at pH ∿ 6.5.

TABLE I

Spin lattice relaxation times T_1(ms) of the tryptophan protons of melittin.
Signal 1, around 7.5 ppm, originates essentially from the proton in C-4. Signal II is due mainly
to the proton in C-2, and signal III from protons in C-5 and C-6 (ref. 15).
The numbers in brackets indicate the extreme range of T_1 values obtained with different melittin
samples.

SOLVENT	pH	Signal I	Signal II	Signal III
D_2O	Neutral	226 (202-249)	247 (225-260)	239 (217-265)
NaCl 0.15 M	6	167 (130-200)	116 (80-135)	170 (144-200)
NaCl 0.15 M	10	260	295	245 (230-260)
NaCl 2 M	7	280	360	260
Na_2SO_4 0.15 M	6	283 (270-300)	332 (315-350)	277 (270-290)
Na_2SO_4 0.15 M	10.5	320	390	350
Phosphate 0.15 M	7.4	170 (150-190)	142 (95-190)	195 (150-240)
Phosphate 1 M	7.4	360	355	310

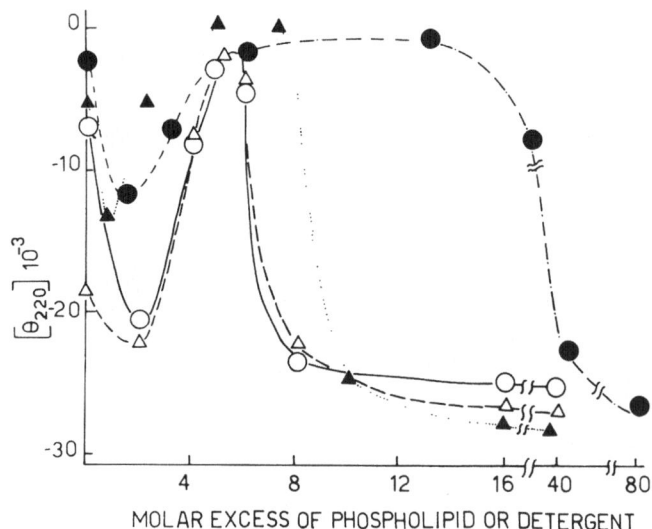

Fig. 7 – Effects of phospholipids or of anionic detergents on the ellipticity at 220 nm of melittin; o—o : egg lecithin in 0.15 M NaCl, pH ~ 6.5; Δ--Δ: egg lecithin in 0.15 M Na₂SO₄, pH ~ 6.0; ▲···▲: sodium dodecylsulphate in 10 mM TES-HCl buffer pH 7; ●—·—·—●: sodium deoxycholate in H₂O, pH ~ 6.

DISCUSSION

On the basis of the CD spectra melittin appears to exist in two main conformational states: random coil in pure H₂O, α-helix in the presence of high salt concentration, or at high pH or when it interacts with anions having a valence higher than 1, or with amphipatic molecules. The NMR results indicate however a more complex situation. As summarized in Table II no direct correlation can be found between the α-helix content from the CD spectra and the presence of upfield shifts of the methyl protons appearing in the region around 0.5 ppm, nor with the relaxation times of the tryptophan C-2 proton. In particular no upfield shift occurs in dilute NaCl even at alkaline pH, or when melittin interacts with phospholipids. On the other hand the T_1 values in dilute NaCl at neutral pH and in dilute phosphate buffer are definitely shorter not only than those found upon addition of Na₂SO₄, of concentrated NaCl or at alkaline pH, but also than those found in pure water. It appears therefore that the structure of melittin can hardly be reduced to two states only, and in particular that the presence of a high α-helical content does not imply absolute

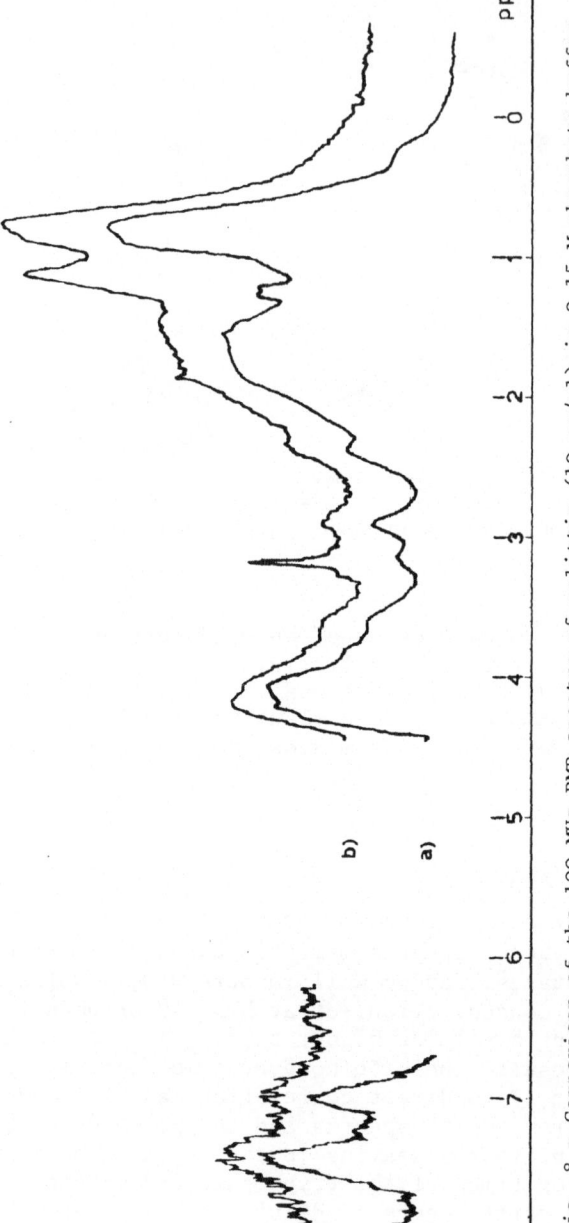

Fig. 8 - Comparison of the 100 MHz PMR spectra of melittin (10 mg/ml) in 0.15 M phosphate buffer, pK 7.2 (a) and upon addition of egg phosphatidylcholine (b) at a lipid-to-protein molar ratio of 1:4.

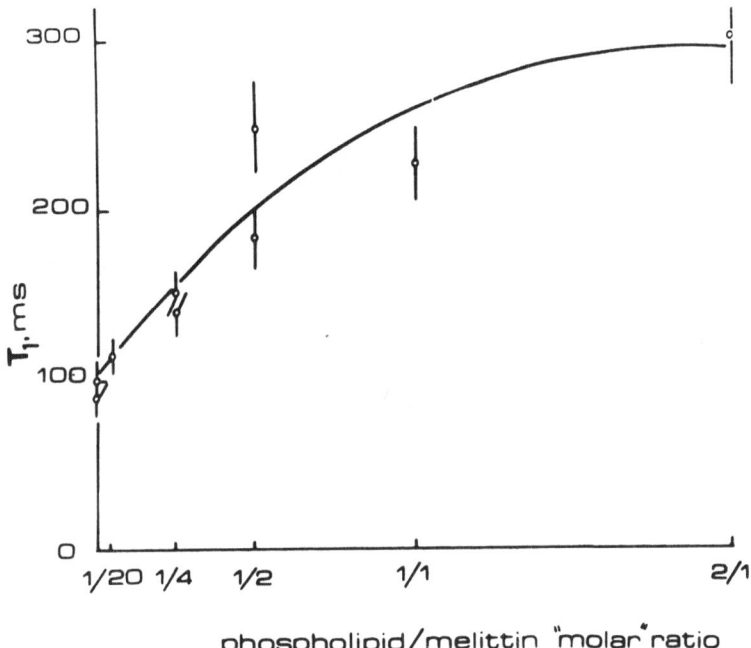

Fig. 9 - Dependence, on the phospholipid-to-melittin molar ratio, of
        the values of the spin lattice relaxation times of the
        triptophan protons giving rise to signal II of Table I
        (taken from ref. 15).

identity of structure.

    In the last column of Table II are reported data from the litera-
ture concerning the aggregation state of the peptide under different
conditions. The correlation, suggested by Talbot et al. (7), between
the ordering of the secondary structure and aggregation in a tetramer
is far from being an absolute one.

    The presence of upfield shifts of methyl resonances seems to
correlate better with the tetrameric state of melittin than with the
α-helical conformation of the molecule - - a notable exception occurring however in dilute NaCl at alkaline pH.

    T_1 values provide information about the mobility of the trypto-
phan residue. At variance with the results from fluorescence polarization studies (9) the dependence of T_1 values in phosphate and
sulphate are markedly different, despite similarities in the other
CD and NMR characteristic features.

TABLE II

Comparison of the information of melittin structure resulting from this paper and from literature data. In order to allow an easier comparison among the different sets of data, they are grouped on a qualitative basis: "neutral" pH covers the range from 5.5 to 7.5; "alkaline" pH is above 10; T_1 measurements are considered as "short" if below 200 ms; "average" when around 250 ms, and "long" if above 290 ms.

Solvent	pH	Secondary structure from CD spectra.	Appearance of upfield-shifted CH_3 resonances (in the region around 0.5 ppm).	T_1 of tryptophan C_2 proton.	Association state.
Water	Neutral	random coil[a,b,c,d]	No[a]	average[a]	monomer[c]
NaCl~0.15M	Neutral	predominantly random coil[a,e]	No[a]	short[a]	intermediate[e]
NaCl~0.15M	Alkaline	α-helical[a]	No[a]	long[a]	tetramer[h]
NaCl~2M	Neutral	α-helical[a,e]	Yes[a]	long[a]	tetramer[e]
Na$_2$SO$_4$~0.15M	Neutral	α-helical[a]	Yes[a]	long[a]	n.d.(presumably tetramer)
Na$_2$SO$_4$~0.15M	Alkaline	α-helical[a]	Yes[a]	long[a]	
Phosphate ~0.15 M	Neutral	α-helical[a,f,d,g]	Yes[a]	short[a,g]	tetramer[h]
Phosphate ~1M	Neutral	α-helical[a]	Yes[a]	long[a]	n.d.(presumably tetramer)
Detergent micelle	Neutral	α-helical[',d,i]	No[d]	n.d.	monomer[d,i]
Phospholipid dispersion	Neutral	α-helical[a,d,f,g]	No[a]	long[g]	n.d.

References: a) this paper; b) ref. 13; c) ref. 5; d) ref. 6; e) ref. 7; f) ref. 14; g) ref. 15; h) ref. 10; i) ref. 8.

The pH dependence of melittin conformation indicates the critical role played by two groups titrating at pH 8.8. This is confirmed by the α-helical structure of formaldehyde-treated melittin. The ordering effect of multivalent anions or of high salt concentration may also be related to a shielding effect of these groups (Fig. 3, 5).

The precise role played by the various ionizable aminoacid residues in determining the structure(s) of the peptide is yet to be defined. The random coil-to-α-helix transition occurring in poly-lysine at alkaline pH is well known (11). On the other hand, poly-arginine undergoes a similar transition upon addition of divalent or polyvalent anions, of dodecylsulphate or of perchlorate, maintaining instead a random-coil structure with Cl⁻ or OH⁻ (12).

In melittin, the transition to α-helical conformation(s) could therefore involve both the lysine and the arginine residues. On the other hand, that a crucial role in allowing or not the α-helical conformation be played by the protonation state of Lys 7 is indicated by the high content of α-helix in melittin interacting with non ionic detergents (where the hydrophobic envoriment is likely to prevent Lys 7 from protonation), and from the consideration that a nucleation center for the formation of an α-helix can be identified in the primary sequence region around this residue.

REFERENCES

1) Sessa, G., Freer, J.H., Colacicco, G. and Weissmann, G., J.Biol. Chem. 244, 3575-3582 (1969).

2) Habermann, E. and Kowallek, H., Hoppe Seyler's Z. Physiol. Chem. 351, 884-890 (1970).

3) Williams, J.C. and Bell, R.M., Biochim. Biophys. Acta, 288, 255-262 (1972).

4) Habermann, E. and Jentsch, J., Hoppe Seyler's Physiol. Chem. 348, 37-50 (1977).

5) Dawson, C.R., Drake, A.F., Helliwell, J. and Hider, R.C., Biochim. Biophys. Acta, 510, 75-86 (1978).

6) Lauterwein, J., Bösch, C., Brown, L.R. and Wüthrich, K., Biochim. Biophys. Acta, 566, 244-264 (1979).

7) Talbot, J.C., Dufourcq, J., de Bony, S., Faucon, J.F. and Lussan, C., FEBS Letters, 102, 191-193 (1979).

8) Knöppel, E., Eisenberg, D., and Wickner, W., Biochemistry, 18, 4177-4181 (1979).

9) Faucon, J.F., Dufourcq, J., and Lussan, C., FEBS Letters, 102, 187-190 (1979).

10) Gauldie, J., Hanson, J.M., Rumjanek, F.D., Shipolini, R.A. and Vernon, C.A., Eur. J. Biochem. 61, 369-376 (1976).

11) Applequist, J. and Doty, P., In: "Polyamino Acids, Polypeptides and Proteins", M.A. Stahmann, Ed., Univ. Wisconsin Press, Madison, Wisc., 1962, p. 161.

12) Ichimura, S., Mita, K. and Zama, M., Biopolymers 17, 2769-2782
     (1978).
13) Jentsch, J., Naturforsch. Z., 24b, 33-35 (1969).
14) Drake, A.F. and Hider, R.C., Biochim. Biophys. Acta, 555, 371-
     373 (1979).
15) Strom, R., Crifò, C., Viti, V., Guidoni, L., Podo, F., FEBS
     Letters, 96, 45-50 (1978).

TEMPERATURE DEPENDENCE OF MATTER TRANSPORT IN <u>Valonia utricularis</u>

D.G. Mita, M. Bianco, P. Canciglia[+], A. D'Acunto,
T. Improta, C. Minatore and F.S. Gaeta

International Institute of Genetics and Biophysics of
C.N.R., via Marconi, 10 - Naples, Italy
+ Istituto di Fisiologia Generale, Facoltà di Scienze,
   via del Vespro 5 - Messina, Italy

1) INTRODUCTION

An intrinsic difficulty in the study of biological membrane transport stems from the multiplicity of transport systems which developed under very strong evolutive pressure due to the capital importance of membrane function for survival. Different membrane transport systems often functionally overlap, thus greatly complicating the investigation of the dependence of transport on physical parameters. A specific complexity appears in the study of temperature effect on membrane transport because of the influence of both average temperature and temperature gradients. Perhaps this circumstance explains why only few studies of the effect of temperature on biological membrane transport have been published in animal (1-4) and plant cells (5-8), in contrast with the keen interest on the role of temperature and temperature gradients on artificial membrane transport (9-25).

It is evident on the other hand that assessment of the temperature dependence is very important in connection with any rate process, in view of the determination of the apparent height of energy barriers. Since most living organisms are poikilotherms, temperature dependence must have put decisive constraints on general trends of biological evolution.

Transmembrane temperature difference is known to be the driving force producing matter transport in both selective dense membranes (14-19) and porous partitions (20-25). In biological systems the different metabolic rates generally occurring in the two compartments separated by a membrane determine the existence of transmembrane

131

temperature gradients and of coupled matter fluxes.

Accordingly we carried out the present investigation of the effects of temperature and of temperature gradients on the rate of water and sulphate ion exchange in the giant coenocitic algal cells of Valonia utricularis, employing a tracer method, based on the use of ^{3}H and ^{35}S.

2) EXPERIMENTAL

2.1 – Materials and Methods

Specimens of Valonia utricularis were collected by care of the Stazione Zoologica of Naples in the sea of Capo Miseno at depths between one and five meters, where the temperature stays between +12°C and +25°C throughout the year. The cells to be used in our experiments were kept in artificial sea water at the temperature chosen for the experiment for at least twelve hours prior to use. Artificial sea water was used to ensure greater uniformity of conditions, even if some authors maintain that this may vary the cell's exchange rates to some extent (8). The artificial culture fluid was prepared according to Harwey's formula (26) in the simplified version of Fleming reported by Pringsheim (27). Valoniae have the advantage of being very large and sturdy cells, easy to manipulate, ideal for separate assay of vacuolar fluid and if needed also of cytoplasm. On the other hand the cell structure presents at least two compartments in series – cytoplasm and vacuole – and four successive barriers interposed between outer medium and vauole: a sturdy wall of cellulose microfibrils, an outer membrane, the plasmalemma, the layer of cytoplasm and an inner membrane, the tonoplast; this is a great complication as we shall see. In Fig. 1 the structure of the cell is schematically represented. We investigated the kinetics of water exchange between vauole and outer medium by measuring both incorporation and efflux rates of ^{3}H$_2$O at various temperatures in the range between +5°C and 45°C. We also performed some preliminary experiments to determine the rate of incorporation of SO_4^{--} into the cytoplasm of Valonia within the same temperature range. Information was thus obtained on the effect of temperature on transport of water – the most abundant and rapidly exchanged cellular material – and of an ion, which some authors maintain to be actively transported in various algal cells (28–29).

To ascertain the effect – if any – of transmembrane temperature differences, these were artificially induced by linearly increasing or decreasing the temperature of cells and surrounding medium at predetermined rates by means of programmed thermostats. If $\Delta T/\Delta t$ (°C·sec^{-1}) is the rate of temperature variation, V and A are cell's volume and outer surface, and ρ and C_p are density and specific heat of cell medium (practically an aqueous saline solution) than a transmembrane heat flux:

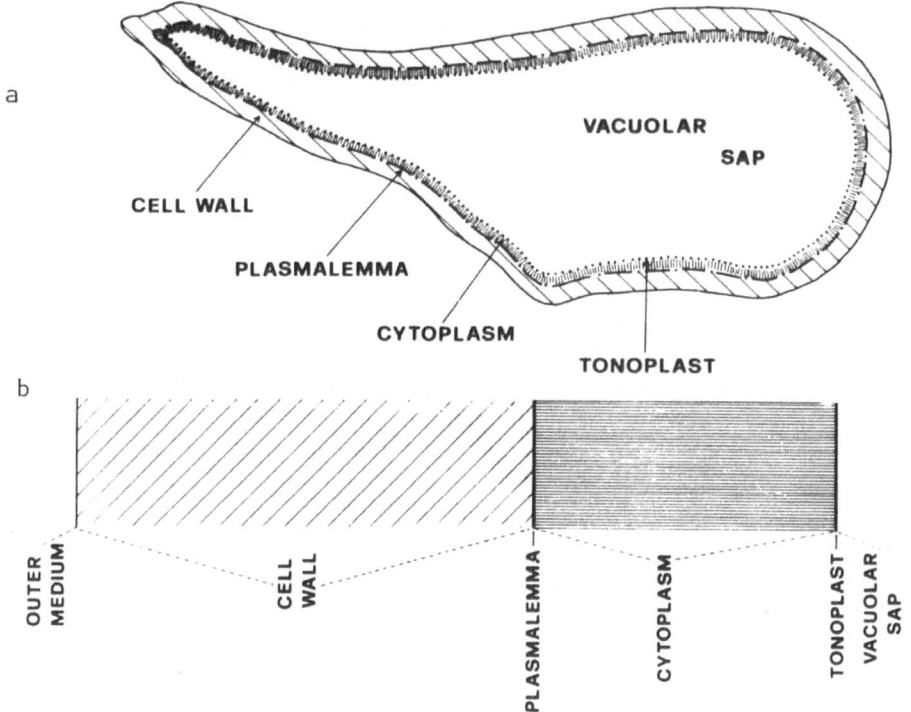

Fig. 1 ⁻ (a) Schematic section (not in scale) through one cell, show-
       ing the succession of layers interposed between outer medium
       and vacuolar fluid. (b) The four layers interposed between
       outer medium and vacuolar fluid, in scale: cell wall $\simeq$ 8 μ;
       plasmalemma $\simeq$ 100 $\overset{\circ}{A}$; cytoplasm $\simeq$ 5 μ; tonoplast $\simeq$ 100 $\overset{\circ}{A}$.

$$J_q = \rho C_p \, \frac{V}{A} \frac{\Delta T}{\Delta t} \tag{1}$$

will be produced. If K is medium thermal conductivity, a radial
temperature gradient $\Delta T/\Delta r$ is established, given by:

$$\frac{\Delta T}{\Delta r} = \frac{\rho C_p}{K} \frac{V}{A} \frac{\Delta T}{\Delta t} \tag{2}.$$

In cooling and heating, this heat flux J_q has equal or, respectively,
opposite sense with the flux of metabolic heat leaving the cell.
Modulation of the rates of matter exchange below and above the isothermal value at the run's average temperature are accordingly expected. Experiments both isothermal and in presence of temperature
gradients were conducted in living Valoniae and in heat-killed cells.
These ones were kept for two hours at a temperature $+70 \pm 2°C$, this
treatment resulting in the disruption of the layer of cytoplasm and
of both cell's membranes, tonoplast and plasmalemma. The heat-killed

cells consist of a small liquid-filled sac bounded by the highly
permeable cell wall, and containing a solution which has ionic concentration and pH values indistinguishable from the ones of the outer
fluid. In addition some particulate matter can be also seen floating
inside. All experiments reported here were conducted in dim light
(intensity of about three foot-candles at cell's surface). The effects of illumination on the rate of water exchange are found to be
rather modest and to affect absolute values but nor relative rates,
so that the present results are well comparable with those already
published by us for water transport in conditions of strong illumination and in the darkness (30-31). Tritium-labelled solution was
prepared by adding ^3H_2O to artificial sea water to obtain a specific
activity of $3.5 \cdot 10^5$ DPM per mgr of solution; ^{35}S-labelled sea water
was prepared by addition of $Na_2{}^{35}SO_4$ to a specific activity of
$2.7 \cdot 10^5$ DPM per mgr.

Activity measurements were carried out by an automatic counter
Tris Carb 3385 by Packard Ltd. Color quenching is not a problem in
the experiments with tritiated water since a practically colorless
vacuolar fluid is introduced in the vials. In the case of sulphate,
where intensely green cytoplasmic material is mixed with the scintillator, a titration curve has been drawn to account for the effect
of color on counting efficiency.

The experimental technique followed was very simple: in incorporation experiments the pre-thermalized cells were shifted into ^3H_2O
or in $Na_2{}^{35}SO_4$ labelled artificial sea water. Then lots of 6 <u>Valoniae</u>
each were extracted at predetermined time intervals. In the experiments with ^3H_2O some vacuolar fluid was extracted by a syringe from
each cell and introduced into a vial containing the scintillator.
The weight increase of each vial after introduction of the sap obtained from 6 algae, allows determination of the DPM/gr in the
vacuolar fluid. When the label content of cytoplasm had to be assessed in the runs with sulphate the following standard procedure
was employed: One end of the cell was cut off and the vacuolar fluid
gently squeezed off; the cytoplasm was then scraped away and mixed
with the scintillator.

In the efflux experiments the cells are kept overnight in labelled artificial sea water, at the same temperature chosen for the run,
this being sufficient in the case ^3H_2O for vacuolar activity to be
in equilibrium with that of the external fluid. Efflux experiments
are then conducted much in the same way as incorporation runs, but
measuring the decrease in time of vacuolar fluid activity, starting
from the moment the cells are immersed in a very large vessel of unlabelled artificial sea water kept in the same thermostated bath.

In the case of non-isothermal runs, aiming to assess the effect
of transmembrane temperature gradients, the shift from unlabelled to
labelled medium or vice-versa takes place in coincidence with the

beginning of the linear temperature variation. Constancy of the
rate of temperature variation in heating and cooling was obtained
by means of a Heto programmable thermostating unit model 02 PG 623.
Values of $\Delta T/\Delta t$ were chosen for these runs between 1.4 and 0.4 centi-
grades per minute. In every other respect these experiments were
conducted as the isothermal runs. The first point was generally
obtained 30 seconds after the shift, successive measurements were
in most cases effected at one-minute intervals in normal runs and
occasionally at longer time intervals in some runs of long duration.
Every experimental curve is constructed from the data of ten runs,
viz. 60 cells per experimental point. Reproducibility of absolute
activity values is poor, on the other hand the relative positions
of curves obtained by changing the values of experimental parameters
are fairly well reproducible.

2.2 - Experimental results

Let us first deal with the isothermal experiments in living
and in heat-killed algae. Water exchange is found to be very rapid
indeed, which is in part a disadvantage, making impossible in fact
to determine the rate of water exchange between outer medium and
cytoplasm and between cytoplasm and vacuole. Already the earliest
measurements practicable with our manual technique, namely the ones
effected 30 seconds after runs initiation yield in the cytoplasm
results near to steady-state conditions. We had to restrict ourselves
therefore to the investigation of tritiated water exchange between
outer medium and vacuole and viceversa. Even this phenomenon is
completed in about two hours in living cells, tracer equilibrium
distribution being quickly attained in both incorporation and efflux
runs. In Figs. 2a and 2b results of such experiments are reported
in graphical form. The dependence of the rate of the process on
temperature is evident. The general trend to higher rates of water
exchange with increasing temperature appears in both families of
curves. When heat-killed cells are employed, the results summarized
in Figs. 3a and 3b are obtained. Here again the tendency to higher
exchange rates with increasing temperature appears, but significative
differences with the behaviour of living cells are also evident.
Tracer incorporation and efflux in this case both proceed at a more
than doubled rate and at the same time the temperature dependence
is much less pronounced. Incorporation and efflux rates are mutually
comparable, the small asymmetry which appears under closer scrutiny
in the living cells is a consequence of difference in water concentration in outer and internal medium. When the effect of this variable is accounted for, complete symmetry between incorporation and
efflux can be observed.

Incorporation of $^{35}SO_4^{--}$ is much slower, allowing determination
of the rate of uptake in the cytoplasm. This study is also greatly
simplified by the circumstance that very little of this substance
- if any - seems to penetrate into the vacuole within the duration

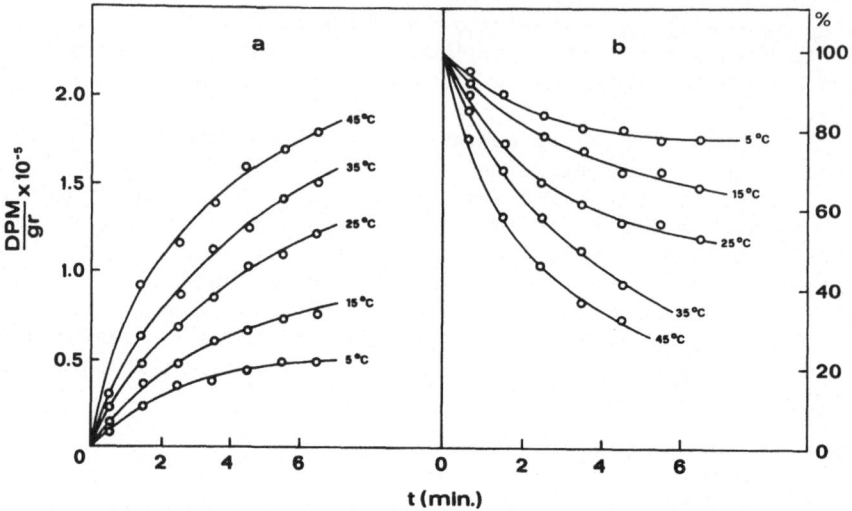

Fig. 2 - Time course of tritium activity in the vacuolar fluid in
(a) incorporation and (b) efflux experiments in living
cells at various temperatures.

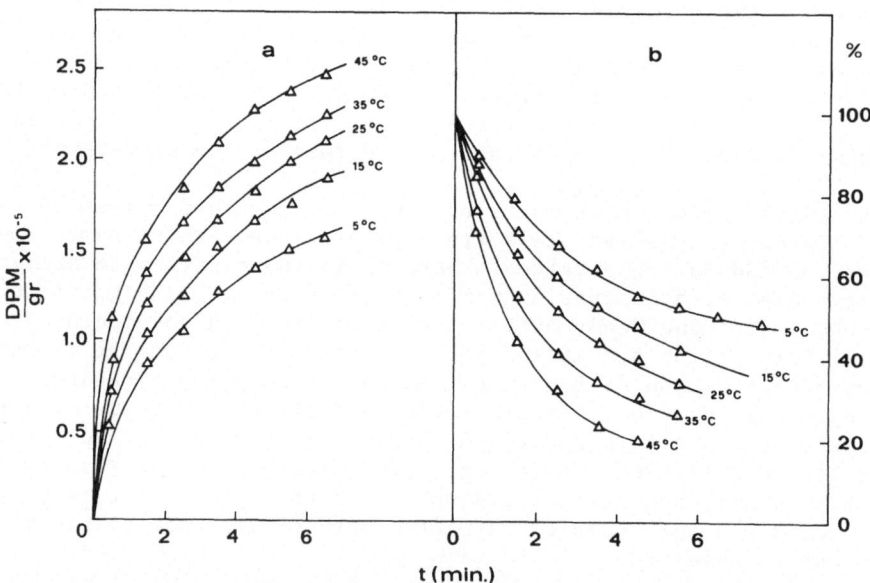

Fig. 3 - Time course of tritium activity in the vacuolar fluid in
(a) incorporation and (b) efflux experiments in heat-killed
cells at various temperatures.

of a run. Also with sulphate, incorporation in heat-killed cells
is found to take place at a higher rate than in the living algae.
A systematic study of sulphate transport in Valonia is presently
under way.

In all our experiments we took care to select for each run cells
of similar weight and not too different shapes. Volume, V and outer
surface A, were evaluated for one typical cell of each lot and then
from the data of Figs. 2 and 3 relative to the case of water, it was
possible to calculate the corresponding water flux J_w by means of
the equation:

$$J_w = \frac{V}{A\{[Q(t)]_o - [Q(t)]_v\}} \cdot \frac{d[Q(t)]_v}{dt} \tag{3}.$$

Here $Q(t)_o$ and $Q(t)_v$ are external and vacuolar tritium activities
at time t, while d $Q(t)_v$/dt is the rate of increase (or decrease)
of tritium activity in the vacuole at the same instant. Once V and
A have been obtained for each sample, the other quantities figuring
in eq. (3) are obtained from the incorporation or efflux curves,
respectively. Owing to the tendency of the radioactive label to
become uniformly distributed between internal and extracellular
fluids, the initial slopes of the isothermal curves are best employed
for flux evaluation. Calculations yield values of the order of
$5 \cdot 10^{-4}$ cm sec^{-1} in the living cells and fluxes about three times as
big in heat-killed cells. These fluxes are plotted against 1/T,
T being Kelvin temperature of individual runs. In Fig. 4 H_2O incorporation results in living and heat-killed cells are represented, while
Fig. 5 refers to efflux experiments. Absolute values of water fluxes
are high, in good agreement with results of other authors (32-34)
for the same organism.

Let us now proceed to survey experimental results obtained with
artificially induced transmembrane temperature gradients in the cases
of water and sulphate transport in Valonia. Tritiated water uptake
in living cells of Valoniae during heating and cooling is modified
relative to isothermal incorporation in the way typically illustrated
in Fig. 6a. Similar efflux experiments with artificial temperature
gradients gave the results reported in Fig. 6b. If water transport
would be temperature-dependent but insensitive to temperature gradient, the non-isothermal runs should yield incorporation and efflux
curves lying entirely between the corresponding limiting isotherms.
If the cells were to adapt immediately to external temperature, the
two non-isothermal curves should intersect in the middle of the time
interval over which the temperature transient is applied. If vice-
versa the cells were to adapt to ambient temperature with a delay
greater than the duration of the temperature shift, then the non-
isothermal curves should each coincide with the respective isotherms
at initial temperature. The experimental curves do not reflect

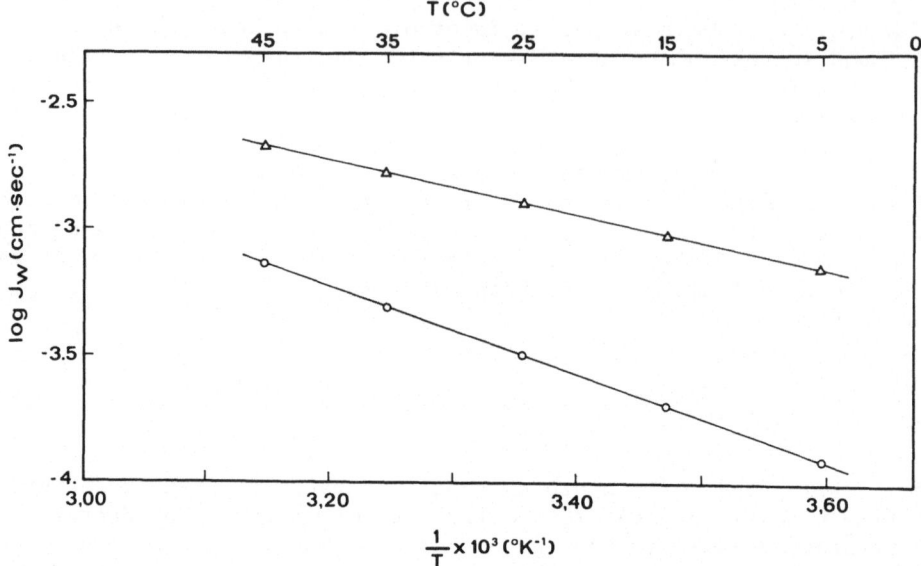

Fig. 4 - Arrhenius plots for tritium incorporation experiments: living cells (o); heat-killed cells (Δ).

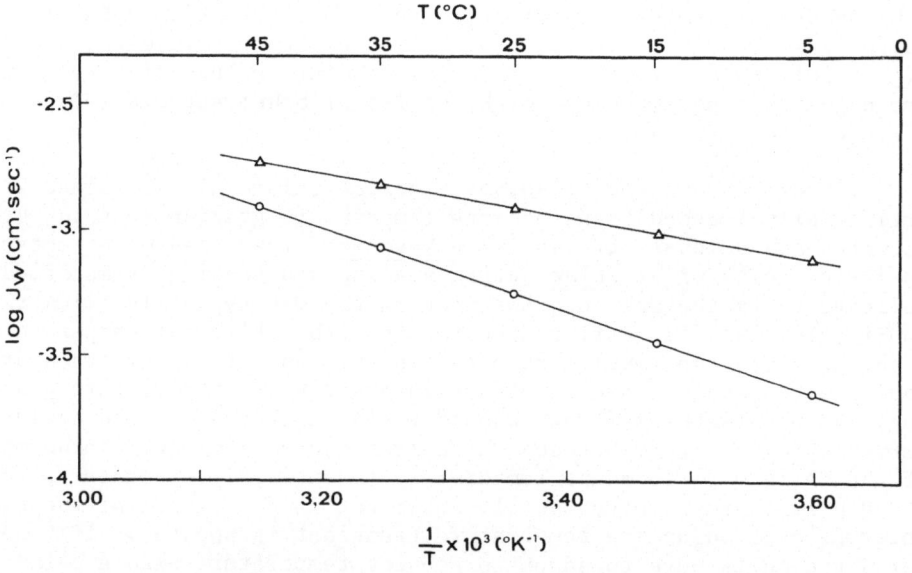

Fig. 5 - Arrhenius plots for tritium efflux experiments: living cells (o); heat-killed cells (Δ).

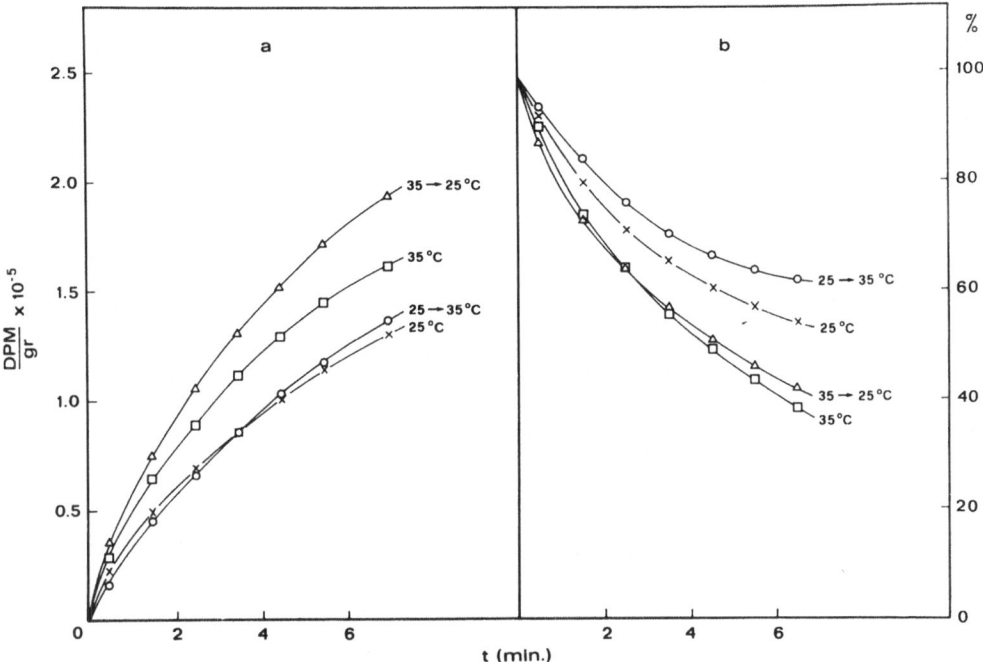

Fig. 6 - Tritiated water uptake (a) and efflux (b) in non-isothermal
        runs during which temperature is linearly raised or lowered
        in the +25°C - +35°C interval.

Table I - Water and sulphate fluxes - calculated from influx experi-
         ments - in isothermal and non-isothermal runs.

PERMEANT	FLUXES (cm·sec^{-1})
Water 25°C	$(3.1 \pm 0.2) \cdot 10^{-4}$
" 35°C	(4.9 ± 0.2) "
" 25°C — 35°C	(2.6 ± 0.2) "
" 35°C — 25°C	(5.7 ± 0.2) "
Sulphate 20°C	$(1.6 \pm 0.2) \cdot 10^{-7}$
" 40°C	(2.0 ± 0.2) "
" 20°C — 30°C	(0.9 ± 0.2) "
" 40°C — 20°C	(2.3 ± 0.2) "

either of these two situations, nor reflect any intermediate behaviour. The two non-isothermal curves indeed are much more widely
separated one from the other than the two limiting isotherms, notwithstanding the fact that their relative difference in temperature
is always smaller than the one between the former; in particular
after 3,5 minutes the temperatures of algae in the heating and in
the cooling cycles are equal. Water fluxes calculated from the
non-isothermal curves markedly differ from the ones obtained from
the respective limiting isotherms. Table I allows comparison of
the respective fluxes, calculated at t = 30 sec.

When similar non-isothermal experiments are performed with the
heat-killed cells the experimental points are found to fall between
the two limiting isotherms, as if only the average temperature mat-
tered in this case for the establishment of the rate of water exchange. Experiments analogous to the ones just described, performed
with labelled sulphate have shown that uptake of this ion in the
cytoplasm takes place in the way tipically illustrated by Fig. 7.
The influence of the artificial temperature gradient is much more
pronounced here notwithstanding the much smaller value of ΔT/Δt than
in the case of water. When heat-killed cells are used, the artificial
temperature gradient does not seem to produce detectable alterations
in sulphate uptake, as already found for dead cells using water as
permeant.

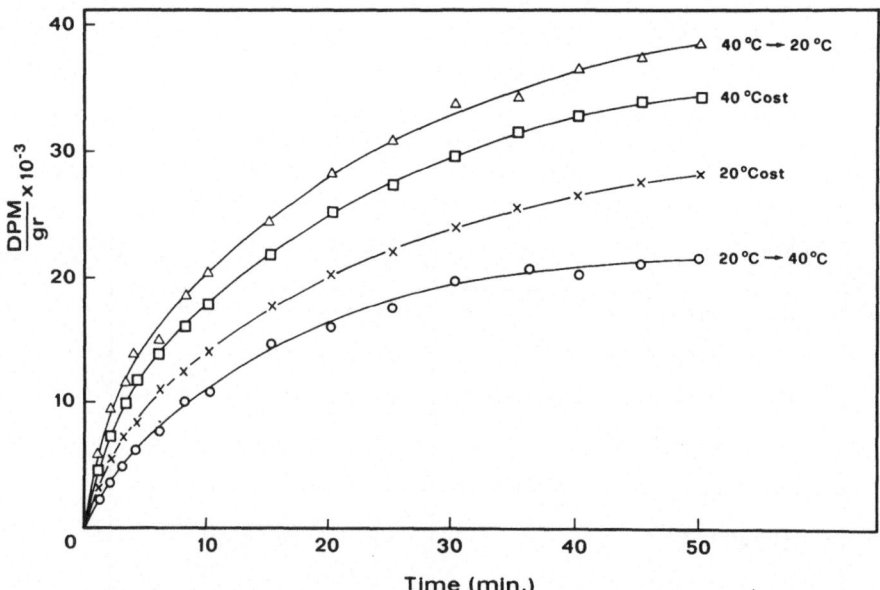

Fig. 7 - Time course of cytoplasmic sulphate incorporation in iso-
         thermal and non-isothermal runs. The latter were effected
         by linearly raising or lowering the temperature in the
         +20°C - +40°C interval.

3) DISCUSSION AND CONCLUSIONS

The significance of our experimental results will be now discussed with reference to the structure of Valonia and to the specific nature of the permeants employed in this study.

Let us start with the temperature-dependence of water transport: what we have determined here is unidirectional tracer flux in presence of a vanishingly small net flux of water. These fluxes are proportional to the number of water molecules effecting the transition through the various energy states characteristic of the energy barriers encountered along a path including the layer of cytoplasm, two biological membranes and the outer cell wall. Calling E the energy of the highest energy state in the series - the bottleneck of the whole process - one may assume the flux to be proportional to the fraction of the population of molecules, or ions, possessing an energy equal or higher than this value. Accordingly:

$$\lim_{t \to 0} \left\{ \frac{1}{C} \ln \left[J_w(t) \right] \right\} = - \frac{E}{RT} \tag{4}$$

where R is the gas constant. From the plots of Figs. 4 and 5 the appearent height of the energy barrier opposing water traffic between vacuole and external medium can be determined. Values of E thus calculated are given in Table II. Incorporation and efflux runs yield practically identical values, while a marked difference is evident between living and heat-killed cells. In particular the value of the activation energy found for water transport in the dead cells is coincident with the one characteristic of passive diffusion in an aqueous saline solution. Remembering that heat-killed algae are bounded only be the sturdy, but highly permeable outer cell wall, this is a very plausible result. The appearent height of the energy barrier in living cells is instead much higher and turns out to be strikingly similar to the one found in non-isothermal water transport through permeable barriers (process of thermodialysis) as can be seen by the data reported in the same table. The higher fluxes found in heat-killed cells reflect the destruction of the two plasma membranes and of the cytoplasmic layer; the value of 4.5 Kcal·mole^{-1} found for the activation energy in this case, shows that the very permeable outer cell wall cannot constitute the bottleneck of water transport in dead - or in living - cells. Of the remaining barriers two are localized in space while the other - the layer of cytoplasm - can be treated as diffuse. A mathematical analysis of the experimental results for water exchange allows to conclude that the bottleneck of this process cannot reside in the layer of cytoplasm (31).

The conclusion is accordingly reached that the temperature--dependence of water fluxes found in our experiments is characteristic of one (or both) plasma membranes: plasmalemma and tonoplast. The activation energy of about 8 Kcal mole^{-1}, which suggests analogy

Table II - Appearent activation energies in water exchange between
vacuolar fluid and extracellular solution in living and
heat-killed <u>Valoniae</u>. Values of activation energies for
self diffusion of water (35,36) and for thermodialysis
of water through artificial porous partitions are included
for comparison.

PROCESS		ACTIVATION ENERGY E $(\text{Kcal} \cdot \text{mole}^{-1})$
Water exchange	living cells	7.7
in <u>Valonia</u> <u>utricularis</u>	heat-killed cells	4.3
Self-	H_2O in H_2O	4.9
-Diffusion	^3H_2O in H_2O	4.4
Water transport by thermo- -dialysis (AP-20 porous partition	in pure water	5.0
	in 0.15 M NaCl aqueous solution	7.5

with a process of thermodialysis, must then reflect a physical property of membrane transport. This conclusion is suggestive but cannot be considered to be conclusive, since the coincidence of the
temperature-dependence of the rate of transport of a permeant <u>in vivo</u>
and in the process of thermodialysis might be accidental. On the
other hand, many analogies have been already found between transport
by thermodialysis and biological membrane transport, there including
the very high selectivity among different permeants exhibited by
unselective porous partitions when crossed by an heat flux (20-25)
the demonstrated existence of uphill transport (23,24) and the
generation of transmembrane differences of electric potentials by
thermodialysis (25). Accordingly the detection in <u>Valonia</u> of fluxes
of water and sulphate produced by an artificially induced heat flux,
superimposed to the one due to metabolic activity, constitutes an
expected effect, which directly proves the influence of transmembrane
temperature gradients on permeants transport through the membranes

of a living cell. The circumstance that water flux J_w is comparatively less affected by the induced heat flow than sulphate flux J_s,
is not surprising, since the very large water flux is simultaneously
influenced by all transport processes – of molecules and ions –
affecting the osmolality of endocellular fluids. For instance,
acceleration of Na^+ and K^+ fluxes in equal amounts would not affect
– per se – water transport.

The results of non-isothermal experiments summarized in Figs.
6a and b for water, and in Fig. 7 for sulphate, are interesting also
in view of their quantitative values. The artificially generated
heat fluxes indeed produce a very sizeable variation of permeant
exchange rates and of the fluxes.

Now, one of the most serious objections which can be opposed to
the assumed significance of heat flow in biological transport stems
from the smallness of the thermodynamic efficiency of the process
as compared with the high efficiency in the use of metabolic energy
often found in biological membrane transport (H.H. Ussing – private
communication). To this fundamental aspect of the problem very careful attention must be doubtlessly paid in future. Anyway, a direct
indication of the ability of heat flow to significantly contribute
to membrane transport in a living system is already provided by the
present results on non-isothermal transport of water and sulphate.
We have seen indeed that notwithstanding the low thermodynamic efficiency of matter transport induced by the flux of thermal energy,
a quite sizeable contribution to water and sulphate transport is
provided by this effect, at least in Valonia.

REFERENCES

1. Good,A.: Biochim. Biophys. Acta, 44:130 (1960).
2. Coldman,M.F. and Good,W.: Biochim. Biophys. Acta, 183:346 (1969).
3. Wright,E.M. and Diamond,J.M.: Proc. Roy. Soc. B., 172:227 (1969).
4. Diamond,J.M. and Wright,E.M.: Proc. Roy. Soc. B., 172:273 (1969).
5. Blinks,L.R.: J. Gen Physiol., 25:905 (1942).
6. Hogg,J., Willimas,E.J. and Johnston,R.J.: Biochim. Biophys.Acta,
       150:640 (1968).
7. Thorhavg,A.: Biochim. Biophys. Acta, 225:151 (1971).
8. Thorhavg,A.: in "Charged Gels and Membranes", Eric Sélégny (Ed.)
       Vol. II, 63, D. Reidel Publishing Company, Dordrecht-Holland
       (1976).
9. Crank,J. and Park,G.S.: "Diffusion in Polymers", Academic Press,
       London and New York (1968).
10. Tuwiner,S.B.: "Diffusion and Membrane Technology", Reinhold Publ.
       Corp., New York and London (1962).
11. Lonsdale,H.K.: "Properties of Cellulose Acetate Membrane in
       Desalination by Reverse Osmosis", Edited by U.Merten, the M.I.T.
       Press (1966).

12. Prigogine,I.: "Thermodynamics of Irreversible Processes".
    Int. Publ. New York (1954).
13. De Groot,S.R. and Mazur,P.: "Non-equilibrium Thermodynamics",
    North Holland Publ. Co., Amsterdam (1962).
14. Rastogi,R.P., Blokhra,R. and Agarwal,R.K.: Trans. Faraday Soc.,
    60:1386 (1964).
15. Rastogi,R.P. and Singh,K.: Trans. Faraday Soc., 62:1754 (1966).
16. Rastogi,R.P., Skukla,P.C. and Yadava,B.: Biochim. Biophys. Acta
    249:454 (1971).
17. Dariel,M.S. and Kedem,O.: J. Phys. Chem., 79:336 (1975).
18. Goldstein,W.E. and Verhoff,F.H.: A. I. Ch. E. J., 21:229 (1975).
19. Vink,H. and Chishti,S.A.A.: J. Membr. Sci., 1:149 (1976)
20. Gaeta,F.S., Mita,D.G. and Perna,G.: "Process of thermal diffusion
    across porous partitions and relative apparatuses". Patents:
    Italy N°928656 of 25/6/71. U.K. n°23590 of 19/5/72. France n°
    72-19189 of 29/5/72. URSS n°1798775/23-26 of 21/6/72. USA n°
    260497 of 7/5/72.
21. Gaeta,F.S. and Mita,D.G.: J. Membr. Sci., 3:191 (1978).
22. Bellucci,F., Bobik,M., Drioli,E., Gaeta,F.S., Mita,D.G. and
    Orlando,G.: Can. J. Chem. Eng., 56:698 (1978).
23. Bellucci,F., Drioli,E., Gaeta,F.S., Mita,D.G., Pagliuca,N. and
    Summa,F.: Trans. Faraday Soc. II, 75:247 (1979).
24. Gaeta,F.S. and Mita,D.G.: J. Phys. Chem., 83:2276 (1979).
25. Mita,D.G., Asprino,U., D'Acunto,A., Gaeta,F.S., Bellucci,F. and
    Drioli,E.: "Heat-flow-induced mass transport through porous
    partitions" in press on Gazzetta Chimica Italiana (1979).
26. Harvey,H.W.: "Recent Advances in the Chemistry and Biology of
    Seawater" p. 29, Cambridge (1945).
27. Pringsheim,E.G.: "Methods for the Cultivation of Algae". In
    "Manual of Phycology, G.H. Smith Ed., Waltham, Mass., USA (1951).
28. Robinson,J.B.: J. Exp. Bot., 20:201 (1969).
29. Robinson,J.B.: J. Exp. Bot., 20:212 (1969).
30. Mita,D.G., Bianco,M., Canciglia,P., D'Acunto,A., Minatore,C.,
    and Gaeta,F.S.: "Kinetics of Water Exchange in Valonia utri-
    cularis". In press on Gazzetta Chimica Italiana (1979).
31. Gaeta,F.S., Mita,D.G., D'Acunto,A., Bolis,L. and Canciglia,P.:
    "Physical Basis of Water Transport in Valonia utricularis" to
    be published in Proceedings of Internat. Conf. on Biological
    Membranes: "Membranes and the Environment" Crans-sur-Sierre
    (Valais) June 11-16 (1979). K.Bloch, L.Bolis, D.C.Tosteson
    editors, Raven Press, N.Y.
32. Gutknecht,J.: J. Gen. Physiol., 50:1821 (1967).
33. Gutknecht,J.: Science, Washington, 158:787 (1967).
34. Gutknecht,J.: Biochim. Biophys. Acta, 163:20 (1968).
35. Simpson,J.H. and Carr,H.Y.: Phys. Rev., 111:1201 (1958).
36. Wang,J.H., Robinson,C.V. and Edelman,I.S.: J.A. Chem. Soc.,
    75:466 (1953).

ACTIVATION ENERGIES IN ISOTHERMAL AND NON-ISOTHERMAL MEMBRANE

TRANSPORT

F.Bellucci[*], F.S.Gaeta, N.Pagliuca, D.G.Mita and
D.Tomadacis[*]

International Institute of Genetics and Biophysics of
C.N.R.
via Marconi, 10, Naples, Italy
[*] Istituto di Principi di Ingegneria Chimica, University
   of Naples, Piazzale Tecchio, Naples, Italy

1. INTRODUCTION

Transport phenomena induced by temperature difference across
a porous partition separating two liquid phases (effect of thermodialysis) has been only recently studied in its general aspects (1-7).
Clear-cut differences appear between thermodialysis and the closely
related effect of thermoosmosis (8-14), in various quantitative as
well as qualitative aspects. On the other hand various significant
analogies with non-isothermal matter transport in the bulk liquid
(effect of thermal diffusion) have already emerged (3, 6, 7).

An investigation of the dependence of matter fluxes upon temperature gradient and average temperature, may yield interesting information on the physical nature of the underlying transport mechanism.

Accordingly we experimentally investigated the influence of
either one of these two parameters on membrane transport. The system
studied was particularly simple, consisting of a synthetic membrane,
the AP-20 Millipore porous partition, sandwitched between two containers filled with distilled water maintained at different temperatures. Isothermal measurements of water transport in the same
system, induced by an hydrostatic pressure difference rather than
by a temperature gradient, were also effected to compare the relative
energy barriers.

Results thus obtained can be discussed in the frame of reference
of non equilibrium thermodynamics (15, 16) or in terms of inherent
molecular transport mechanisms.

145

2. EXPERIMENTAL

2.1 - Apparatus and methods

The apparatus used in non isothermal transport experiments is of a type already employed by us in previous researches (4,5,7). Since its design and performance have already been discussed in detail in the literature cited above, a very coincise description will suffice here.

Fig. 1 is a schematic representation of the apparatus, consisting of two adjacent semi-cells, communicating through a rectangular window over which the porous partition is stretched in the vertical

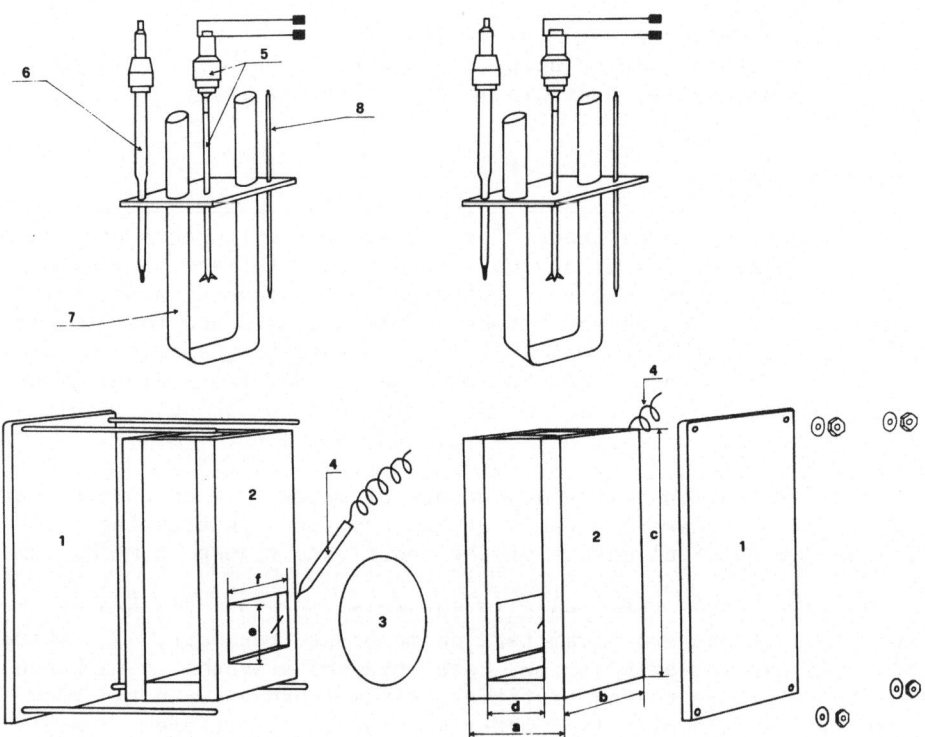

Figure 1. Schematic representation of apparatus and actual dimensions. The numbers indicate: (1) metallic holding frame; (2) semicells; (3) porous septum; (4) thermocouples; (5) motor, shaft and propeller; (6) Vertex; (7) U-shaped glass tube for thermostatting fluid; (8) Thermometer.
Dimensions: a = 35; b = 115; c = 145; d = 15; e = 60; f = 70 mm. Perspex walls are 1 cm thick.

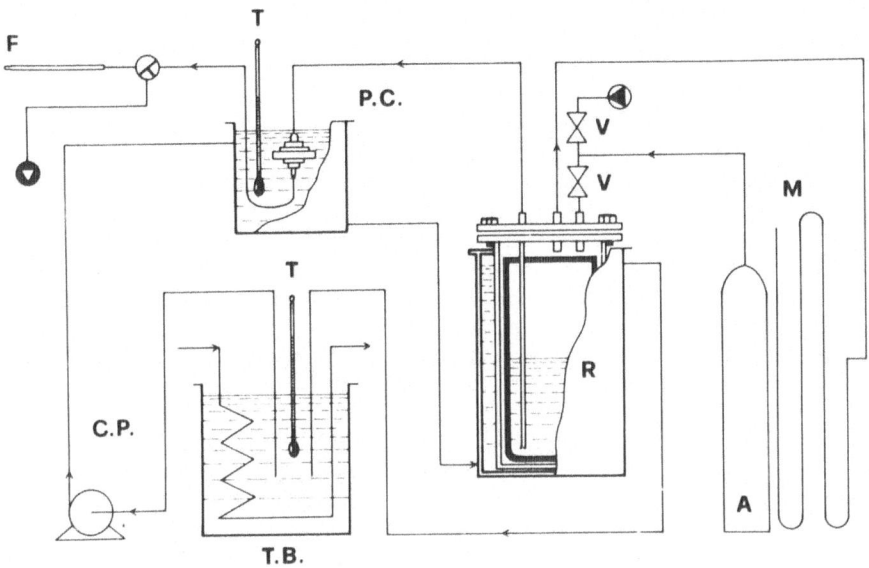

Figure 2.   Schematic representation of the permeability equipment:

P.C. , permeability cell;   T.B. , thermostatic bath;
R   , reservoir;   C.P. , circulation pump;
A   , air tank;   M   , manometer;
F   , capillary.

plane. Each semicell is kept at a predetermined temperature by
circulating an externally thermostated fluid through a U-shaped tube
immersed in the liquid. The fluid in both containers is stirred
continuously, to ensure uniform temperature distribution. While
the temperature in each semicell is measured by thermometers immersed
in the fluid, thermocouples positioned adjacent to the two membrane
faces read directly the transmembrane temperature difference. The
perspex walls allow precise level measurements to be effected in
each semicell during the runs, by means of a cathetometer. A porous
partition AP-20 Millipore, made of borosilicated glass microfibers
held together by an acrylic glue separates the warm distilled water
contained in one semicell, from the cooler water on the other side.
Mechanical and thermal resistence of these partitions in the range
of pressures and temperatures employed in this study is fair. The
reproducibility of the observed water fluxes, both upon repeated use
of the same partition in successive runs and in runs performed with
different AP-20 filters is satisfactory. Overall reproducibility
of results was found to be well within 5% (to be compared with a
10% overall variability observed in the runs with phenol and acetic
acid solutions (4, 5, 7)).

Isothermal measurements of hydraulic permeability at various
temperatures were performed with the simple apparatus schematically
represented in fig. 2. It essentially consists of a permeability
cell (P.C.), a reservoir (R) containing distilled water, both kept
at the same constant temperature T by the bath (TB), from which the
thermostating fluid is circulated through circulation pump (C.P.).
The transmembrane driving pressure is provided by an air tank (A),
and is measured by an open-air, low pressure intermediate fluid
(Hg-H_2O) manometer (M) having an accuracy of ± 1 mm Hg. The flow
rates are measured by means of the flow-pipe capillary (F) and a
stop watch.

2.2 - Results

Non isothermal experiments are conveniently summarized in fig.3,
where the observed water fluxes J_1 (gr/cm^2 min) are plotted against
temperature difference ΔT across the partition.

Actually in each run one measures the rate of level variation
of the water contained in each semicell. Constant rates have always
been found within the first 120 minutes of running. Since apparatus
geometry is well defined (see Fig.1) from these constant rates, the
water flux can be immediately calculated. These fluxes are plotted
in the figure against ΔT for various average temperatures T_{av} over
the range between + 18°C and + 40°C. Water is always observed to
flow from the warm to the cold semicell. Various interesting
features are discernible in these plots:
a) - Water fluxes are proportional to transmembrane temperature

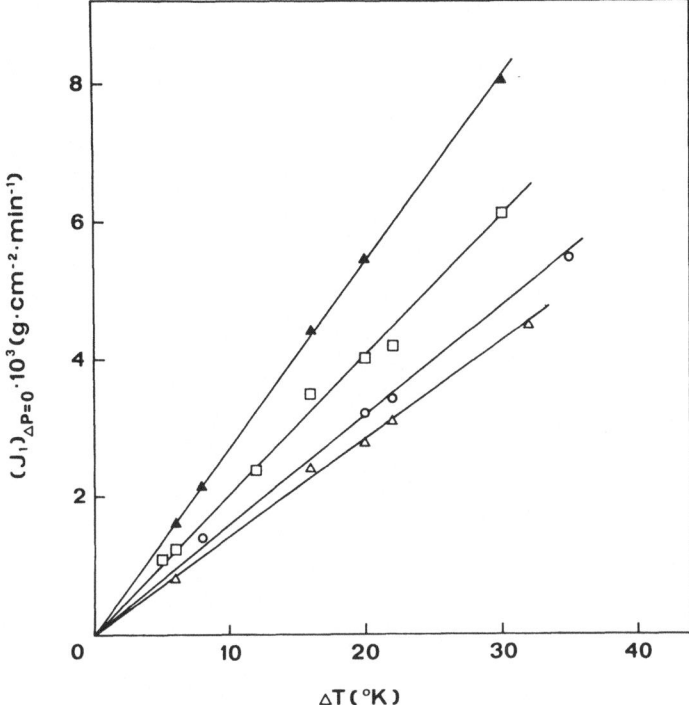

Figure 3. Water flux J_1 in g cm^{-2} min^{-1} at $\Delta P = 0$ as a function of
transmembrane temperature difference ΔT in °K, at differ-
ent average temperatures:

$\triangle$ T_{av} = 18°K; $\bigcirc$ T_{av} = 23°K; $\square$ T_{av} = 30°K; $\blacktriangle$ T_{av} = 40°K.

gradient. This linear relationship holds within all the explored
temperature range, and for temperature gradients ranging from zero
up to almost 10^3(°C/cm) (AP-20 partition thickness is 0.3 mm).
b) - Fluxes increase with average temperature (at constant ΔT).
c) - All plots extrapolate to coordinates origin, this circumstance
showing that the temperature difference measured by thermocouples
adjacent to the membrane faces is coincident with the effective
transmembrane temperature drop. Furthermore this same circumstance
also shows that there are no appreciable contributions to the observed flux from hydrodynamic agitation produced by stirring, nor
from the hydrostatic pressure head produced by the volume flow.
Isothermal steady-state water fluxes measurements performed with
the apparatus of fig.2, are reported in fig.4 as a function of hydrostatic pressure difference ΔP. The following main characteristics
emerge from examination of these plots:
a) - fluxes are proportional to transmembrane pressure difference

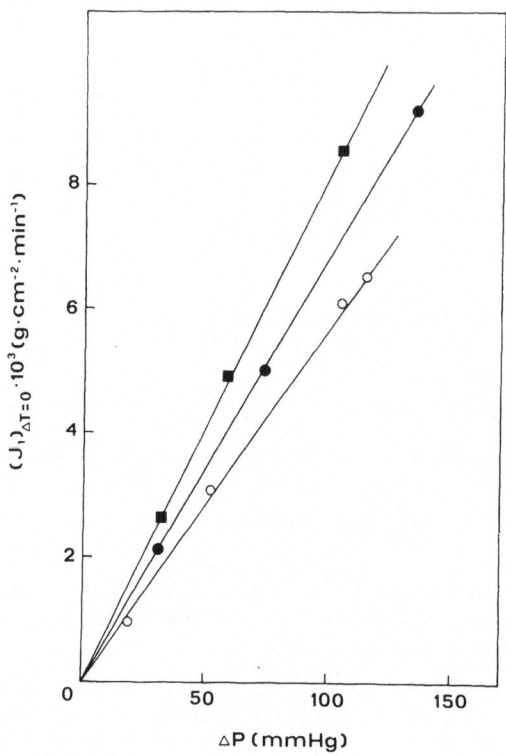

Figure 4. Water flux J_1 in g cm^{-2} min^{-1} at $\Delta T = 0$ as a function
           of hydrostatic pressure difference ΔP at different
           temperatures:
      o T = 30°K; ● T = 45°K; ■ T = 60°K.

throughout the explored pressure range (zero to 150 mm Hg).
b) - Fluxes increase with increasing temperature (at constant ΔP).
c) - All plots extrapolate to coordinates origin, thus confirming
that measured ΔP's accurately represent the effective transmembrane
pressure difference.

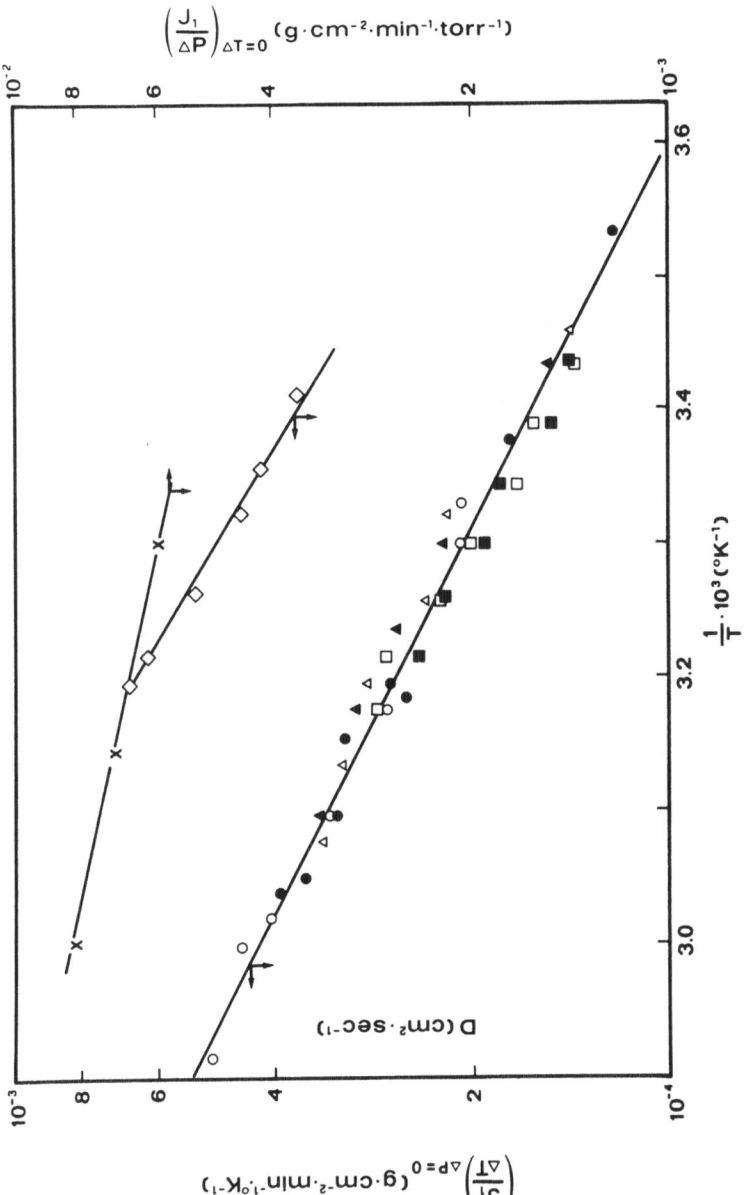

Figure 5. $(J_1/\Delta T)_{\Delta P=0}$ in g cm^{-2} $min^{-1}°K^{-1}$ as a function of T^{-1} (external left hand scale) at different ΔT:

○ $\Delta T = 6°K$; ● $\Delta T = 8°K$; △ $\Delta T = 12°K$; ▲ $\Delta T = 16°K$; □ $\Delta T = 20°K$; ■ $\Delta T = 22°K$.

X : $(J_1/\Delta P)_{\Delta T=0}$ in g cm^{-2} min^{-1} $torr^{-1}$ as a function of T^{-1} (right hand scale)

◇ : Diffusion coefficient data (inner right hand scale) from ref.(25).

In fig.5 both $(J_1/\Delta T)_{\Delta P=0}$ and $(J_1/\Delta P)_{\Delta T=0}$ are plotted in the classical Arrhenius form. As can be seen both sets of runs exhibit a linear behaviour with an high correlation coefficient. The somewhat higher correlation of isothermal results is in great part due to the better relative precision of measurement which can be achieved in those cases.

3. DISCUSSION AND CONCLUSIONS

These experimental results can be conveniently discussed within the frame of reference of non-equilibrium thermodynamics, relating the flux of the i-th species, J_i and the heat flux J_q through a permeable phase boundary, with the generalyzed forces X_q and X_k by the well-known phenomenological equations:

$$J_i = \sum_{k=1}^{N} \alpha_{ik} X_k' + \alpha_{iq} X_q$$

$$ (i=1,2,\ldots N) \tag{1}$$

$$J_q = \sum_{k=1}^{N} \alpha_{ql} X_k + \alpha_{qq} X_q$$

N being the number of species present, and α_{ik}, α_{iq}, α_{ql} and α_{qq} being the phenomenological coefficients. These coefficients, independent from the generalized forces and fluxes can on the other hand depend in an arbitrary way upon the mean values of the system's state variables (temperature, pressure, composition etc.) as well as on the nature of the permeable partition (porous, dense, etc.).

Equation (1) applies near equilibrium where X_k and X_q are small. What "small" means cannot be specified theoretically; relative orders of magnitude having to be experimentally assessed in each case.

In our system the linear dependence of water flux on temperature and pressure gradients (Figs.3 and 4) verifies the applicability of eq.(1). It will thus be possible to write the two following relations:

$$\left(\frac{J_1}{\Delta T}\right)_{\Delta P=0} = f\ (T_{av}); \quad \left(\frac{J_1}{\Delta P}\right)_{\Delta T=0} = g\ (T) \tag{2}$$

T being the Kelvin temperature at which each run was performed (average temperature in the case of the non-isothermal experiments). Equations (2) apply throughout the explored interval of the driving forces, down to the limiting case $\Delta T = \Delta P = 0$. The linearity of the plots of Fig.5 shows that our experimental data are well correlated by classical Arrhenius functions, that is:

$$f\ (T_{av})\ =\ f_o\ \exp\ (-\ E_{P,T}/R\ T_{av})$$

$$g\ (T)\quad =\ g_o\ \exp\ (-\ E_{P,P}/R\ T) \tag{3}$$

where $E_{P,T}$ is the apparent activation energy for water permeation under a temperature gradient, while $E_{P,P}$ is the activation energy for the pressure-driven process. These energies are directly related to the molecular mechanisms of water transport in either process (17, 18).

From the present results it appears that Eqs (3) apply throughout the explored temperature intervals (+ 30°C ÷ + 65°C in the hydraulic permeability experiments and + 10°C ÷ +70°C in the non-isothermal runs).

The respective values of $E_{P,T}$ and $E_{P,P}$ calculated from Fig.5 are about 4.8 Kcal/mole and 2.3 Kcal/mole respectively; the low activation energy of the isothermal process being indicative of a viscous flow mechanism while the relatively higher value of $E_{P,T}$ is comparable with those generally found in diffusive processes (17-19). The concept of apparent activation energies for transport in a temperature gradient arises from the different dissolution enthalpies of the permeant on opposite partition sides. From the definition of enthalpy it follows that

$$\Delta H_1\ +\ \Delta H_2\ =\ \int_{T_1}^{T_2} (C_{p,b}\ -\ C_{p,m})\ dT\ \simeq\ (C_{p,b}\ -\ C_{p,m})\ (T_2-T_1) \tag{4}$$

where ΔH_1 and ΔH_2 are dissolution enthalpies at temperatures T_1 and T_2, while $C_{p,b}$ and $C_{p,m}$ are the molar specific heats of water in the bulk liquid and in the membrane phase respectively. It is known that in dense membranes $C_{p,m}$ is different from $C_{p,b}$ (20). Our porous partition having higher water contents should presumably present much smaller differences between $C_{p,b}$ and $C_{p,m}$.

Equation (4) represents an overall energetic balance through which it is possible to assess that the value of the sum of dissolution enthalpies is rather small, while nothing can be said about their individual values.

Let us develop a simple kinetic argument enabling us to assess the influence of either one ΔH_1 or ΔH_2 on the overall process. Introducing a direct and an inverse process rate r_d and r_i, the net rate of water transport r will be:

$$r = r_d - r_i = K_d T_1 - K_i T_2 \tag{5}$$

where K_d and K_i are proportionality factors dependent upon T_1 and ΔH_1 and T_2 and ΔH_2 respectively. On the other hand the experimental results reported in fig.3 show that $r = J_1 = B(T_1 - T_2)$, where B is the permeability coefficient in the non-isothermal runs. By comparing the result with eq.(5) one finds that $K_d \simeq K_i$, otherwise a non linear dependence on ΔT should have been observed. This argument shows that the temperature variation of ΔH does not significantly affect water transport within the explored linear range; phase transition on the other hand should only account for moderate ΔH values. Even in the case of dense cellulose acetate membrane, ΔH is found to be 1.3 Kcal/mole (21). Now $E_{P,T}$ shall be:

$$E_{P,T} = E_{D,T} + \Delta H_T \tag{6}$$

where $E_{D,T}$ is the activation energy for the merely diffusive process within the membrane. From the above value of $E_{P,T}$ and the estimated order of magnitude of ΔH one finds that $E_{D,T}$ must be some 5 to 6 Kcal/mole. Incidentally we observe that the value of the activation energy for the self-diffusion of isotopically labelled water is about 5 Kcal/mole (22-24). Data on water transport in cellophane membranes from the literature (25) have been reported in graphical form in Fig.5 for comparison sake. The computed activation energy from these data is 5.5 Kcal/mole.

Non-isothermal water transport in AP-20 porous partition thus appears to be of a diffusive type. Porous structure and dimensions should enable us to specify the importance of membrane matrix influence on such a transport mechanism.

Poiseuille equation for viscous flow in our case is:

$$J_1 = \frac{\varepsilon \, R^2}{8\eta \, \tau \, \delta} \, \Delta P \tag{7}$$

where ε is pore fractional surface, τ a tortuosity factor, δ and R membrane thickness and equivalent pore radius. When applied to the experimental results of fig.4, for T = 30°C gives $R \simeq 3 \cdot 10^{-5}$ cm having assumed $\varepsilon = 0.5$ and $\tau = 2$. This value of R is several orders of magnitude greater than water hydration diameter, thus water membrane diffusion is seen to occur in a basically homogeneous liquid phase.

We can now express the mass flux J_1 as volume flux $\overline{J}_1 = J_1/\rho$ and define an integral phenomenological coefficient D'^* as:

$$D'^* = \frac{\overline{J}_1}{\Delta T/\delta} \tag{8}$$

having the dimensions óf an ordinary coefficient of thermal diffusion D' $cm^2 sec^{-1}°K^{-1}$. From the nature of the transport phenomenon described above, D'^* appears representative of self-thermal
diffusion of water in membrane water. Interestingly enough D'^*
turns out to be of the expected order of magnitude for such a parameter, that is : $D'^* \simeq 1.0·10^{-7}$ cm^2/sec °K at + 30°C.

From the analysis of the experimental results discussed above,
the following conclusions can be drawn:

a) The difference between the activation energies for the isothermal
   and non-isothermal processes $E_{P,P}$ and $E_{P,T}$ are significative.
   They clearly indicate the occurrence of two different transport
   mechanisms.
b) The higher value of $E_{P,T}$ is indicative of a mass transport process
   taking place by movement of single particles (diffusion) or by
   translocation of small groups of particles. The relatively lower
   value of $E_{P,P}$ is assumed to be indicative of a viscous flow
   mechanism.
c) The ΔH value seems to be indicative of a somewhat different
   structural organization of bulk water and membrane water, involving breaking - to some extent - of water molecules hydrogen
   bonding, in the transition from the bulk into the membrane phase.

On the other hand having recently suggested that thermodialysis
might play a significant role in biological membrane transport (2),
(26), (27) we find interesting to compare the results obtained here
for non-isothermal water transport through synthetic porous partitions with the analogous known data of water transport in biological membranes.

REFERENCES

1) Gaeta F.S., Mita D.C. and Perna G.: "Process of thermal diffusion
   across porous partitions and relative apparatuses". Patents:
   Italy N°928656 of 25/6/71. U.K.n°23590 of 19/5/72. France n°
   72-19189 of 29/5/72. URSS n°1798775/23-26 of 21/6/72. USA n°
   260497 of 7/5/72.
2) Gaeta F.S. and Mita D.G., Proc.IV Int. Winter School of Biophysics
   of Membrane Transport", Wisla (Poland), Febr.18-28, part III,
   197 (1977).
3) Gaeta F.S. and Mita D.G., J.Membr.Sci., 3:191 (1978)
4) Bellucci F., Bobik M., Drioli E., Gaeta F.S., Mita D.G. and
   Orlando G.: Cand.J.Chem.Eng., 56:698 (1978).
5) Bellucci F., Drioli E., Gaeta F.S., Mita D.G., Pagliuca N., and
   Summa F.: Trans.Faraday Soc., II, 75:247 (1979)
6) Gaeta F.S. and Mita D.G.: J.Phys.Chem., 83:2276 (1979)
7) Mita D.G., Asprino U., D'Acunto A., Gaeta F.S., Bellucci F., and
   Drioli E., "Heat-flow-induced mass transport through porous

        partitions", in press on Gazzetta Chimica Italiana

8) Haase R., Z.Phys.Chem. (N.F.) 21:244 (1959).

9) Haase R. and De Greiff H.J., Z.Naturforsch., 26:1773 (1971).

10) Rastogi R.P., Blokhra R. and Agarwal R., Trans.Faraday Soc.
        60:1386 (1944).

11) Dariel M.S. and Kedem O., J.Phys.Chem., 79:336 (1975).

12) Goldstein W.E. and Werhoff F.H., A.I.Ch.E.J., 21:229 (1975).

13) Vink H. and Chishti A.A., J.Membr.Sci., 1:149 (1976).

14) Rastogi R.P. and Mishra B., J.Membr.Sci., 4:1 (1978).

15) Prigogine I., "Thermodynamics of Irreversible Processes", Inter-
        science; New York (1954).

16) De Groot S.R. and Mazur P., "Non-equilibrium Thermodynamics",
        North Holland, Amsterdam (1962).

17) Crank J. and Park G.S., "Diffusion in Polymers", Academic Press,
        London and New York (1968).

18) Tuwiner S.B., "Diffusion and Membrane Technology", Reinhold
        Publ.Corp., New York and London (1962).

19) Lonsdale H.K., "Properties of cellulose acetate membrane in
        Desalination by Reverse Osmosis", Edited by U.Merten, the
        M.I.T. press (1966).

20) Burghoff H.G. and Pusch W., Journal of Applied Polymer Sci.,
        23:473-484 (1979).

21) Mauersberger H.R., ed., Textile Fibers, 6th ed., John Wiley &
        Sons, New York, 1954, pp.886-7.

22) Glasstone S., Laidler K.J. and Eyring H., "The theory of Rate
        processes", New York, McGraw-Hill Book Co.Inc., 1941.

23) Simpson J.H. and Carr H.Y., Phys.Rev., 111:1201 (1958).

24) Wang J.H., Robinson C.V. and Edelman I.S., J.A.Chem.Soc., 75:
        466 (1953).

25) Madras S., McIntosh R.L. and Mason S.G., Can.J.Research, 27B:
        764 (1949).

26) Gaeta F.S. and Mita D.G., Chimia, 30:261 (1976).

27) Mita D.G., Bianco M., Canciglia P., D'Acunto A., Minatore G. and
        Gaeta F.S.: "Kinetics of water exchange in Valonia Utricularis",
        in press on Gazzetta Chimica Italiana.

TRANSPORT COEFFICIENTS IN A NONLINEAR SERIES MEMBRANE ARRAY

G. Monticelli and F. Celentano

Istituto di Fisiologia Generale e di Chimica biologica
dell'Università

via Mangiagalli, 32 - 20133 MILANO

Despite its many noticeable successes, several limitations of
the linear thermodynamic description of membrane transport, as proposed by Kedem and Katchalsky[1], have been encountered. These arise
as a consequence both of the various experimental non linear force-
-flow relationships found to date and of some intrinsic limits of
the linear treatment[2,3,4]. Suggestions intended to overcome the
mentioned difficulties have been made by Patlak et al.[5] and by Kedem
and Katchalsky[6], utilizing series membrane arrays. Series arrays are
particularly interesting as they approximate the structure of the
epithelia and allow a correlation between the experimentally accessible global phenomenological coefficients of the array and the intrinsic characteristics of the single membranes. Some of these corre-
lations will be analyzed in the following emphasizing the problems
they pose in experiments.

THE MODEL SYSTEM

Let consider an isothermal system composed by two parallel
membranes a and b of thickness δ^a and δ^b, separating an internal
compartment 2 of finite volume from two external infinite phases 1
and 3. The three compartments contain an aqueous solution of the
nonelectrolyte s with concentration C_i and hydrostatic pressure p_i
(i=1,2,3). A solute active transport J_A takes place across the
membrane a.

In order to avoid the difficulties arising from the introduction
of a mean solute concentration, independent of the local flow conditions, the practical equations for steady state flow 1 may be
written in local form[5], obtaining for membrane a

$$J_v = - L_P^{'a} \left(\frac{dp}{dx} - \sigma^a RT\phi \frac{dC}{dx} \right)$$

$$J_s - J_A = (1 - \sigma^a) C(x) J_v - \omega^{'a} RT \frac{dC}{dx}$$

and for membrane b

$$J_v = - L_P^{'b} \left(\frac{dp}{dx} - \sigma^b RT\phi \frac{dC}{dx} \right)$$

$$J_s = (1 - \sigma^b) C(x) J_v - \omega^{'b} RT \frac{dC}{dx}$$

where J_v and J_s are the volume and solute flows, $L_P^{'i}$, σ^i and $\omega^{'i}$ are respectively the hydraulic permeability, the reflection coefficient and the solute permeability for the membrane i, $\emptyset$ is the osmotic coefficient and x is the flow direction, perpendicular to the membranes and positive from compartment 1 towards 3.

The equations above are integrated across the a and b membranes with the boundary conditions $C^a(0) = C_1$, $C^a(\delta^a) = C^b(0) = C_2$, $C^b(\delta^b) = C_3$ and the similar for pressure. The integrated equations may yield J_v, J_s, C_2 and p_2 as a function only of the accessible variables pertaining to the external compartments 1 and 3 [5,7,8]. We will write here only the volume flow equation, involving three important transport coefficients.

$$J_v = - \Lambda \left\{ \Delta p - RT\phi \left[\sigma^b + \frac{(\sigma^a - \sigma^b)(1-\sigma^b)(1-e^{\alpha J_v}) e^{-\beta J_v}}{(1-e^{\alpha J_v})(1-\sigma^b) + (e^{-\beta J_v}-1)(1-\sigma^a)} \right] \Delta C \right.$$

$$+ RT\phi \frac{(\sigma^a - \sigma^b)(1-e^{\alpha J_v})(e^{-\beta J_v}-1)}{(1-e^{\alpha J_v})(1-\sigma^b) + (e^{-\beta J_v}-1)(1-\sigma^a)} \frac{J_A}{J_v} \tag{1}$$

$$\left. - RT\phi \frac{(\sigma^a - \sigma^b)^2 (1-e^{\alpha J_v})(e^{-\beta J_v}-1)}{(1-e^{\alpha J_v})(1-\sigma^b) + (e^{-\beta J_v}-1)(1-\sigma^a)} C_1 \right\}$$

where $\Delta p = p_3 - p_1$, $\Delta C = C_3 - C_1$, $\Lambda = L_P^a L_P^b / (L_P^a + L_P^b)$, $\alpha = (1-\sigma^a)/P^a$, $\beta = (1-\sigma^b)/P^b$, with the positions $L_P^i = L_P^{'i}/\delta^i$ and $P^i = \omega^{'i} RT/\delta^i$.

Equation (1) is implicit in J_v and composed by four additive terms in Δp, ΔC, J_A and C_1, the first three of them being analogous to the ones found in the linear J_v equation [9,11]. The fourth addendum

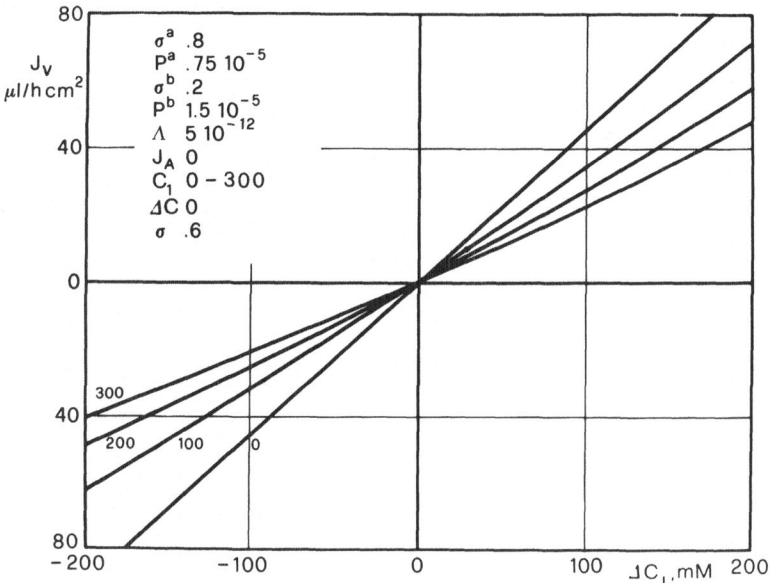

Fig. 1 - Volumetric flow calculated by means of eq. (1) as a function
of hydrostatic pressure difference, expressed as osmolarity
difference of an impermeant solute. For C_1 = 0 the curve is
reduced to the straight line representing the classical filtration law by D'Arcy.

accounts for the concentration of the permeating solute taken as a
reference for calculating ΔC. This term introduces a non negligible
correction to the classical linear filtration law, to which eq. (1)
is reduced when no solute is transported, even when ΔC = 0, fig. 1.

TRANSPORT COEFFICIENTS

The filtration coefficient L_p is defined in a non linear situation as the slope of the J_v (Δp) curve at zero solute osmotic pressure difference $\Delta \pi$ = RT $\emptyset$ ΔC = 0. As eq. (1) is implicit in J_v but
can be readily solved for Δp, $1/L_p$ can be calculated. This is a
rather complicated expression whose limit for vanishing J_v is shown

$$\lim_{J_v \to 0} \left(\frac{\partial \Delta p}{\partial J_v}\right)_{\Delta C=0} = \frac{1}{\Lambda} + RT\phi \frac{(\sigma^a - \sigma^b)^2}{P^a + P^b} \left\{ C_1 + \frac{J_A}{2(P^a + P^b)} \right\} \qquad (2)$$

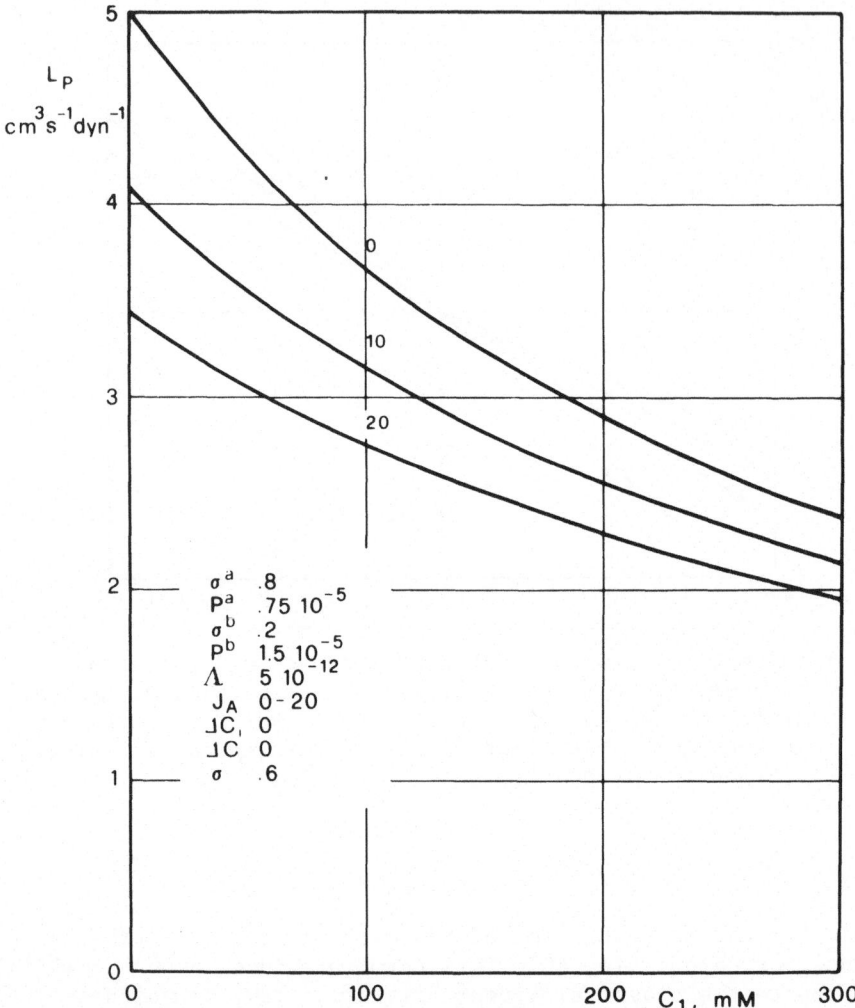

Fig. 2 - Filtration coefficient as a function of C_1 for different
values of the solute active transport. The hydraulic permeability of the membrane array decreases with increasing
solute concentration and active transport.

It follows that the resistance to filtration is not merely the sum
of the single membranes resistances, as a consequence both of active
transport and of the very presence of a permeating solute (Fig. 2).
In a similar way $1/L_{PD}$ can be obtained:

$$\lim_{J_v \to 0} \left(\frac{\partial \Delta \Pi}{\partial J_v}\right)_{\Delta p=0} = \frac{p^a + p^b}{\sigma^a p^b + \sigma^b p^a} \left\{ \frac{1}{\Lambda} + RT\phi \frac{(\sigma^a - \sigma^b)^2}{p^a + p^b} \right. \cdot$$

$$\left. \cdot \left[C_1 + \frac{J_A(2\sigma^b - 1)}{2(\sigma^a p^b + \sigma^b p^a)} \right] \right\} \tag{3}$$

L_{PD} behaves similarly to L_P in function at C_1 and J_A.

The reflection coefficient is defined in the linear theory as $\sigma = (\Delta p/\Delta \Pi)_{J_v = 0}$. Similarly, from eq. (1)

$$\lim_{J_v \to 0} \left(\frac{\Delta p}{\Delta \Pi}\right) = \frac{\sigma^a p^b + \sigma^b p^a}{p^a + p^b} + \frac{\sigma^a - \sigma^b}{p^a + p^b} \frac{J_A}{\Delta C} \tag{4}$$

This reflection coefficient is not a constant, as its value depends on the experimental conditions and on J_A. It formally corresponds to the apparent reflection coefficient previously defined for a single linear membrane[10,11]. For $J_A = 0$ the eq. (4) is reduced to

$$\sigma = \frac{\sigma^a p^b + \sigma^b p^a}{p^a + p^b} \tag{5}$$

which, when $\delta^a = \delta^b$ becomes identical to the reflection coefficient determined by Kedem and Katchalsky[6] for two membranes in series.

Eq. (5) can be obtained also from the ratio of eqs. (3) and (2) when $J_A=0$, but eq. 4 cannot be obtained from the same ratio. In effect it is true that

$$\lim_{J_v \to 0} \left(\frac{\partial \Delta p}{\partial J_v}\right)_{\Delta \Pi=0} \Big/ \lim_{J_v \to 0} \left(\frac{\partial \Delta \Pi}{\partial J_v}\right)_{\Delta p=0} \neq \lim_{J_v \to 0} \frac{\Delta p}{\Delta \Pi}$$

From eq. (1) it is also possible to calculate the Δp^* and $\Delta \Pi^*$ values of the hydrostatic and osmotic pressure differences required to zero the volume flow in presence of active transport when, respectively, $\Delta\Pi = 0$ or $\Delta p = 0$:

$$\Delta p^* = RT\phi \; \frac{\sigma^a - \sigma^b}{p^a + p^b} \; J_A$$

$$\Delta\Pi^* = - RT\phi \; \frac{\sigma^a - \sigma^b}{\sigma^a p^b + \sigma^b p^a} \; J_A$$

$$(6)$$

It is readily seen that also the ratio $\Delta p^*/\Delta\Pi^*$ gives the reflection coefficient (4).

DISCUSSION

The procedure proposed by Patlak et al.[5] in order to recover the intrinsic non linearity of transport across thick membranes seems a very useful one, because it recognizes the limits of the integral practical equations derived by Kedem and Katchalsky[1] and utilizes these equations in a local form, a situation where the limits pointed out by several Authors[2,3,4] vanish. As demonstrated by Axel[3], such a treatment leads to an integral equation for solute flow coincident with the one derived by hydrodynamic methods.

The model system described above remains anyhow a simplified model for epithelia, as it neglects diffusive phenomena in the inner compartment and the possibility of coupling between the passive and active solute flows, which can be excluded only in mosaic membranes. In any case the results seem useful in order to explain the experimental data and to obtain a better control of the experiments. In effect, the variability of the filtration, ultrafiltration and apparent reflection coefficients which appear to be a function of both the experimental conditions and of the activity of the epithelia, imposes to be very careful in programming the experiments and when comparing the values existing in literature, often determined in poorly defined conditions.

The $J_V(\Delta p)$ curve in fig. 1 is similar to experimentally determined curves, i.e. for corneal endothelium[12] and frog skin[13]. Unfortunately the fitting of eq. (1) to the experimental data by purely statistical methods seems rather difficult, requiring the determination of five parameters. Anyhow three independent relationships between the parameters can be chosen within eqs. (2) to (6) expressing experimentally measurable quantities as a function of the unknowns to be determined and of C_1 and J_A, which are also measurable. It follows that the independent determination of only two parameters is required, a problem that can be solved observing that

$$\lim_{J_v \to \infty} \left(\frac{\partial \Delta p}{\partial \Delta \Pi}\right) C_1, \ J_A = \sigma^b$$

$$(7)$$

$$\lim_{J_v \to \infty} \left(\frac{\partial \Delta p}{\partial \Delta \Pi}\right) C_3, \ J_A = \sigma^a$$

Thus the reflection coefficients of the two membranes can be determined by means of graphical or numerical derivative of the curve obtained with couples of Δp and $\Delta \Pi$ values producing the same volumetric flow in experiments with C_1 or C_3 constants respectively for positive or negative values of the flow. The values obtained by this method, although estimated by extrapolation to infinite volumetric flow, are rather accurate because the curve obtained by derivation is limited between 0 and 1 and thus becomes sufficiently flat for low, experimentally accessible, flows.

REFERENCES

1. O. Kedem and A. Katchalsky, Analysis of the permeability of biological membranes to non-electrolytes, Biochim. Biophys. Acta 27:229 (1958)
2. G. S. Manning, E.H. Bresler and R.P. Wendt, Irreversible thermodynamics and flow across membranes, Science 166:1438 (1969)
3. L. Axel, Flow limits of Kedem-Katchalsky equations for fluid flux, Bull. Math. Biol. 38:671 (1976)
4. D.C. Mikulecky, A simple network thermodynamic method for series--parallel coupled flows. II The non-linear theory, J.Theoret. Biol. 69:511 (1977)
5. C.S. Patlak, D.A. Goldstein and J.F. Hoffman, The flow of solute and solvent across a two-membrane system, J. Theoret. Biol. 5:426 (1963)
6. O. Kedem and A. Katchalsky, Permeability of composite membranes. III Series array of elements, Trans. Faraday Soc. 59:1941
7. F. Celentano and G. Monticelli, Non Linear force-flow relationships in biological membranes, in:"Proc. 2nd Winter School on Biophysics of Membrane Transport" Agricultural Academy, Wroclaw (1975)
8. F. Celentano and G. Monticelli, A theoretical model for non linear osmosis in biological membranes, in: "Proc. 2nd Natl Conf. Biophys.", Bucarest, (1976)
9. O. Kedem, Criteria of active transport, in "Membrane Transport and Metabolism", A. Kleinzeller and A. Kotyk eds., Academic Press, New York (1960)
10. F. Celentano and G. Monticelli, Un confronto fra alcune descrizioni termodinamiche del trasporto attivo, in: "Atti 1° Riunione Scient. S.I.B.P.A." Camogli (1973)

11. G. Monticelli, F. Celentano, and G. Torelli, Sodium Chloride
    reflection coefficient in rabbit gall bladder, Biochim.
    Biophys. Acta 401:41 (1975)
12. J. Fishbarg, Fluid transport by corneal endothelium, in:
    "Comparative Physiology: Water, Ions and Fluid Mechanics"
    K. Schmidt-Nielsen, L. Bolis and S.H.P. Maddrell eds.,
    Cambridge U.P., Cambridge (1978)
13. F. Celentano, G. Monticelli and M.N. Orsenigo, Non linear
    volume flow dependence on osmotic pressure difference in
    frog skin, J. Physiol. (Paris) 74:365 (1978)

STRUCTURAL ASPECTS OF ENERGY TRANSDUCING MEMBRANES

Giorgio Lenaz, Giovanna Curatola, Laura Mazzanti

Istituto di Biochimica, Università di Ancona

and

Giovanna Parenti-Castelli, Mauro Degli Esposti, Enrico Bertoli

Istituto di Chimica Biologica, Università di Bologna

INTRODUCTION

The elucidation of membrane functions is close to the stage where the properties of a given system are correlated with and explained by the knowledge of its structure. This stage has been already reached for the function of some soluble proteins, e.g. haemoglobin. The functional properties of haemoglobin (sigmoidal shape of the O_2-binding curve, Bohr effect, the effect of diphosphoglycerate etc.) have been correlated with the precise knowledge of the tridimensional structure of the haemoglobin molecule and of its conformational changes when the ligand (O_2) and the allosteric effectors (H^+, diphosphoglycerate etc.) are bound.

This stage has not yet been reached for any membrane function, but the functional investigations and the structural studies have often obtained such a degree of sophistication to attempt a structure to function correlation in many cases.

In general, the functions linked to biomembranes (electron transfer, energy coupling, active transport, hormone action etc.) are known in better detail than the structures to which they are associated.

For example "ionic channels" are better known as biophysical (kinetic and thermodynamic) entities than as real molecular structures. Nevertheless, the knowledge available of the general structural properties of membranes has reached certain important conclusions, so that attempts may be made to suggest how ionic channels can be organized in a natural membrane at the molecular level.

TABLE I

Structural problems in oxidative phosphorylation

Electron transfer

1) Is electron transfer asymmetric?

Elucidation of asymmetry and molecular organization of each electron transfer component within each complex in the membrane and role of "noncatalytic" components.

2) How are electrons transported?

Elucidation of distances between prosthetic groups in a complex and eventually of the complete tridimensional structure in the complex. Orientation and mobility of protein complexeses in the membrane. Orientation and the mobility of quinones (For example chemiosmotic coupling requires that quinones are reduced inside and reoxidized outside).
Existence of a lipid annulus and its effect on interprotein electron transfer or the interactions of quinones.

Coupling

1) Are proteins ejected through asymmetric electron transfer alternated with hydrogen transfer or by proton pumps?

Elucidation of the complete tridimensional structure will give evidence where are prosthetic groups located and if proton channels are present.

2) What is the mechanism of ATP synthesis?

Structure of ATPase and its proton channel
How are protons coupled to ATP synthesis?

Ion movements

Is Ca^{2+} transport mediated by a channel? How is it coupled to the high energy state (H^+ cotransport?)

The membrane structure described by the model of Singer and
Nicolson (1) has the major advantages of considering three universal
properties of membranes, i.e. the organization of lipids as a fluid
bilayer, the penetration of proteins into the bilayer, and the a-
symmetry of the membrane components. This model as such is however
too general and perhaps not accurate in the description of protein
tridimensional structures in the bilayer.

Notwithstanding the dramatic increase of knowledge of the last
few years, there are a number of uncertainties even for simple
problems as the lipid organization in the bilayer. One further
complication is represented by the dynamic character of the membrane,
so that, for example, the same protein may be integral or peripheral
(1), depending on the functional state of the protein.

Reviews covering the main aspects of membrane structure are
available (if.e.g. 2-4).

The final stage in the understanding of a function at the molecular level is represented by the knowledge of the structure to
which it is linked; this will often mean to know the exact position
in space of each atom of an integral protein within the lipid bilayer.

Research is not far from this stage in the case of bacteriorhodopsin, the protein involved in the light-driven proton pump of
Halobacterium halobium (5).

Let us take as an example the problem of coupling in energy-
conserving membranes such as those of mitochondria, chloroplasts and
chromatophores. A critical evaluation of the chemio-osmotic hypothesis of coupling (6) will depend upon the answers to several questions listed in Table I in the case of mitochondrial oxidative
phosphorylation.

Orientation and mobility of Coenzyme Q

A problem which has been investigated in our laboratory concerns
the sidedness of the redox cycle of ubiquinone (Coenzyme Q, CoQ).

In view of the important role of CoQ in the respiratory chain
and of the assumption that it is a mobile electron carrier between
Complexeses I or II and Complex III (7), we have investigated the
effects of different quinone homologs in mitochondrial electron
transport (8). The results we have found have suggested some kind
of compartmentation, so that short-chain quinones reduced by NADH
cannot reach their site of reoxidation, whereas longer-chain quinones
can. Due to the higher polarity of short-chain quinones in comparison with the very hydrophobic CoQ_{10}, such compartmentation could be
the result of sidedness of the redox cycle of the quinone in the
membrane (9). If the reduced quinone must cross the membrane to

TABLE II

Oxidation of ubiquinones in mitochondria and SMP (10)

	Specific activity		Relative efficiency (°)	
	Mitochondria	SMP	Mitochondria	SMP
CoQ-3	0.104	0.202	51	100
CoQ-7	0.167	0.037	82	18

(°) Assuming as 100% the rate of CoQ_3 in SMP

TABLE III

Effects of ubiquinones on the physical state of membrane lipids

A) EPR studies in lipid vesicles (egg lecithin) (11)

Quinone (0.4mM)	5-NS	16-NS
	$T_{,,}$ (gauss)	$(10^{-9} sec)$
-	25.6	0.8
$CoQ_3 ox$	25.2	0.7
$CoQ_3 red$	25.9	1.0
$CoQ_{10} ox$	24.6	0.7
$CoQ_{10} red$	25.0	0.8

B) Fluorescence polarization studies in mitochondria (E.Bertoli, unpublished)

Quinone (0.04mM)	Polarization of perylene fluorescence
-	0.05
$CoQ_3 ox$	0.28
$CoQ_7 ox$	0.06

be reoxidized, it is suggestive to postulate that the more polar
short-chain homologs may not be able to cross the membrane with a
sufficient rate. Additional support to this hypothesis originates
from the difference observed in hydroquinone oxidation in intact
mitochondria and submitochondrial particles, which are known to have
opposite polarity. It was found that reduced CoQ_3 is oxidized at
much higher rate in submitochondrial particles (SMP) than in mitochondria, whereas reduced quinones having longer chains are oxidized
at higher rates in intact mitochondria than in SMP (10) (Table II)

The above studies suggest that the site of oxidation of natural
ubihydroquinone is located near the inner side of the inner membrane.

Structural studies are under way to substantiate this hypothesis
in order to understand the interactions of oxidized and reduced
quinones having different chain lengths in both artificial lipid
bilayers and in the inner membrane of mitochondria. The low mobility
of short chain CoQ homologs is compatible with spin and fluorescence
probe data (11) indicating that they have an ordering effect on lipid
bilayers, particularly when in the reduced state; on the other hand,
long-chain quinones have a disordering effect (Table III). The
ordering effect of short-chain quinones may be related to structural
effects on membrane protein leading to functional alterations, such
as the reversible loss of oligomycin sensitivity of mitochondrial
ATPase (Table IV).

The problem of the boundary lipids

On the basis of spin label evidence, Jost et al (12) have suggested that a layer of lipid molecules is immobilized in contact with
integral protein molecules; these "boundary lipids" exchange with
surrounding lipids at much lower rate than lipids in the bulk bilayer.

The demonstration of the existence of boundary lipids appears
very important for the properties of mitochondrial proteins. A
boundary layer would represent a dynamic outer surface of each protein, capable of affecting conformational plasticity and hence catalytic activity of the protein itself, and also its interactions with
other proteins or different molecules (such as quinones!), besides
affecting passive permeability in the areas surrounding the proteins.
Evidence in favor and against boundary lipids is listed in Table V.

We have found that general anesthetics (that are lipid-soluble
molecules) induce some fluidization of the lipid bilayer in liposomes
prepared from mitochondrial lipids, but the effect is rather small
except when high concentrations of anestetics are used. However,
in mitochondrial membranes we have found a large disordering effect
induced by anesthetics (17).

Usually the fluidization is much greater in the core of the

TABLE   IV

Effect of CoQ$_3$ on oligomycin and DCCD sensitivity of mitochondrial
ATPase (M. Degli Esposti, G. Parenti-Castelli, E. Bertoli, unpublished
observations).

Addition	% Inhibition of ATPase activity	
(n moles)	Q added before oligomycin	Q added after oligomycin
1) Oligomycin (50 mg/ml)		
none	41	47
Q$_3$ (342)	9	26
Q$_7$ (352)	38	36
Q$_3$ + Q$_7$	31	27
2) DCCD (0.8 mM)		
none	53	53
Q$_3$ (228)	20	53

TABLE   V

Evidence for and against boundary lipids

1) Yes - Spin labels and fluorescence polarization suggest increase
of order in lipids induced by protein; DSC, X-ray etc. suggest that
in presence of proteins on aliquot of membrane lipids do not undergo
a cooperative transition (2)
2) No - NMR (DMR) evidence suggests that lipids are disordered by
proteins (13)
3) Yes - Possibility that a space-average disorder (detected by NMR)
is accompanied by a time-average increase in viscosity (detected by
EPR). The relaxation times in NMR are several orders of magnitude
slower than in EPR.
4) No - Possibility that immobilized lipids (EPR) are only trapped
between proteins at high protein concentrations (evidence with
gramicidin A) (14)
5) Yes - In cytochrome oxidase immobilized lipids (EPR) are found
also at very low protein concentrations. At high protein concentrations experimental spectra are less "mobile" than theoretical spectra
suggesting trapping (15).
6) Yes - Anesthetics abolish immobilized lipids. Their effects on
particle distribution, enzyme kinetics and protein conformation are
quivalent to the effects of delipidation, suggesting that anesthetics
affect primarily lipid protein interactions (16,17).

TABLE VI

Effects of anesthetics on the mobility of stearic acid spin labels in phospholipid vesicles and in mitochondrial membranes

Anesthetics	Concentration mM	5-NS		16-NS	
		Vesicles	Mitochondria	Vesicles	Mitochondria
		Decrease of $2T_{,,}$(gauss) (°)		Decrease of τ_c (sec 10^{-10}) (°°)	
Butanol	5	–	0	–	-1.2
	10	0	-0.5	-0.3	-2.2
	50	-3	-2	-3	-6.5
Halothane	0.37	–	-1	–	-1.1
	1	–	-2	–	-1.1
	1.0	0	-2.5	0	-4
Ketamine	0.08	–	-0.5	–	-1.3
	0.48	–	-1	–	2.7
	2.0	-0.5	-1	-3	-4.5

(°) $2T_{,,}$ was 50.D gauss in control vesicles and 55.5 gauss in control mitochondria

(°°) τ_c was 13×10^{-10} sec in control vesicles and 20×10^{-10} sec in control mitochondria

membrane (as probed by decrease of the rotational correlation times (τ_c) of 16-doxylstearate) than near the membrane surface (as probed by decrease of the hyperfine splitting ($2T_{11}$) of 5-doxylstearate) (Table VI and Fig. 1).

Anesthetics appear to abolish the short-range immobilization induced by proteins on lipids (tentatively ascribed to boundary lipids): in other words, there is a labilization of lipid proteins interactions induced by the membrane perturbation. This appears to us as an indirect proof in favor of the existence of boundary lipids. The disordering effect in membranes is induced at low anesthetic concentrations, compatible with those known to cause general anesthesia.

The effects of anesthetics in particle distribution in the fracture faces after freeze-etching microscopy, on enzyme kinetics and on membrane protein conformation, are equivalent to the effects of lipid removal, suggesting that they affect primarily lipid-protein interactions (17-21) (Table VII).

In particular, conformational changes in membrane proteins have been observed by means of circular dichroism upon addition of anesthetics. Anesthetics induce a decrease of intensity of the negative dichroic bands in the region of 224-208 nm (indicative of loss of

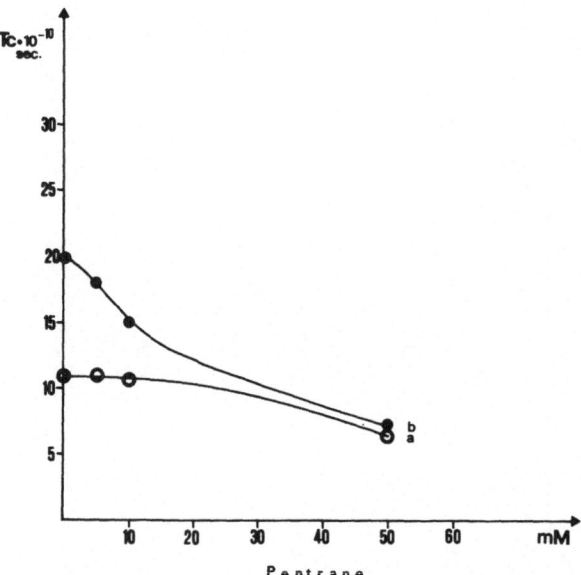

Fig. 1 - Effect of pentrane on the mobility of 16-doxylstearate in mitochondrial membranes (b) and in vesicles obtained from mitochondrial phospholipids (a).

TABLE VII

Effects of anesthetics and lipid removal on some kinetic properties of mitochondrial ATPase and on conformation of membrane proteins (16, 19-21)

Parameter studied	Anesthetics	Delipidation
ATPase		
Arrhenius plots		
discontinuity	abolished or flattened	maintaned* or abolished**
activation energy	increased above the discontinuity	slightly increased* or increased*
Substrate concentration	uncompetitive inhibition	uncompetitive inhibition*
oligomycin sensitivity	abolished	decreased***
	Hill coeff. from-2to-1	Hill coeff. from-2to-1*
Conformation of proteins (byCD)(°)	loss of α-helix	loss of α-helix
Aggregation of intramembrane particles	increased	increased

Delipidation by means of phospholipase A_2(*), Triton.X.100(**) or acetone(***)

(°) Effect of anesthetics on the CD spectra of purple membrane from H.Halobum (G.Parenti-Castelli, G. Lenaz, M.M.Long, unpublished).

	Decrease in ellipticity (%)	
	210 nm	224 nm
Butanol 0.35 M	-32	-39
Ketamine 1.7mM	-32	-39
Ketamine 4.3mM	-50	-36

α-helical structure) in several membranes (16-17); similar changes were found reversibly by delipidation of mitochondrial membranes (19).

The α-helix decrease induced by delipidation or perturbation of lipid-protein interactions by anesthetics has a theoretical explanation in the fact that an increased accessibility of water to peptide bonds induces a competition of H_2O-peptide hydrogen bonds with peptide-peptide hydrogen bonds that are required to stabilize secondary structure such as α-helix.

It is therefore justified to advance the hypothesis that anesthetics primarily alter lipid protein interactions at the level of the boundary lipids that exert their role on catalytic activity and conformation.

REFERENCES

1 - Singer S.J. and Nicolson G.L. Science 175, 720, 1972.
2 - Lenaz G., "Membrane Proteins and their Interaction with Lipids" (R.A. Capaldi ed.) M. Dekker, New York, 1977, p. 47.
3 - Lenaz G., SubCell Biochem. 6, 233, 1979.
4 - Vanderkooi G., Biochim. Biophys. Acta 344, 307, 1974.
5 - Stoeckenius W., Lozier R.H. and Bogomolni R.A., Biochim. Biophys. Acta 505, 215, 1979.
6 - Mitchell P. in "Electron Transfer and Oxidative Phosphorylation" (E. Quagliarello et al., eds.) North Holland, Amsterdam, 1975, pp. 305.
7 - Hill M.W., Biochim. Biophys. Acta 356, 117, 1974.
8 - Lenaz G., Pasquali P., Bertoli E., Parenti-Castelli G., Folkers K.,Arch. Biochem. Biophys. 169, 217, 1975.
9 - Lenaz G., in "Molecular Biology of Membranes" (S. Fleischer et al. eds.) Plenum, 1978, p. 137.
10 - Lenaz G., Landi L., Cabrini L., Pasquali P., Sechi A.M., Ozawa T., Biochem. Biophys. Res. Commun. 85, 1047, 1979.
11 - Spisni A., Masotti L., Lenaz G., Bertoli E., Pedulli G.F., Zannoni C., Arch. Biochem. Biophys. 190, 454, 1978.
12 - Jost P.C., Griffith O.H., Capaldi R.A., Vanderkooi G., Biochim. Biophys. Acta, 311, 141, 1973.
13 - Seelig A., Seelig J., Hoppe-Seyler's Z. Physiol. Chem. 359, 1747, 1978.
14 - Chapman D., Cornell B.A., Eliase A.W., Perry A., J. Mol. Biol. 113, 517, 1977.
15 - Marsh D., Watts D., Mashke W., Knowles P.F., Biochem. Biophys. Res. Commun. 81, 397, 1978.
16 - Lenaz G., Curatola G., Mazzanti L., Parenti-Castelli G., Mol. Cell. Biochem. 22, 3, 1978.
17 - Curatola G., Mazzanti L., Bertoli E., Lenaz G., Bull. Mol. Biol. Med. 3, 1235, 1978.
18 - Lenaz G., Curatola G., Mazzanti L., Parenti-Castelli G., Bertoli E., Biochem. Pharmacol. 27, 2835, 1978.

19 - Masotti L., Lenaz G., Spisni A., Urry D.W., Biochem. Biophys. Res. Commun. 56, 892, 1974.
20 - Parenti-Castelli G., Sechi A.M., Landi L., Cabrini L., Mascarello S., Lenaz G., Biochim. Biophys. Acta 547, 161, 1979.
21 - Pasquali Ronchetti I., Curatola G., Mazzanti L., Lenaz G., Bertoli E., XII Congress Ital. Soc. Electron Microscopy, Ancona 20-22 Sept. 1979.

KINETIC ASPECTS OF ENERGY COUPLING IN BACTERIAL PHOTOPHOSPHORYLATION UNDER CONTROLLED THERMODYNAMIC CONDITIONS

Giovanni Venturoli, Aurelio De Santis and Bruno Andrea Melandri

Institute of Botany, University of Bologna
Bologna (Italy)

The most essential feature of the chemiosmotic hypothesis for electron transport-linked phosphorylation is the postulation of a vectorial translocation of protons mediated by the redox carriers of the electron transport system (1). According to the original proposal of the hypothesis this translocation gives rise to an electrochemical gradient of protons in diffusion equilibrium with the aqueous bulk phases capable of driving ATP synthesis through an asymmetric ATPase. A widespread agreement exists at present about the general idea of a protonic coupling operating in energy transducing membranes of mitochondria, chloroplasts and bacteria (2). On the basis of theoretical considerations and of experimental results, it has been however suggested that rapid diffusion phenomena of protons not in equilibrium with the bulk phases could be involved in the energy transduction mechanism (3,4); these processes, localized at the interphases (4), could lead to short range interactions between the photosynthetic units and the ATPase complexes and could give rise to a degree of coupling between redox reactions and phosphorylation higher than that expected on the basis of the original proposal of the chemiosmotic hypothesis (4,5,6). In this situation, it is clear that a systematic kinetic study of the transient events of energy transduction (7,8) can be utilized to integrate the informations obtained in experiments based on a classic thermodynamic approach, which are generally based on a comparison of the phenomenological forces involved in the process under stationary conditions (5,9).

The most relevant and direct parameter in the study of transients of ATP synthesis is the affinity of the phosphorylation reaction (A_p). In fact the onset of phosphorylation can be identified as an inversion in the process of energy transduction due to ATP hydrolysis. During this inversion, caused by the onset of a coupled force (the protonic

gradient) opposing the input flow (ATP hydrolysis), the system proceeds through a state in which no net flow of the ATPase reaction
occurs. Taking into account the stoichiometry and the degree of
coupling, the protonic potential difference balances, under this
condition, the input force for ATP hydrolysis. The free energy
change for ATP hydrolysis therefore sets the energetic threshold to
be overcome when the onset of phosphorylation takes place (10).

The photophosphorylating system of photosynthetic bacteria is
especially suitable for the study of the transient events of ATP
synthesis under controlled thermodynamic conditions. Single turnovers of the cyclic electron transport can in fact be induced by
10 μs xenon flashes of saturating light: this allows the resolution
of the induction kinetics into the elementary processes which contribute to energy transduction. Flash-induced phosphorylation can
be evaluated coupling ATP formation to the luciferin-luciferase
reaction (11) and monitoring the luminescence of the system subsequent
to the enrgization of chromatophores. This method offers a very high
sensitivity, which allows the detection of the ATP synthesized even
in a single turnover of the redox system, and makes it possible to
evidentiate threshold effects eventually due to small changes in the
affinity of the phosphorylation reaction.

The results of a set of experiments in which the ATP produced
by chromatophores of Rhodopseudomonas capsulata during sequences of
single turnover flashes was measured using the luciferin-luciferase
assay, are shown in Fig. 1. In these assays the adenylate kinase
activity present in bacterial chromatophores and in the crude firefly luciferase was utilized as a buffer enzyme system for the control
of ATP/ADP concentration ratios. The addition of AMP or of exogenous
ATP enabled us to change the affinity for ATP synthesis in a range
of the order of 3 Kcal·mole^{-1}. At the maximal value of A_p attainable
with the described methodology (-11.6 Kcal·mole^{-1}), photophosphorylation is not immediately induced, but requires several turnovers of
the electron transport. On the contrary, when the energetic threshol
which opposes the onset of the protonic force is lowered to -8.2 Kcal
mole^{-1}, phosphorylation starts after the first single flash with an
yield which is 50% of the maximal yield, reached already after the
second flash at this A_p. The data presented in Fig. 1 are therefore
fully consistent with the general concepts of chemiosmosis: they
clearly indicate that the formation of the driving force for ATP
synthesis is a cooperative process to which more than one turnover
of the same photosynthetic unit can contribute.

A comparison between the electrical capacitance and the intrinsi
buffer capacity of the membrane leads to believe that during the firs
turnovers of the electron flow, the driving force is mainly formed by
the electrostatic component of the protonic gradient and that a
measurable ΔpH is built up only by a prolonged energization of the
system (12,7). The response of flash-induced phosphorylation to

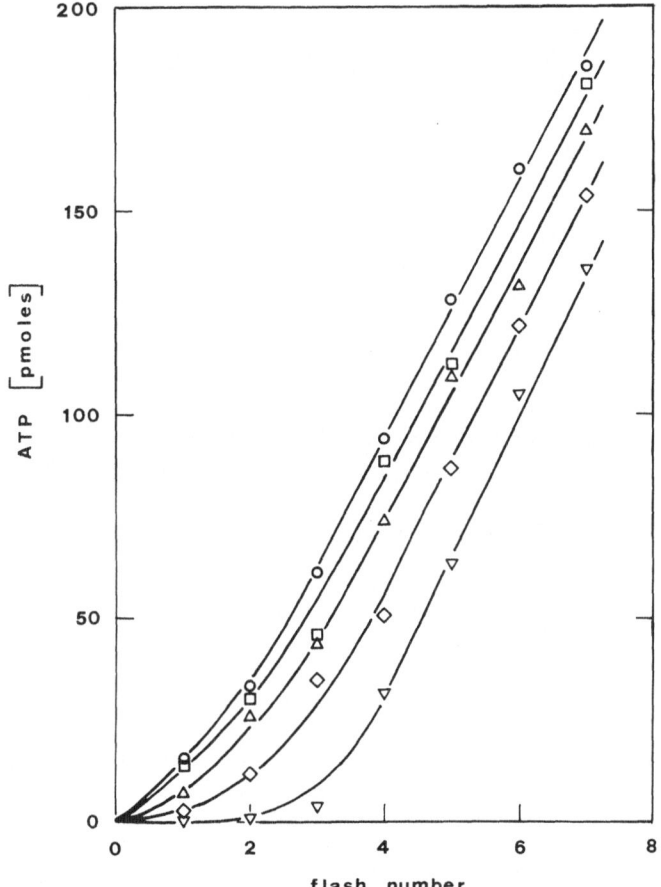

Fig.1 - The dependence of the induction kinetics of phosphorylation
upon the magnitude of the affinity for ATP synthesis (A_p)
in chromatophores from Rhodopseudomonas capsulata.
A_p Kcal·mole^{-1} : (○-○) -8.2; (□-□) -9.6; (△-△) -9.8;
(◇-◇) -11.1; (▽-▽) -11.5.
The assay medium contained in 2 ml, at 30°C: 100 mM sodium
glycylglycine (pH 7.75), 10 mM magnesium acetate, 0.1% bovine
serum albumin, 8 (or 1) mM inorganic phosphate, 0.2 mM sodium
succinate, 20 µM ADP, 10 mM KCl, 4 µM nigericin, 60 µM luci-
ferin (Sigma), 1.2 mg of crude firefly lantern extract (Sigma
FLE-50) and chromatophores corresponding to 10 µl BChl.
Photosynthetic reactions were induced by xenon flashes
(10 µs half width) of saturating intensity, fired 160 ms
apart. The actinic light was filtered through two layers
of 88A wratten filters plus a 665 nm cut of Schott glass
filter. Light-induced ATP synthesis was assayed measuring
the luciferin-luciferase luminescence by means of a photo-
multiplier screened by a Corning L-96 filter. The luminescence was calibrated by addition of known amounts of ATP.

uncoupling by ionophores was consistent with these considerations
and confirmed the energetic interpretation of the results of Fig. 1.
In the presence of 10 mM KCl, nigericin, which catalyzes a $K^+- H^+$
exchange dissipating ΔpH, did not affect the ATP yield during the
first single turnover flashes; on the other hand, the K^+-conductor
valinomycin, which dissipates the electrostatic component of the
protonic gradient, causes a considerable lag in the onset of ATP
synthesis, which takes place only after 20 single turnover flashes
(7).

These data, as a whole, indicated that during the first turnovers of the electron flow, the ATP yield - at critical and known
values of the phosphate potential - was strictly dependent on the
energetic state of the membrane and could therefore give direct informations about the electric fields involved in energy transduction.
This observation suggested the possibility of an accurate analysis
of the decay kinetics of the light-induced high energy state of the
membrane through measurements of the ATP yield after one single flash
during a postillumination period. In this set of experiments (whose
results are shown in Fig. 2) a high membrane potential was elicited
by a 125 ms flash; a single turnover flash was then fired after a
dark time of variable length and ATP monitored. The results showed
that a maximum ATP yield was reached when the single short flash was
fired 200-300 ms after the long flash. If the flash was delayed further, the ATP yield decreased steadily and vanished after about 30 s.
This behaviour indicated that the energetic state induced by 125 ms
flash relaxed very slowly since it could contribute to the synthesis
of ATP for several seconds after phosphorylation in postillumination
had stopped. The results of similar experiments performed in the
presence of different concentrations of valinomycin clearly demonstrated the destabilizing action on the high energy state promoted by
the ionophore: at $3 \cdot 10^{-8}M$ valinomycin and 10 mM KCl the ATP yield
induced by the single flash during postillumination was already
negligible after 1 s of darkness.

A spectroscopic method based on the light-induced absorption
band shift of endogenous carotenoids is widely utilized to evaluate
the membrane potential in photosynthetic systems (13). The spectral
response of these pigments has been extensively studied in natural
and model systems, and a good agreement exists that this shift of
carotenoids can be considered as an "in situ" indicator of electric
fields involved in the processes of energy transduction (14). Doubts
have been, however, expressed either on the quantitative use of this
method which is based on calibrations with KCl pulses plus valinomycin in the dark and, particularly, on the actual localization of
the electric fields detected by this approach (15). We have performe
an analysis of the carotenoid shift signals under the same experimental conditions in which the relaxation of the high energy state of
the membrane was directly assayed by measuring the flash-induced ATP
yield (Fig. 2). The values of the carotenoid signals observed before
and after the single turnover flash, as a function of the dark time

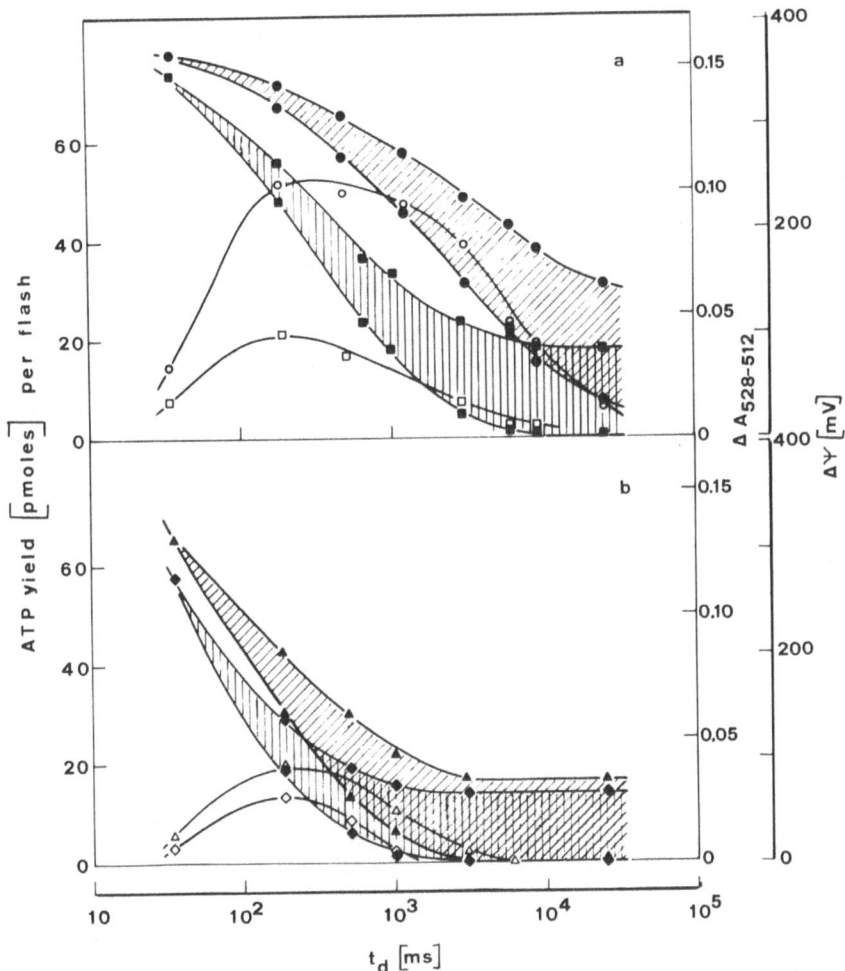

Fig. 2 - The ATP yield (open symbols) and the carotenoid band shift
(closed symbols) induced by a single turnover flash during
a postillumination period as a function of the dark time
(t_d) between the 125 ms and the 10 μs flashes.
Conditions as described in Fig.1. Carotenoid band shift
was monitored using a dual wavelength spectrophotometer at
528-512 nm. The oscillations between the values of the
carotenoid signals observed before and after the single
flash are indicated by the shaded areas.
In addition to a control experiment (Fig.2a,○,●) the effect of valinomycin is also shown: (Fig.2a,□,■) 10mM valinomycin; (Fig.2b,△,▲) 20 mM valinomycin; (Fig.2b,◇,◆)
30 mM valinomycin. The phosphate potential was controlled
throughout the experiments and was about -10 Kcal mole[-1].

between the long and the short flashes fired in sequence are shown
in Fig. 2. The decay of the absorption band shift is relatively slow
in the absence of valinomycin; when the ionophore is present this
decay is strongly accelerated and, as shown for the ATP yield, no
postillumination effect is observed after 1 s of darkness at $3 \cdot 10^{-8}$M
valinomycin. The strict correlation between electrochromic signals
of carotenoids and ATP yield during the relaxation of the light-
-induced high energy state evidentiated by these data, was confirmed
by the results of further experiments in which the yield of ATP
synthesis and the corresponding carotenoid signals, observed in pulsed
light after a few initial turnovers, were studied as a function of
flash frequency.

The correlation between ATP yield and carotenoid shift emerging
from the results of these last experiments and from the data obtained
in postillumination (see Fig. 2) is represented in Fig. 3. On the
basis of purely phenomenological observations, this plot clearly
demonstrates the validity of the carotenoid shift as a quantitative
indicator of an electrostatic energetic state of the membrane driving
ATP synthesis. This conclusion is independent of any assumption
about the actual profile of the membrane potential and, of course,
of the acceptance of the calibration of the carotenoid signals with
K^+ diffusion pulses in the dark.

CONCLUSIONS

The set of experiments described in this paper demonstrates that
the formation of a high energy state of the membrane can require more
than one turnover when photophosphorylation operates against highly
negative affinities for ATP synthesis; this high energy state is
essentially electrostatic and it is quantitatively measured by the
extent of the carotenoid signal.

The experiments demonstrate, on the other hand, that ATP synthesis takes place only as a consequence of a turnover of the photosynthetic electron transport, even if a high membrane potential is
present and is potentially able to drive ATP synthesis upon excitation
of electron flow by additional flashes, as clearly shown in Fig. 2.
This study therefore confirms, and resolves in signle turnover events,
the strict coupling existing between redox reactions and ATP synthesis
in chromatophores of photosynthetic bacteria, a coupling which, however, requires also the concurrent existence of a large transmembrane
difference in protonic potential.

Acknowledgments

This work was supported by a grant from Consiglio Nazionale
delle Ricerche (n. 78.02186.04).

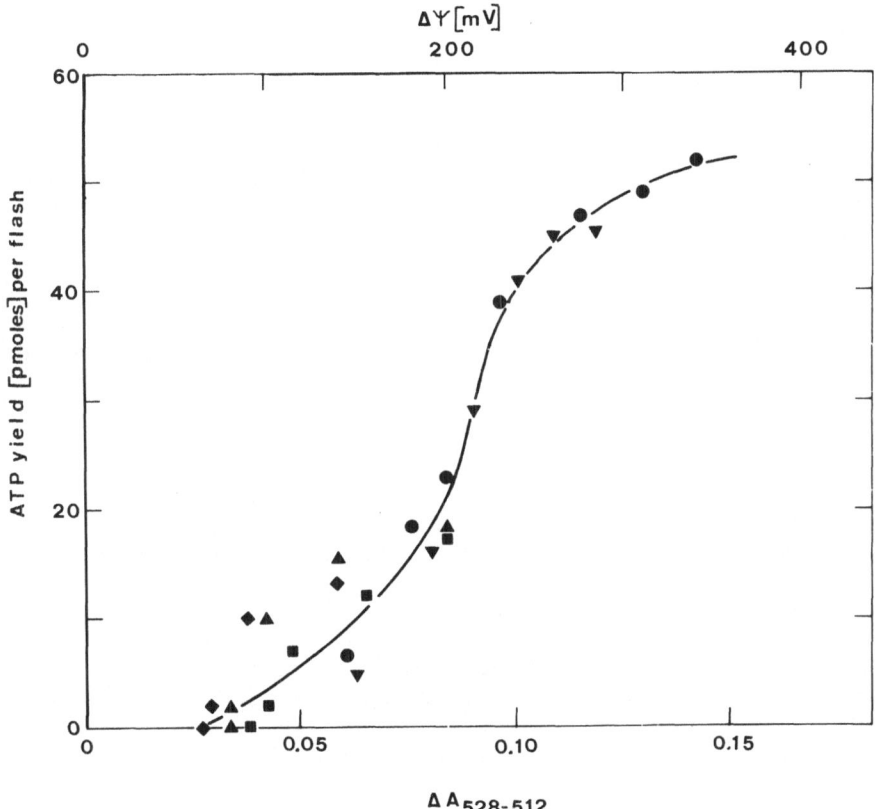

Fig. 3 - Correlation between the ATP yield per flash and the carotenoid band shift obtained from the data shown in the
descending parts of the curves of Fig.2 and from the
results of experiments in which the ATP yield and the
carotenoid shift were studied as a function of flash
frequency.

REFERENCES

1) P.Mitchell, Chemiosmotic coupling and energy transduction, Glynn Research, Bodmin (1968).

2) P.D.Boyer, B.Chance, L.Ernster, P.Mitchell, E.Racker and E.C. Slater, Oxidative phosphorylation and photophosphorylation, Annu.Rev.Biochem. 46:955 (1977).

3) R.J.P.Williams, The multifarious couplings of energy transudction Biochim.Biophys.Acta 505:1 (1978).

4) D.B.Kell, On the functional proton current pathway of electron transport phosphorylation, Biochim.Biophys.Acta 549:55 (1979).

5) H.V.Westerhoff and K.Van Dam, Irreversible thermodynamic description of energy transduction in biomembranes, in: Current Topics in Bioenergetics, vol.9, D.R.Sanadi ed., Academic Press, New York (1979).

6) B.A.Melandri and A.Baccarini-Melandri, Bioenergetics of the early events of bacterial photophosphorylation, in: Cation flux across biomembranes, Y.Mukohata and L.Packer eds., Academic Press, New York (1979).

7) B.A.Melandri, A.De Santis, G.Venturoli and A.Baccarini-Melandri, The rates of onset of photophosphorylation and of the protonic electrochemical potential difference in bacterial chromatophores, FEBS Lett. 95:130 (1978).

8) K.M.Petty and J.B.Jackson, Kinetic factors limiting the synthesis of ATP by chromatophores exposed to short flash excitation, Biochim.Biophys.Acta 547:474 (1979).

9) H.Rottenberg, S.R.Caplan and A.Essig, A thermodynamic appraisal of oxidative phosphorylation with special reference to ion transport by mitochondria, in: Membranes and ion transport, vol.1, E.E.Bittar ed., Wiley Interscience, London (1970).

10) A.Baccarini-Melandri, R.Casadio and B.A.Melandri, Thermodynamics and kinetics of photophosphorylation in bacterial chromatophores and their relation with the transmembrane electrochemical potential difference of protons, Eur.J.Biochem. 78:389 (1977).

11) A.Lundin and M.Baltscheffsky, Measurement of photophosphorylation and ATPase using purified firefly luciferase, in: Methods in Enzymology, vol.57, A.M.DeLuca ed., Academic Press, New York (1978).

12) B.A.Melandri, R.Casadio and A.Baccarini Melandri, Energy levels and rates of photophosphorylation in bacterial chromatophores, in: Photosynthesis '77, D.O.Hall, J.Coombs and T.W.Goodwin eds., The Biochemical Society, London (1978).

13) J.B.Jackson and A.R.Crofts, The high energy state in chromatophores from Rhodopseudomonas sphaeroides, FEBS Lett. 4:185 (1969).

14) H.T.Witt, Energy conversion in the functional membrane of photosynthesis. Analysis by light pulse and electric pulse methods, Biochim.Biophys.Acta 505:355 (1979).

15) H.Rottenberg, The measurement of transmembrane electrochemical Proton gradients, Bioenergetics 7:61 (1975).

ROLE OF A "CRITICAL" FATTY ACID CONCENTRATION ON PHOSPHOLIPID

EXCHANGE AND SIZE ENLARGEMENT OF SONICATED VESICLES

S. Massari and R. Colonna

Centro per lo studio della Fisiologia dei mitocondri -

Istituto di Patologia Generale - Università di Padova

INTRODUCTION

In recent times fatty acids were used not only as probes for the phospholipid bilayer structure, but also as inducers of important biological processes such as membrane fusion (1), membrane permeability changes (2), phospholipid exchange (3,4) and phospholipid vesicle size enlargement (3-7). These observations point out that fatty acids, having a higher concentration in heart and brain than in other tissues, may play an important role in regulating many biological functions. In the present work the role of fatty acids in phospholipid exchange and in vesicle size enlargement is investigated.

RESULTS

The capacity of release of PL molecules from sonicated vesicles was measured by means of the translocation rate of PL molecules across a dialysis membrane. DPPC vesicles, about 2 μmol Pi, were added to the upper cell of the dialysis apparatus. The amount of PL translocated was measured by determining the inorganic phosphate content in the lower cell and it was found to increase linearly with the time. Fig. 1 shows the PL translocation rate from vesicles with different chain length when variable amounts of myristic acid were added to the upper cell of the dialysis apparatus. Temperature was 50°C. The translocation rate abruptly increased at a "critical" myristic acid concentration, dependent on the phospholipid chain length. Similar results were obtained when myristic acid was co-sonicated with PL molecules to form mixed vesicles. The abrupt increase occurred with vesicles either in the solid or in the liquid--crystalline state. However the "critical" value of myristic acid

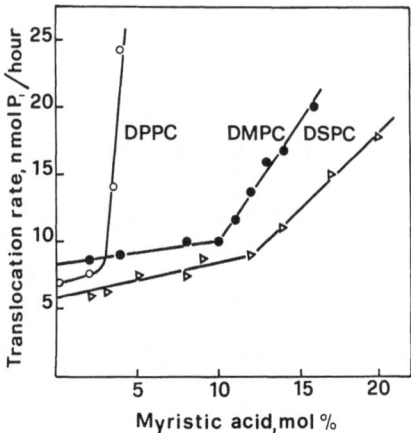

Fig. 1 - Phospholipid translocation rate in vesicles with different
phosphatidylcholine chain length.

responsible of the marked increase of the translocation rate decreased
as the incubation temperature was increased.

Fig. 2 shows the fluorescence response of ANS and pyrene and
the permeability properties of the vesicle membrane to K^+ when a
DPPC vesicles suspension maintained at 25°C was supplemented with
myristic acid. The addition of myristic acid decreased the ANS
fluorescence, due to a release into the solution of bound ANS molecules or to an increase of polarity or viscosity of the ANS microenvironment. Electrostatic repulsion between membrane bound ANS
molecules and negatively charged fatty acid molecules should not
occur since the apparent pK_a of myristic acid bound to the membrane
is approximately 8, while the experimental pH was 5. The excimer-
-monomer ratio of pyrene (2.5 mol% cosonicated with PL), obtained
by dividing the fluorescence intensity at 470 nm by that at 394 nm
(excitation wavelength 340 nm) increased by increasing the fatty
acid concentration. Similar results, obtained by Usher et al. (8)
were explained by the pyrene clusters formation. The fluorescence
responses of both ANS and pyrene abruptly changed at the same "critical" myristic acid concentration causing the abrupt increase of
the PL translocation rate. Furthermore, the marked pyrene fluorescence change compared to ANS response indicated that membrane
structural modifications occurred probably in the hydrocarbon region.
K^+ trapped into the vesicles was measured with a K^+ electrode after

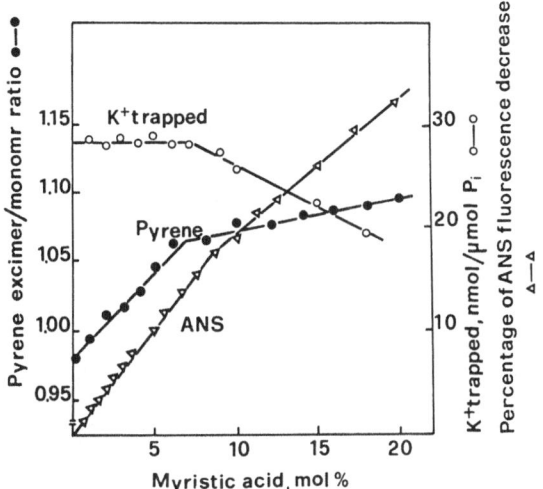

Fig. 2 - ANS and Pyrene fluorescence response and K$^+$ trapped at
various myristic acid concentration in DPPC vesicles.

lysis of the vesicles with TRITON-X 100. Fig. 2 shows the amount
of K$^+$ trapped in the vesicles after 2 hours of incubation in the
presence of variable amounts of myristic acid. At a "critical" fatty
acid concentration (the same causing the abrupt increase of the PL
translocation rate) an increased K$^+$ efflux was observed.

The rate of size enlargement of PL vesicles was followed by
measuring the time dependent light scattering increase. In the presence of relatively high amounts of myristic acid, DPPC vesicles
showed a temperature dependence growth in size (Fig. 3A). ΔI represents the light intensity increase during the first 10 minutes,
I_0 represents e light intensity at time zero. Excitation and emission
wavelength were 400 nm. The maximum rate of size enlargement occurred
at 37-38°C, two degrees below the critical temperature of the vesicles.
The size enlargement was irreversible for several reasons: 1) the
light scattering values of the size enlarged vesicles remained stable
at 38°C even after cooling or warming the sample. 2) A transformation
of the smaller vesicle into more extended structures was confirmed
by analysis of the eluate of a Sepharose 4 B chromatography column.
Fig. 3B shows that a marked increase of the light scattering response
was induced at a "critical" fatty acid concentration. However this
value was higher than that inducing the abrupt increase of PL translocation rate. When measured at various PL concentrations, the rate

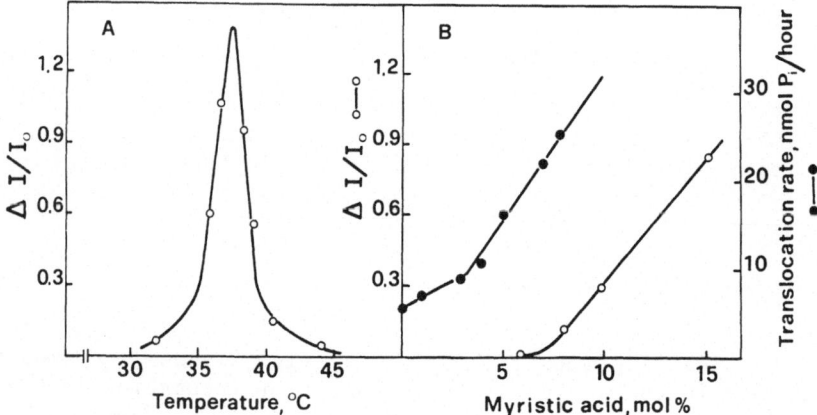

Fig. 3 – A. Light scattering increase induced by the temperature in DPPC vesicles. B. DPPC vesicles: light scattering increase ●——● and phospholipid translocation rate ○——○ at various myristic acid concentrations.

of light scattering response increased following a second-order kinetics, suggesting that the size increase could occur by a collision
process. This interpretation was confirmed by experiments in which
the surface potential of the vesicles was modified either by changing
the pH or by adding low amounts of negative PL. As the vesicle surface becomes charged, the size increase process is strongly inhibited.
By using fatty acid – DPPC cosonicated vesicles, it was shown that
the "critical" concentration value of fatty acid inducing the abrupt-
-increase of the PL translocation rate and of light scattering response was dependent on the chain length of the fatty acid used
(myristic, palmitic and stearic). The longer the fatty acid chain
length, the higher concentration of fatty acid responsible of the
abrupt changes of the phenomena was necessary. Local anesthetics,
at concentrations of $10^{-4} - 10^{-3}$ M strongly inhibited the rate of
size increase.

When DMPC and DPPC vesicles were incubated together with myristic
acid, a progressive incorporation of DPPC molecules into DMPC vesicles
occurred. In fact the amplitude of the light scattering change at
the phase transition (ΔI in Fig. 4) increased for the DMPC vesicles
and decreased for the DPPC vesicles. Moreover the transition temperature of DMPC vesicles slightly increased, whereas the transition
temperature of DPPC vesicles remained stable. Lipid exchange was
analyzed in terms of changes of the ratio between amplitudes of the
transition intensity changes after a constant incubation time.

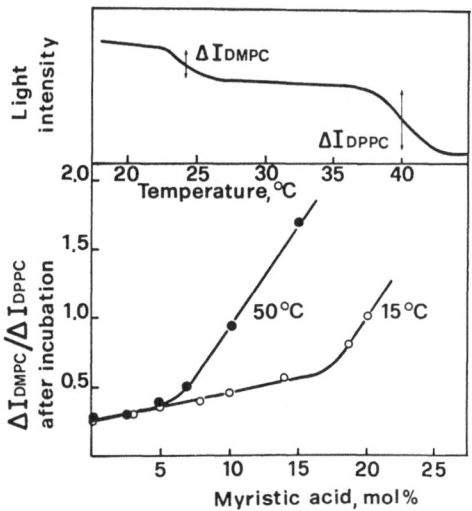

Fig. 4 – Lipid phase transition curves of an equimolar mixture of
        DPPC and DMPC vesicles (upper part), and effect of myristic
        acid on the amplitude of the lipid phase transition changes.

Fig. 4 shows that this ratio abruptly increased at a "critical"
fatty acid concentration, with vesicles both in solid and liquid
crystalline state. As shown in Fig. 1, 10 mol% of myristic acid
was sufficient fro reaching the "critical" concentration for DPPC
vesicles, but not for DMPC vesicles. The destabilization effect
of the fatty acid above its "critical" concentration may induce a
release of PL molecules more pronounced in DPPC than in DMPC vesicles.
Lipid exchange may occur: 1) by free lipid molecules diffusion;
2) by vesicle collision, 3) by vesicle merging. In the collision
or merging mechanisms the rate of PL exchange is expected to show
a dependence on the concentration of both donor and acceptor vesicles.
In the case of diffusion of free molecules, the rate limiting reaction is presumably the PL release from donor vesicles. We have
found that, in the presence of a "critical" fatty acid concentration,
the rate of PL exchange depended on the concentration of donor
vesicles and not on the concentration of acceptor vesicles. Moreover increase of net charges on the vesicle surface did not inhibit
the PL exchange process. These observations indicate that a PL
diffusion of free molecules seems the more probable mechanism of
the PL exchange.

DISCUSSION

NMR experiments using mixed fatty acid-synthetic diacylphospha-
tidylcholine bilayers were interpreted in terms of an increased rate
of isomeric fluctuations of the methylene groups around the C-C bands,
leading to an average increase in the number of gauche isomers along
the alkyl chain (9). The degree of motional anisotropy of the chain
proton groups was gradually reduced by progressively increasing the
chain length disparity of the fatty acid and PL molecule. The gauche
isomeric state shortens the lipid chain thereby creating "holes"
around the lecithin terminal CH_3 units, as shown in the model presented in Fig. 5.   The holes created by the chain shortening are
eliminated through a local or an overall thinning of the bilayer.

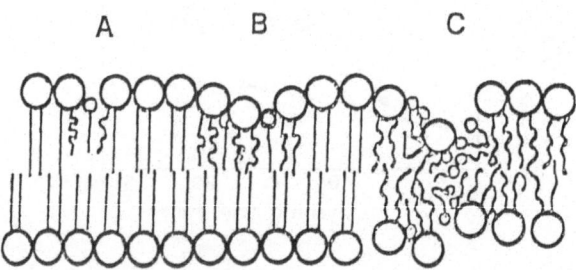

Fig. 5 - Membrane model illustrating the effect of fatty acids on
         the bilayer structure.

1) Effect of the "critical" fatty acid concentration on the bilayer structure

a) Below the "critical" value, the hole formation allows the insertion and the aggregation of pyrene molecules. Therefore the excimer/monomer ratio of pyrene increases.
b) Above the "critical" value, the probability that two or more fatty acid molecules interact with the same PL molecules is not negligible: the consequent shortening of the lipid chain progressively extends in the bilayer and the holes created continue to be eliminated through a thinning of the membrane. Consequently the excimer/monomer ratio of pyrene increases less than before. The bilayer thinning explains also the increase of the membrane permeability to K^+.

2) PL-exchange

Above their "critical" concentration, fatty acids may induce a decrease of hydrophobic attraction between fatty acid-phospholipid complexes and neighbouring phospholipid molecules. This may be due to: 1) a decrease of the Van der Waals interaction among the terminal methyl and methylene groups of the lipid hydrocarbon chain (10);
2) a decrease in area occupied by the hydrocarbon chain of the lecithin-fatty acid complex consequent to the gauche isomeric state formation. The diminished attraction between phospholipid molecules, along with an increase of water solubility of the lecithin-fatty acid complex, may considerably increase the concentration of phospholipid molecules or phospholipid-fatty acid complexes free in the aqueous solution.

3) Vesicle size enlargement

When the temperature is in the range of the solid-fluid transition temperature, solid and fluid domains of phospholipids coexist, with the dimension and the number of these domains dependent on the temperature (11). It is well known that fatty acids distribute preferentially in the fluid region of the bilayer (12,13). Therefore, the fatty acid concentration in the fluid domains is higher than in the solid domains. When the fatty acid concentration reaches a "critical" value, the bilayer structure of these domains is extensively modified (cf. Fig. 5,C). Calorimetric and light scattering experiments in fact, indicate that fatty acid, above a threshold concentration, induces a destabilization of the bilayer structure (3,14). Fatty acids may form a micellar-like structure in these destabilized areas, thereby increasing the probability of vesicle sticking after collision. Since the work for adducting two electrically neutral zwitterionic phospholipid bilayers is high, LeNevey et al. (15) suggested that a destabilization or a transient rearrangement of the phospholipid molecules may account fro moving away the polar head groups. This hypothesis is supported by the inhibitory effect of local anesthetics which, by increasing membrane fluidity (16,17), allow a more homogeneous partitioning of fatty acids on the bilayer membrane and they decrease the fatty acid concentration in the fluid domains.

ABBREVIATIONS USED

ANS: 8-anilino-1-naphtalene sulfonic acid.
Pi: inorganic phosphate
DPPC: dipalmitoylphosphatidylcholine
DMPC: dimiristoylphosphatidylcholine
DSPC: distearoylphosphatidylcholine
PL: phospholipids

REFERENCES

1. J.I. Howell, D. Fisher, A.H. Goodall, M. Verrinder and J.A. Lucy, Interactions of membrane phospholipids with fusogenic lipids, Biochim. Biophys. Acta 332:1-10 (1973).
2. R. Hori, Y. Kagimoto, K.Kamiya and K.J. Iuni, Effects of free fatty acids as membrane components on permeability of drugs across bilayer lipid membranes. A mechanism for intestinal absorption of acidic drugs, Biochim. Biophys. Acta 509:510-518 (1978).
3. J.M.H. Kremer and P.H. Wiersema, Exchange and aggregation in dispersions of dimiristoyl phosphatidylcholine vesicles containing myristic acid, Biochim. Biophys. Acta 471:348-360 (1977)
4. D. Papahadjopoulos, S. Hui, W.J. Vail and G. Poste, Studies on membrane fusion. 1. Interaction of pure phospholipid membranes and the effect of myristic acid, lysolecithin, proteins and dimethylsulfoxide, Biochim. Biophys. Acta 448:245-264 (1976).
5. H.L. Kantor and J.H. Prestegard, Fusion of fatty acid containing lecithin vesicles, Biochemistry 14: 1790-1795 (1975).
6. H.L. Kantor, S. Mabrey, J.H. Prestegard and J.M. Sturtevant, A calorimetric examination of stable and fusing lipid bilayer vesicles, Biochim. Biophys. Acta 466:402-410 (1977).
7. H.L. Kantor and J.H. Prestegard, Fusion of phosphatidylcholine bilayer vesicles: role of free fatty acid, Biochemistry 17: 3592-3597 (1978).
8. J.R. Usher, R.M. Epand and D. Papahadjopoulos, The effect of free fatty acids on the thermotropic phase transition of dimiristoylglicerophosphocholine, Chem. Phys. Lipids 22:245-253 (1978).
9. F. Podo and J.K. Blasie, Proton magnetic relaxation studies of mixed phosphatidylcholine fatty acid and mixed phosphatidyl-choline bimolecular bilayers, Biochim. Biophys. Acta 419:1-18 (1976).
10. A.W. Eliasz, D. Chapman and D.F. Ewing, Phospholipid phase transitions. Effects of n-alcohols, n-monocarboxylic acids, phenylalkyl alcohols and quaternary ammonium compounds, Biochim. Biophys. Acta 448:220-230 (1976).

11. A.G. Lee, Functional properties of biological membranes. A physical-chemical approach, Prog.Biophys. Molec. Biol. 29: 3-56 (1975).

12. K.W. Butler, N.H. Tattrie and I.C.P. Smith, The location of spin probes in two phase mixed lipid systems, Biochim. Biophys. Acta 363:351-360 (1974).

13. C.L. Bashford, C.G. Morgan and G.K. Radda, Measurement and interpretation of fluorescence polarizations in phospholipid dispersions, Biochim. Biophys. Acta 426:157-172 (1976).

14. S. Mabrey and J.M. Sturtevant, Incorporation of saturated fatty acids into phosphatidylcholine bilayers, Biochim. Biophys. Acta 486:444-450 (1977).

15. D.M. LeNeveu, R.P. Rand and V.A. Parsegian, Measurement of forces between lecithin bilayers, Nature 259:601-603 (1976).

16. W.L. Hubbell, J.C. Metcalfe, S.M. Metcalfe and H.M. McConnell, The interaction of small molecules with spin labelled erythrocyte membranes, Biochim. Biophys. Acta 219:415-427 (1970).

17. D. Papahadjopoulos, K. Jacobson, G. Poste and G. Shepard, Effects of local anesthetics on membrane properties. 1. Changes in the fluidity of phospholipid bilayers, Biochim. Biophys. Acta 394: 504-519 (1975).

DEPOLARIZATION CURRENT ANALYSIS AND ELECTRICAL BIREFRINGENCE IN

WATER-IN-OIL MICROEMULSIONS

Donatella Senatra, Cecilia Gambi, Antonio Neri, Matteo Vannini

Liquid State Physics Laboratory, Institute of Physics
University of Florence

Largo E.Fermi 2 (Arcetri), 50125 Florence, Italy

INTRODUCTION

Microemulsions, consisting of water-in-oil (dodecane plus hexanol) stable and optically clear dispersions with an ionic surfactant as emulsifier, demonstrate several peculiar properties depending on the amount of water present (1-3).

A qualitative phase diagram or phase map study performed on these systems over the temperature interval $(-20 \div +80)°C$, keeping constant both the surfactant/hydrocarbon and the alcohol/hydrocarbon ratio, showed that, as the water content is increased, four macroscopically different structure-regions may be identified: I°- an optically transparent region, where the system is a stable and isotropic dispersion of water spherical droplets in a continuous oily phase (1-4-5); II°- a Gap region, where the system separates into two different phases: an optically clear and isotropic w/o microemulsion and an optically birefringent, slightly translucid, liquid-crystalline mesophase (6); III°- a birefringent region, where the system displays mesomorphic liquid-crystalline phases (9); IV°- a second optically transparent region, corresponding to an oil-in-water type of dispersion. In the liquid-crystalline region, depending upon the water content, both a "nematic" and a lamellar mesophase are supposed to develop. Although each of the two structures had been observed separately by many authors (6-8) in different w/o and o/w systems, none has ever detected both of them within a given system.

The main purpose of the present paper was to find an experimental evidence which could substantiate the presence, in a given w/o micro-

emulsion, of both a nematic and a lamellar mesophase. Starting with
the "microemulsion state" of the system, we investigated, upon the
water addition, the early structural modifications that precede the
liquid-crystalline phase as well as the physical characteristics
distinguishing one mesophase from the other. Finally we analyzed
the nematic-lamellar transition which has hitherto been overlooked.

From a general point of view, microemulsions are extremely useful systems for studying the role of water adjacent to charged surfaces including lipid interfaces of biologic interest. In addition,
it should be pointed out that, transitions from the bilayer configuration to the so-called "water hose" have been reported for some
phospholipidic membranes during dehydration processes (9), and the
liquid-crystalline nature of phospholipids has been recognized long
since (10). See (Fig. 1). Structures as water cylinders and water
lamellae have also been demonstrated by many authors for various
lipid-water systems (11).

From the above considerations it follows that microemulsions
could become a powerful tool for analyzing the structural properties
of biological membranes. Unfortunately, although lyotropic liquid-
-crystals occur abundantly in nature, particularly in living systems,
their structures and properties are only just beginning to be eluci-
dated.

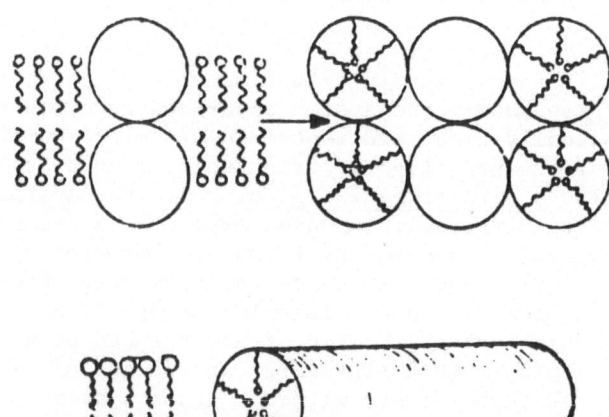

Fig. 1 - Transition of phospholipid from bilayer to the concentric
        water hose pattern

Depolarization current analysis

    The thermally stimulated release of dielectric polarization
was investigated in w/o microemulsions by means of the "ionic thermo-
conductivity method" (ITC) (12-15). The most important feature of
this experimental approach is that through a single measurement it
is possible to obtain a complete picture of the temperature-dependent
relaxation-processes induced into a dielectric by an impressed electric field, as well as to evaluate their characteristic parameters,
the activation energy ξ and the relaxation time τ. The experimental
sequence of the ITC method is the following: (See Fig. 2), an electric
field (10 V) $E_p(\underline{1})$, is applied to a condenser filled with the sample
($\underline{2}$) at a polarizing temperature T_p (20°C). After a time t_p (3 min.)
the polarization induced by the impressed field is "frozen-in" by
lowering the temperature of the sample to that of liquid nitrogen ($\underline{3}$).
At this point the E_p field is removed and the sample connected to
an electrometer ($\underline{4}$): while the distortion effects in the sample relax,
the orientational polarization that was induced in the latter at the
T_p temperature does not. The temperature of the sample is then
raised in a linear manner with a constant heating rate $b=dT/dt$
(0.1°C/s) ($\underline{8}$), and the ionic thermocurrent $J(T)$ is recorded ($\underline{9}$) as
the oriented dipoles of the sample lose their orientation and revert
to a random distribution. In the case of complex relaxations, i.e.
if more kinds of dipolar species are present in the system under
test, a series of current peaks are observed extending from the freezing temperature (-203°C) to higher temperatures (50°C) to which the
sample may be raised. The resultant curve is called "the thermally
stimulated depolarization spectrum" (TSDS), each of the current peaks
corresponding to a given depolarization process. Assuming a first
order kinetic for dipole orientation and introducing for the relaxation time the relation

$$\tau = \tau_o \exp (\xi/KT) \tag{1}$$

the $J(T)$ current is expressed by:

$$J(T) = \frac{N\, p^2 E_p\, \alpha}{KT\tau_o} \exp\left[-\frac{\xi}{KT} - (b\tau_o)^{-1} \int_{T_0}^{T} \exp\left(-\frac{\xi}{KT}\right) dT \right] \tag{2}$$

with: N dipole concentration, p dipole moment, K Boltzmann's constant,
α a geometrical factor depending on the possible dipolar orientation,
b heating rate, T absolute temperature, ξ activation energy and
τ_o relaxation time constant.

    In Eq. 2 $J(T)$ does first increase exponentially, reach a maximum
and then drop to zero; Eq. 2 describes the depolarization of a
single dipolar species. If a complete peak separation is achieved,
for a given depolarization process, the activation energy and the

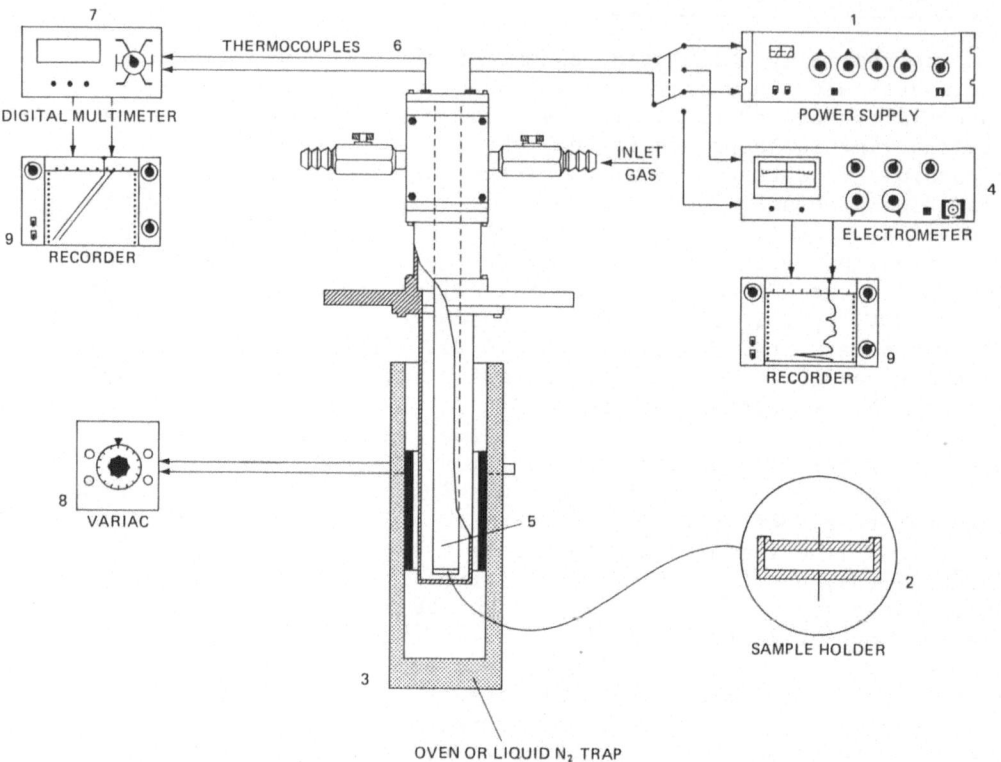

Fig. 2 - Experimental set up for ITC measurements on w/o micro-
        emulsions.

The sample is contained into a teflon cylindrical cell (2) with
gold plane parallel electrodes 3 mm apart from each other. The
sample holder is then placed at the bottom of a stainless steel
cylinder (5) which is in turn set within a second concentric cylinder of the same material. The gap between the two cylinders is
vacuum-controlled by a flow of nitrogen gas. The temperature of
the sample is followed with a thermocouple (6) placed as close as
possible to the sample itself; the electromotive force is measured
with a digital multimeter (7). The heating device is realized by
means of an electric owen regulated through a Variac (8). The
cooling process is performed by immersing the whole apparatus (5)
into a Dewar filled with liquid nitrogen (3). The ITC depolarization current and the sample temperature are simultaneously
recorded on a double-pen instrument. The condenser charge current
is also monitored on a recorder (9).

relaxation time constant can be obtained by utilizing the whole experimental J(T) curve, i.e., by integration of the area delimited by the ITC depolarization band, and the following equation, first developed by Bucci and Fieschi (13), applies:

$$\ln\tau(T) = \ln\tau_o + \frac{\xi}{KT} = (\ln\int_{t(T)}^{\infty} J(t')dt') - \ln J(T) \qquad (3)$$

Therefore the above parameters may be evaluated by computing the slope and the intercept of the regression line expressing the least squares fit to the experimental points; substitution of ξ and τ_o in Eq. 1 gives the relaxation time τ.

ITC Results

Typical TSD spectra of w/o microemulsions with different water contents are given in Figs. 3 and 4.

In Fig. 3: samples in the "microemulsion state"; TSD spectra show only one depolarization peak at T = ≃ 20°C.

In Fig. 4-a): samples exhibiting birefringence. Two main depolarization bands are recorded, one still occurring around – 20°C, the other one occurring at T = + 35°C; the latter peak was found to be characteristic of the first liquid-crystalline mesophase. It is worthy to note that the peak at + 35°C was detected at the begin-

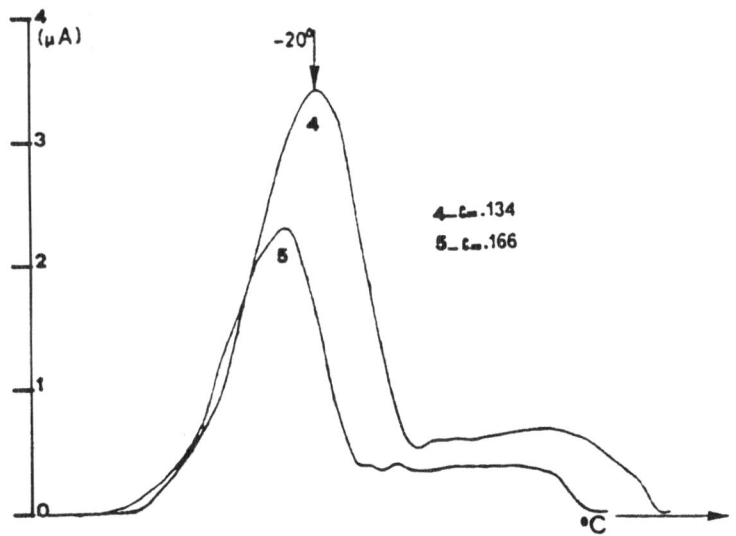

Fig. 3

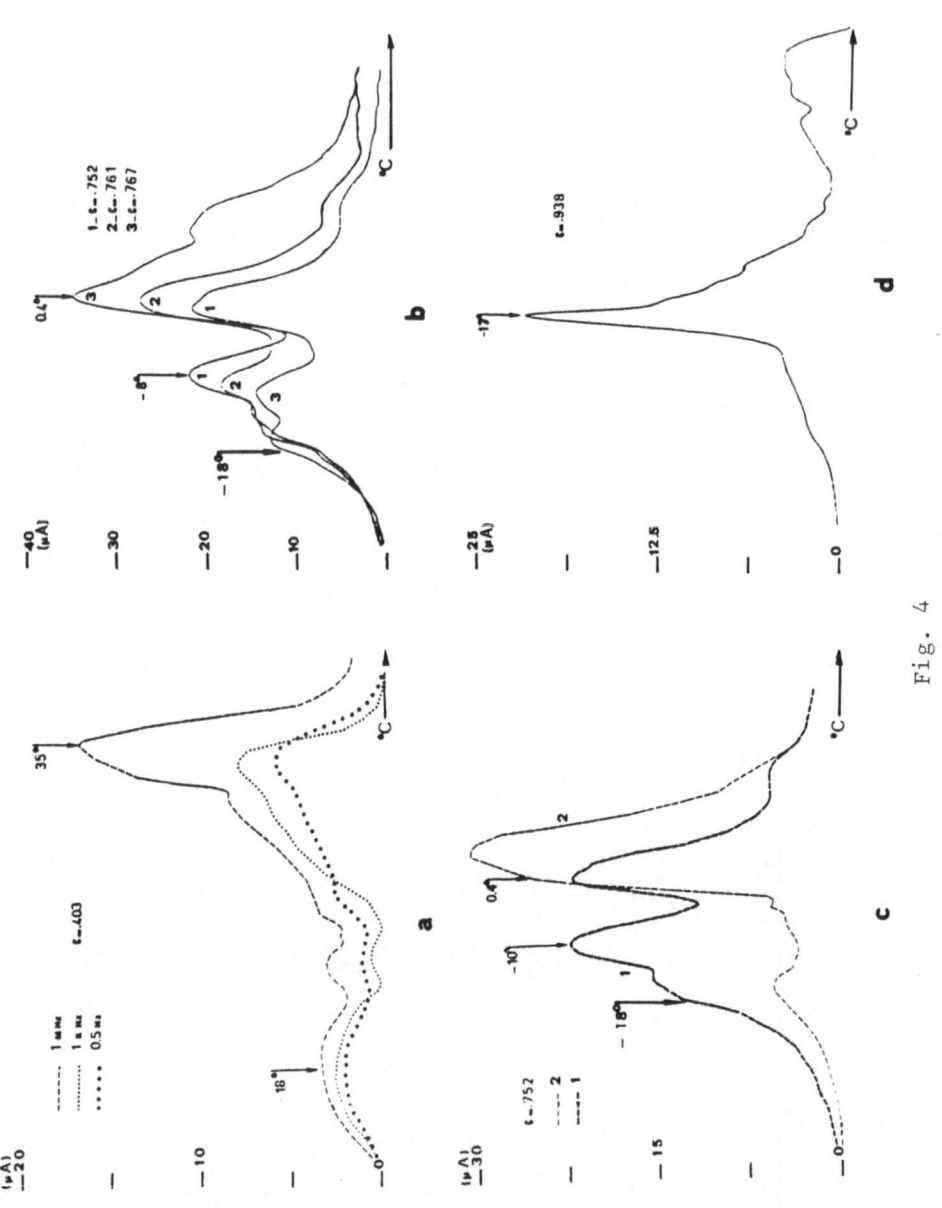

Fig. 4

ning of the Gap region at a concentration at which the system still appears as homogeneous and transparent. Since peak cleaning in the liquid-crystalline "band" was impossible with the presently available experimental equipment, a separation was attempted by polarizing the sample with a positive (+ 10 V) square pulse of 0.5 Hz, 1 KHz and 1 MHz. The + 35°C peak displayed the greatest absorption in the MHz region, while the first depolarization band, (T $\simeq$ - 20°C), did not show any significant difference. The result may offer some information about the order of magnitude of the oriented structure.

In Fig. 4-b): Strongly birefringent samples with water contents beyond the first liquid-crystalline region (c > 0.6). Three main depolarization bands are recorded. TSDS demonstrate that, upon increasing the water content, the peak occurring at T = + 0.4°C increases while that at T = - 8°C decreases. Note the first peak at T $\simeq$ - 20°C is still present without any appreciable variation. Since, for a given system, a lamellar mesophase should always occur at a temperature below the nematic domain, we believe the above findings indicate that, depending on the w/o ratio, both a lamellar and a nematic mesophase may develop in water-in-oil microemulsions. The lamellar phase, as it should be, occurring at a lower temperature (0.4°C) than the nematic (35°C).

In Fig. 4-c): Typical TSD spectra for a sample exhibiting viscoelastic properties at the end of the lamellar region. Curve 1 depicts the TSD spectrum corresponding to an extremely viscous gel sample; Curve 2 regards the spectrum performed on the same sample 30 days later: the system is collapsed into a perfectly fluid, slightly translucid sample. The depolarization peak at T = + 0.4°C shows that the structure is not yet destroyed.

In Fig. 4-d): Depolarization spectrum of an oil-in-water type of dispersion. Note that only one peak is detected as for the samples in the w/o microemulsion "state" plotted in Fig. 3.

At the present stage of research, the Bucci and Fieschi method was applied to the first depolarization peak occurring at T $\simeq$ - 20°C.

In Fig. 5 the trend is reported of the maximum current intensity of the latter vs. concentration (mass fraction). The different structure-regions encountered are also depicted.

The activation energy ξ and the relaxation time τ, calculated by means of Eqs. 3 and 1, are plotted in Fig. 6 as a function of concentration. Both ξ and τ were evaluated with a 2% accuracy.

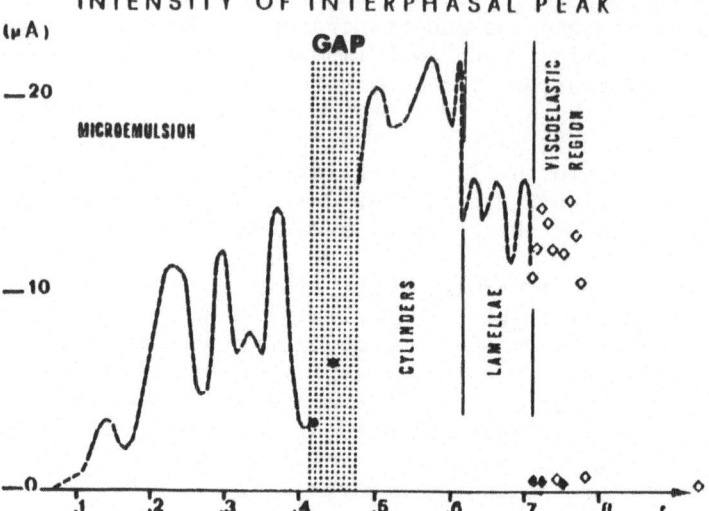

Fig. 5

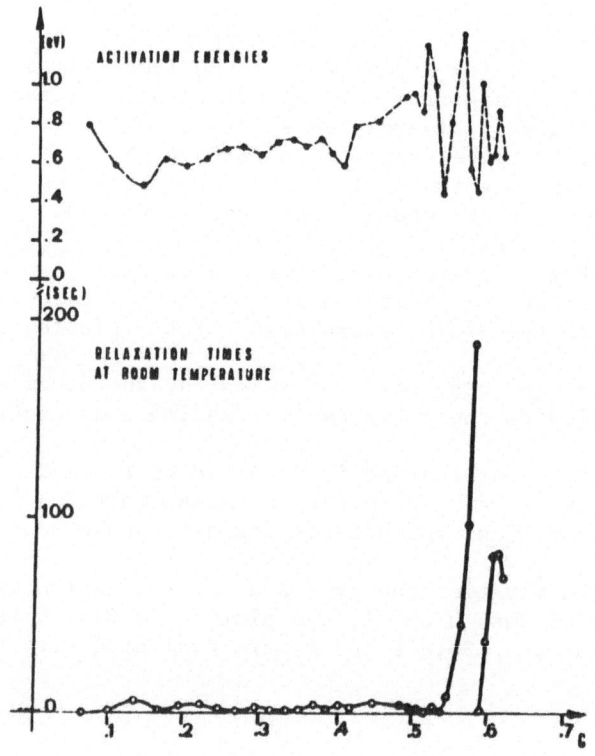

Fig. 6

Electrical birefringence

Electric anisotropy studies were carried out on w/o microemulsions at 20°C. The modifications induced in the system by a low-
-level (10 V) electric field were analyzed through optical anisotropy
measurements, using, as light source, an He-Ne Laser with a 6328 Å
wave length. The experimental set up was that usually adopted for
Kerr effect studies. Laser fluctuations were monitored by means
of a beam splitter placed before the entrance of the sample holder
device. The cell was a fused sylica parallelepipedon with rectangular gold plane parallel electrodes at 3 mm distance. Both the
splitted and the transmitted beams were monitored by means of two
linear detectors and the signals (mV) were simultaneously recorded;
the current during the charging process of the condenser was recorded too. In this way it was possible to evaluate the time required by the system for reacting to the impressed field.

Electrical birefringence results

Electrical birefringence may be induced in w/o microemulsions
with unusually low-level electric fields in a concentration interval
falling within the Gap region, where the system appears macroscopically as a·transparent and homogeneous dispersion. The amount of
induced birefringence grows to a maximum at a concentration corresponding to the Gap region middle point (c=0.48), then it drops to
zero at the end of the Gap (c≈0.5). In order to localize the various
concentration intervals, see Fig. 5. The typical behavior of birefringence measurements in the Gap region is shown in Fig. 7 where
the dotted line indicates the field removal. In order to compare
the different trends, each of experimentally recorded curves was
normalized to its starting value (mV/mV_i). Further addition of
water, as it is well known, leads to a spontaneously birefringent
system, in this case the imposition of the field decreases the
endogenous birefringence. The latter result is reported in Fig. 8
where: the dashed wurve corresponds to a measure performed on a
sample exhibiting the first liquid-crystalline phase (c=0.545);
the full line curve represents the trend of a measure on a sample
exhibiting the second liquid-crystalline mesophase (c=0.623).

The behavior was found to characterize the samples with a w/o
ratio within the concentration interval 0.6 < c < 0.7. Arrows indicate the field removal; symbols have the same meaning as in Fig. 7.

It should be pointed out that usually microemulsions samples
in the lamellar region, if observed through two crossed polaroids,
using a white light source, look strongly birefringent, whereas they
do appear practically no-birefringent at the 6328 Å wave length of
the Laser beam employed in the present work.

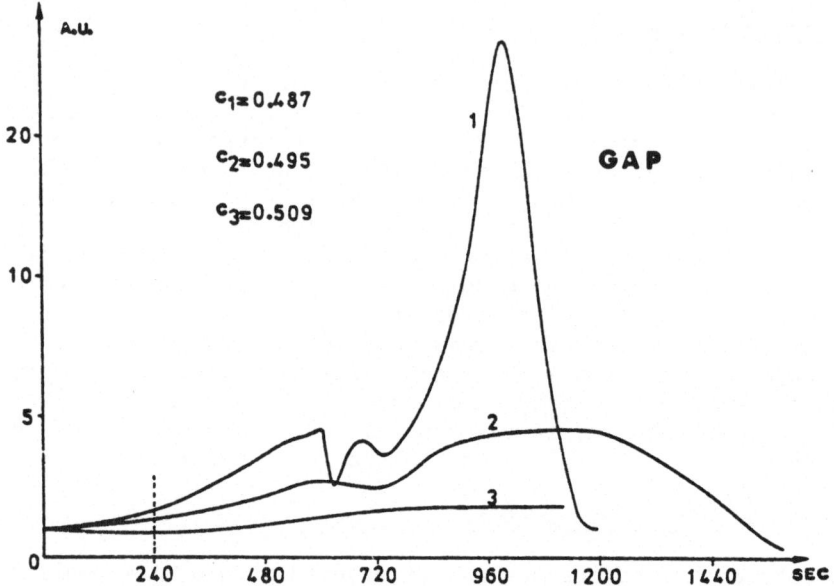

Fig. 7

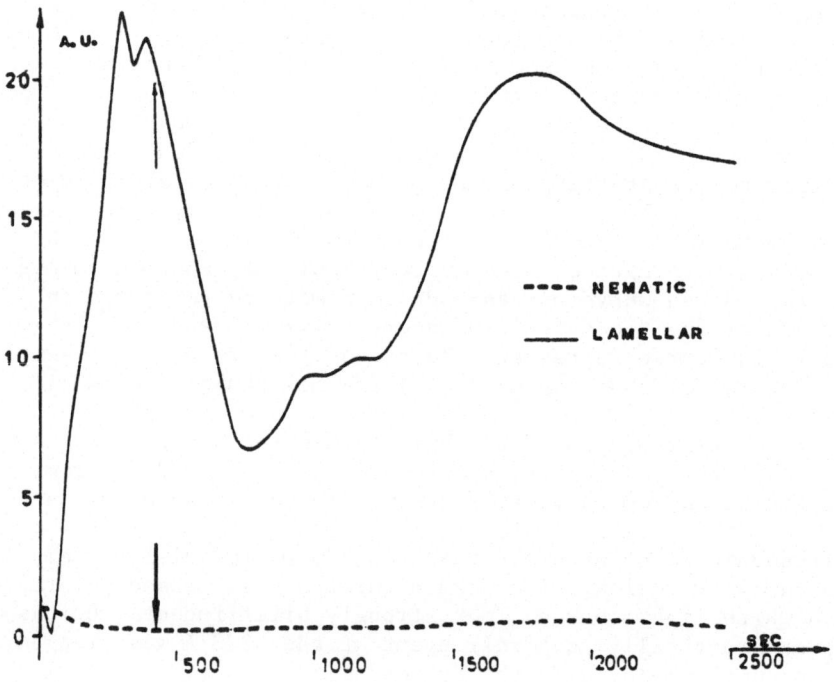

Fig. 8

DISCUSSION

The results gathered by means of the ITC approach demonstrate that, independently from the w/o ratio, in each of the depolarization spectra a peak always occurs at the same temperature ($\approx -$ 20°C). The intensity of this peak strongly depends on the water content of the sample; its behavior against concentration offers a general picture of the different structure-regions that develop in w/o micro-emulsions upon water addition (Fig. 5). Thus the aforementioned peak should be the expression of a polarizable entity, always present within the system, able to follow the various structures through which the system organizes itself upon increasing water contents.

Considering the low intensity of the electric field imposed to the sample as well as the peak-characteristics, we believe this oriented entity be the electrified interface between the dispersed water and the continuous oily phase. For the above reasons the peak recorded at T = $-$ 20°C was labeled "the interphase peak" (16).

Unlike the interphase peak, the other main depolarization peaks observed, respectively, at + 35°C and + 0.4°C, do occur within a well defined concentration interval and are no more detectable out-side a given liquid-crystalline region; we interpreted these peaks as "structure-peaks" due to the presence of either liquid-crystalline structure. It should be noted that, since liquid-crystalline meso-phases of w/o microemulsions are both lyotropic and thermotropic, the structure peaks depend on the polarizing temperature, whereas the interphase peak does not. The above result indicates that each liquid-crystalline phase has a threshold temperature above which that given structure is destroyed. Considering the behavior of both the activation energy and the relaxation time of the interface--polarization process (Fig. 6), we may conclude that, as far as the interphase peak is concerned, a transition occurs at c = 0.580. Taking into account the "structure-peaks" as well as the concen-tration intervals within which they manifest themselves, it follows that the aforementioned transition develops at the end of the nematic region where the interface changes from a cylindrical to a lamellar configuration. It is anyhow surprising that ξ and τ do not show any peculiar singularity in the concentration interval where the system from the "microemulsion state" passes into a liquid-crystal-line state. The first structure-peak observed at 35°C, (Fig. 4-a), and mainly the behavior of optical anisotropy measurements (Fig. 7), on macroscopically homogeneous and clear samples, suggest that the transition from the w/o isotropic microemulsion to a lyotropic liquid crystal, takes place gradually. A possible mechanism could be the following: as the water content increases (c $\approx$ 0.37), the system organizes itself in larger structures which may be regarded as the building blocks that, upon further water addition (c $\approx$ 0.5), will give rise to a spontaneously birefringent liquid-crystalline phase.

At the beginning of the Gap the water content of the system is not
enough to allow a definite structural liquid-crystalline configuration, but the structured entities, within the system, may be indeed
oriented and spatially ordered by means of an impressed electric
field.   Therefore a Kerr-like effect can be observed, due to the
optical anisotropy induced by the field.

REFERENCES

1. Schulman,J.H., and Riley,D.P.,   J.Colloid Sci. 3:383 (1948)
2. Schulman,J.H., Stoeckenius,W., and Prince,L.M.,   J.Phys.' Chem.
    63:1677 (1959).
3. Prince,L.M.,   J.Colloid Interface Sci. 52:182 (1975).
4. Senatra,D. and Giubilaro,G.,   J.Colloid Interface Sci. 67:448
    (1978).
5. Senatra,D. and Giubilaro,G.,   J.Colloid Interface Sci. 457
    (1978).
6. Ballaro',S., Mallamace,F., Wanderlingh,F., Senatra,D. and
    Giubilaro,G.,   J. of Physics (C) in press (1979).
7. Sjöblom,E. and Friberg,S.,   J.Colloid and Interface Sci. 67:16
    (1978).
8. Friberg,S. and Burasczenska,I.,   Progress in Colloid and Polymer
    Sci. 63:1 (1977).
9. Green,D.E.,   Ann. N.Y. Acad. Sci. 195:150 (1972).
10. Byrne,P. and Chapman,D.,   Nature (London) 202:987 (1964).
11. Ekwall,P., Mandell,L., and Fontell,K.,   Acta Chem. Scand. 22:
    373 (1968).
12. Bucci,C. and Fieschi,R.,   Phys. Rev. Lett. 12:16 (1963).
13. Bucci,C., Fieschi,R. and Guidi,G.,   Phys. Rev. 148:816 (1966).
14. Cappelletti,R. and Fieschi,R., in "Electrets", Perlman Ed.
    The Electrochemical Society Inc., p. 1 (1973).
15. Bini,S. and Cappelletti,R., in "Electrets", Perlman Ed.
    The Electrochemical Society Inc., pp 66 (1973).
16. Senatra,D., Gambi,C., Neri,A. Interlab Comunication pp 1-95
    "Int.Institut für studien-dokumentation und information" I-
    39040, Gossensass (BZ) (1979).

PHOTOBIOLOGY

ON THE NATURE OF THE PRIMARY PHOTOCHEMICAL EVENTS IN BACTERIO-

RHODOPSIN

U. Dinur[*], B. Honig[*≠] and M. Ottolenghi[*]

[*] Department of Physical Chemistry
The Hebrew University of Jerusalem, Israel

[≠] Department of Biological Sciences
Columbia University, New York, N.Y. 10027, U.S.A.

ABSTRACT

The sequence of primary events following light absorption by light-adapted (all-trans) bacteriorhodopsin (BR_{570}^t) is considered by analyzing recent picosecond absorption and emission data. Theoretical calculations are carried out in order to predict the ground state and the excited state (singlet-singlet) absorption spectra of the molecule. It is concluded that excitation leads to a photochemically important non-fluorescent excited state (I) which decays into a photoproduct (J_{625}). In J_{625}, which is most probably a ground state molecule, the chromophore has undergone a structural change, presumably trans→13-cis isomerization. It is suggested that the subsequent process $J_{625} \xrightarrow{11psec} K_{610}$, reflects a relaxation of the protein environment involving proton transfer. The conclusions preclude models which, based on the assumption that J_{625} is a primary excited state, attribute the first photochemical event to proton translocation. Based on a variety of common features, the general picture bears also on the nature of the primary processes in vision.

Introduction

The photocycle of bacteriorhodopsin, the protein pigment of the purple membrane of Halobacterium halobium[1], has generated considerable interest in the field of molecular photobiology, (see Ref. 2-4 for recent views). This is due to the close relationship between its photocycle and the mechanism of the light-driven proton pump[5] which is capable of driving ATP synthesis[6], and also to the striking structural and photophysical analogies between bacteriorhodopsin

and visual pigments[3,4]. The light-adapted form of bacteriorhodopsin which absorbs at 570 nm, contains an all-trans retinyl-polyene
chromophore (BR_{570}^t). Its photocycle is initiated with the generation
of a primary red-shifted intermediate absorbing around 610 nm, denoted as K_{610}[2]. This species is in many respects analogous to
bathorhodopsin (BATHO), the early photoproduct of visual pigments
[3,4]. In both systems elucidation of the structural changes associated with the generation of the respective batho intermediates is
crucial for understanding the primary event as well as the subsequent
steps in the photocycles.

Models for the primary event

    As in the case of visual pigments, three principal classes of
models have been proposed to account for the molecular changes associated with the primary photochemical event in all-trans bacteriorhodopsin. a) Trans-cis isomerization, probably about the C_{13-14}
double bond[7-9]. This first event induces a secondary environmental
relaxation involving proton translocation in the protein[8,9]. (It
has recently been proposed that the double bond isomerization is
accompanied by a concomitant rotation about the C_{14-15} single bond
[10]). b) Proton translocation to the Schiff base nitrogen[11-14].
c) Proton translocation between acid-base residues in the opsin[15-
18].

    In all of the above models proton transfer must be assumed so
as to account for the transient picosecond phenomena observed by
Applebury et al.[12]. These pioneering experiments have shown that
the K_{610} generation kinetics exhibit a deuterium isotope effect and
a temperature dependence, suggesting that at low temperatures the
process is controlled by proton tunneling. The situation is completely analogous to the generation of bathorhodopsin from bovine
rhodopsin, though in the latter system the deuteration effect is
considerably larger[11]. Although the existence of an ultrafast
proton translocation reaction[11-12] is now well accepted, it is
evident that the various models differ in defining the exact nature,
as well as the sequence, of primary events. The two major equivocal
problems are the identification of the groups involved in the proton
transfer reaction and the question as to what is the first photochemical step.

    While disagreeing on the nature of the groups associated with
proton translocation, both approached, b and c share the essential
feature of proton-transfer as the primary excited state reaction.
The process is presumed to be induced by the shift in electron density accompanying excitation. In such schemes, K_{610} is formed from
a red shifted precursor (denoted here as J_{625}) which is assigned as
the lowest singlet excited state, S_1. On the other hand, in the
isomerization model (a)[7-9,19] both for visual pigments and for
bacteriorhodopsin, proton transfer is a ground-state process which
follows isomerization of the chromophore. Accordingly, J_{625} is

assigned as a ground state precursor of K_{610}. The $J_{625} \rightarrow K_{610}$ transformation involves a thermal protein relaxation (proton transfer) process induced by the preceding photochemical isomerization: $BR_{570}^t \xrightarrow{h\nu} J_{625}$. Warshel's model[17,18] also assumes that proton translocation occurs on the ground state (S_0) electronic surface. It differs, however, from the sequence of Hurley et al.[8] and Honig et al.[9] in that it involves excited vibrational levels of S_0 and that it does not require the occurrence of trans-cis isomerization as an essential feature in the formation of J_{625} and K_{610}. In conclusion, the various approaches differ not only with respect to molecular models but also with regard to the spectroscopic assignments upon which they are, to a great extent, based.

It is evident that a proper assignment of J_{625} is of central importance in understanding the photochemistry of BR_{570}^t. In the present work we have attempted to formulate a sequence for the primary photochemical events in BR_{570}^t, referring also to the basic question as to whether J_{625} corresponds to a ground state or to an excited state. With this in mind, we have analyzed all of the available data on the ultrafast kinetics following the excitation of BR_{570}^t and have carried out theoretical calculations of ground and excited singlet-singlet absorption maxima and oscillator strengths. The conclusions lead to the suggestion of a scheme of primary processes in which the formation of J_{625} involves an isomerization of the chromophore. This primary act induces a subsequent ground state proton transfer reflected by the $J_{625} \rightarrow K_{610}$ transition.

Picosecond absorption kinetics

It seems worthwhile to first examine the picosecond absorption measurements on BR_{570}^t to see if they can be fitted into a consistent picture. Confining ourselves for the moment to room temperature, the major observations are: I) K_{610} is formed from a precursor (J_{625}) in 11 psec, the process being slowed down to 18 psec upon deuteration[12]. II) The ground state of BR_{570}^t is most probably partially repopulated in less than ~ 6 psec[12]. III) By applying a 615 nm subpicosecond excitation and detection technique at room temperature, Ippen et al.[20] have recently time resolved the 1.0±0.5 psec growing-in of a red shifted photoproduct of BR_{570}^t, claimed to be stable for more than 50 psec. This species was identified by them as K_{610}.

The apparent inconsistency between observations I and III is readily resolved by recalling that the 615 nm monitoring wavelength of Ippen et al.[20] is close to the isosbestic point between the spectra of K_{610} and J_{625}, as measured by Applebury et al.[12]. This explains the failure of the former investigators to observe the 11 psec growing-in of K_{610}. We therefore suggest that their 1 psec process should be attributed to the formation of J_{625} rather than to that of K_{610}. The most reasonable sequence of events is then:

$$BR_{570}^t \xrightarrow{h\nu} I \xrightarrow{1 \text{ psec}} J_{625} \xrightarrow{11 \text{ psec}} K_{610}$$

where I is the precursor of J_{625} and is presumably an excited state species. In this scheme J_{625} may be another excited state formed from I (Fig. 1A), or, as suggested previously, a ground state species (Fig. 1B). In either case there is a "dark reaction" $I \to J_{625}$ preceeding the deuteration-sensitive formation of K_{610}. In the next sections, we discuss evidence arguing against the assignment of J_{625} as an excited state. Especially, it is ruled out that J_{625} may be identified with the photochemically important "common excited state" [7,8] which determines the yield of photocycling, both in rhodopsin and in bacteriorhodopsin. Such a state is tentatively identified with the I intermediate.

On the rate of ground-state repopulation

Extensive arguments, based on absolute quantum yield values as well as on their temperature and wavelength dependence, have been previously presented showing that the primary event in both rhodopsin and bacteriorhodopsin is controlled by the quantitative population of a common excited state minimum along the coordinate between the pigment and its bathophotoproduct[7,8]. From such a minimum parti-

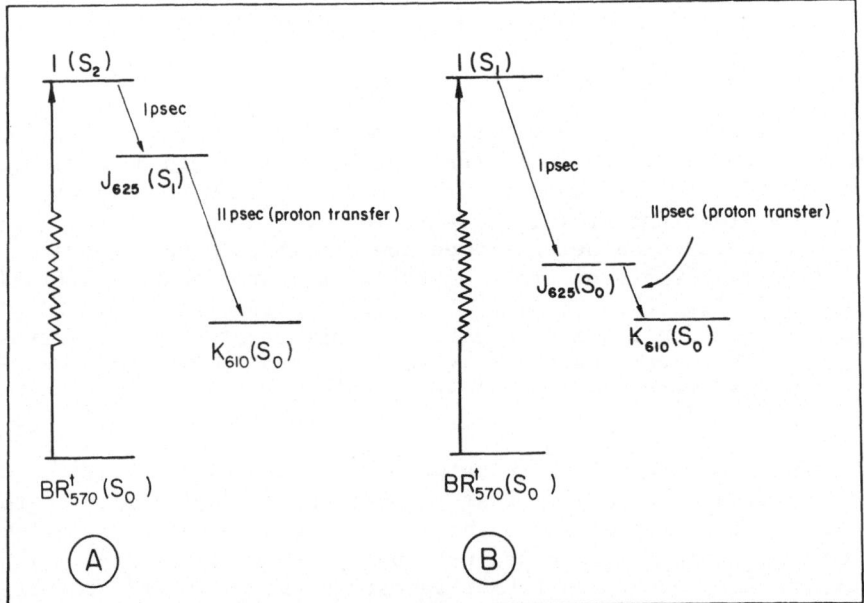

Fig. 1 - Two possible schemes for the sequence of events in Bacteriorhodopsin.

tion takes place to the corresponding ground states. In the case
of bacteriorhodopsin the partition ratio is 0.75 to BR^t_{570} and 0.25
to K_{610}. In the bovine rhodopsin system the fraction decaying to
bathorhodopsin (BATHO) is 0.7 while that repopulating the original
ground state pigment is 0.3.

In both systems Rentzepis and coworkers[11,12] (and also Monger
et al., for rhodopsin[21]) failed to observe the ultrafast repopulation process of the primary ground state pigment. Given the significant quantum yields for ground state repopulation, it should thus
be concluded that this process is completed within times which are
shorter than the ~6 psec resolution time of the above measurements.
The "common excited state" hypothesis[7,8] then requires that the
partition leading to the formation of K_{610} also occurs in less than
~6 psec. However, since (especially at low temperatures) the deuterium sensitive growing-in of K_{610} is much slower than 6 psec[11,12],
it is evident that J_{625} cannot be identified with the photochemically
important "common excited state".[a] The same conclusion is also applicable to the red shifted precursor (BATHO')[11] of bathorhodopsin.
Having excluded J_{625}, it becomes quite reasonable to identify the
"common excited state" with the primary transient I. This obviously
requires that the ground state of BR^t_{570} is repopulated in ~ 1 psec
(the observed rise time of J_{625}[20]). Such a prediction can in
principle be experimentally verified.[b]

On the basis of the lack of a ground-state repopulation, match-
ing the growing-in of K_{610}, an additional, more general, argument
can also be made against the identification of I and J_{625} as the
second (S_2) and the first (S_1) excited states of BR^t_{570}, respectively.

---

[a] Formally, a mechanism may be suggested which retains the "common
excited state" assignment for J_{625} if it is assumed that BR^t_{570} is
regenerated from J_{625} indirectly, via a long-lived intermediate,
X. e.g., $J_{625} \xrightarrow{11psec} K_{610}$, competing with $J_{625} \xrightarrow{11\ psec} X \cdots$
$\xrightarrow{slow..} BR^t_{570}$. The mechanism essentially implies the existence
of an additional photocycle (parallel to the well established
sequence: $K_{610} \rightarrow L_{540} \rightarrow M_{410} \rightarrow O_{640} \rightarrow BR^t_{570}$), a phenomenon which
is not indicated by the accumulated photochemical evidence for
BR^t_{570}[2-4]. Nevertheless, careful experiments looking for a BR^t_{570}
regeneration reaction, other than from the latter sequence, may
have to be carried out in order to definitely rule out the above
alternative.

[b] We note that an alternative feasible assignment for I is that of
vibrationally excited form of J_{625}. According to the present model
(see also below), the $I \rightarrow J_{625}$ process will thus represent a vibrational relaxation process in S_o.

In fact, if J_{625} is identified with S_1 [12,15] a reasonable assignment for I would be that of a higher singlet state, e.g. S_2. However, according to our previous discussion (see also Fig. 1A) this would lead to violation of the general rule of ground state repopulation from S_1 (rather than S_2), mading the S_1 assignment of J_{625} highly unlikely.

Theoretical predictions of ground ($S_1 \leftarrow S_0$) and excited ($S_n \leftarrow S_1$) absorption spectra

A line of reasoning similar to that outlined above can be carried on further on the basis of spectroscopic (rather than kinetic) arguments, in order to test the suggestion [15] that J_{625} is an excited state (S_1) retaining the basic all-transoid configuration of BR_{570}^t. Since J_{625} is not directly populated following light absorption, this would imply that S_1 is optically forbidden while S_2 (i.e. I) is optically allowed. This requirement can be checked by theoretical calculations. Moreover, the spectrum of J_{625} has been determined by Applebury et al. [12] so that a theoretical analysis of the $S_n \leftarrow S_1$ spectral transition should also be relevant to its assignment.

For a clarification of the above questions we have carried out calculations of spectroscopic transition energies and oscillator strengths for a model all-trans bacteriorhodopsin molecule. As required by the Resonance Raman data for both rhodopsin [22-25] and bacteriorhodopsin [26,27], we have assumed that in BR_{570}^t the Schiff base nitrogen is fully protonated. This assumption constitutes the basis of models $\underline{a}$ and $\underline{c}$, but not of model $\underline{b}$ where full protonation is claimed only for the bathophotoproducts, J_{625} and bathorhodopsin.

The method of calculation was basically the CNDO/S method [28] supplemented with CI calculations that included singly and doubly excited configurations. A slight change was introduced into the original CNDO/S parameters in that Ohno formula was used for the repulsion integrals and β for the oxygen was changed to 31.5 eV. The use of Ohno formula is necessary whenever doubly excited configurations are included in the calculation [29], while the change of $\beta(O)$ gives a better fit to aldehyde spectra [30]. Molecular geometries were fixed in each case according to the bond order-bond length relation [31]. Methyl groups as well as the saturated part of the β-ionylidenic ring were excluded from the calculations and replaced by hydrogens. Calculations were first carried out for a free protonated Schiff base. External charges were then added in order to account for the actual ionic interactions of the chromophore with its surroundings. This was carried out using the model of Honig et al. [32a] for visual pigments. Agreement between the calculated and theoretical $S_1 \leftarrow S_0$ spectrum was obtained with one external charge 3Å away from the nitrogen and another one at 5Å perpendicularly to the center of the ring [32b]. The interaction of the e-

lectronic charge distribution with the external charges was calculated with Ohno formula with γ_{aa} = 13.0 eV for the external charges
(i.e., as if the external charges were oxygen atoms).

The 351x351 matrix included all singly and double excited configurations that are constructed from 10 π,π^* MO's. Oscillator
strengths were calculated according to:

$$f_{nm} = k\Delta E_{nm} \; (\underline{R}_{nm})^2 \tag{1}$$

where n,m denotes two electronic states ($E_n > E_m$), $\underline{R}_{nm}$ is the transition dipole and k is a unit-bearing constant that was given here
an arbitrary value (see below). The absorption bands were synthesized according to the following formula:

$$A_i(E) = \sum_{j>1} \; f_{ij} \; \sqrt{\frac{\alpha}{\pi}} \exp \{-\alpha(E-\Delta E_{ji})^2\} \tag{2}$$

where $A_i(E)$ is the absorption of the i'th singlet state. Each transition was thus assigned a gaussian lineshape. As a further approximation we have chosen α (the band width parameter) to be the same
for all transitions, and equal to the experimental value for the
$S_1 \leftarrow S_0$ band. Second, since calculations of oscillator strengths are
reliable only with respect to the relative intensities, we have
normalized the f_{ij} and the f_{01} which was fixed to be 1 (through a
proper choice of k, eq. (1)). In this way, all transitions and band
shapes are compared to the $S_0 \leftarrow S_1$ band. (Note that $A_0(E)$ is a normalized extinction coefficient for the $S_n \leftarrow S_0$ transition, related to
the molar extinction (ε) via $A=\varepsilon \cdot 10^{-5}$).

While $S_1 \leftarrow S_0$ spectral calculations are fairly routine and highly
reliable, calculations of transitions between excited states are
subject to greater uncertainty. In order to test our parameters on
a known system we have also calculated the $S_n \leftarrow S_1$ absorption of all-
-trans retinol, the only derivative of the chromophore for which such
a spectrum has been measured experimentally[33]. Since it appears
that the $^1A_g^-$ state may be the lowest excited state for this molecule
[34-36] we have performed two sets of calculations, one assuming that
S_1 corresponds to the $^1A_g^-$ state and one assuming it corresponds to
the $^1B_u^+$. The results of such calculations are shown in Fig. 2a.
The corresponding data for a free PRSB and for all-trans bacteriorhodopsin are shown in Fig. 2b assuming that S_1 is $^1B_u^+$.

Let us first consider the $S_1 \leftarrow S_0$ optical transition. There is
now considerable experimental and theoretical evidence[29,34-36]
showing that in retinols, retinals and other polyenes the optically
allowed $^1B_u^+$ state corresponds to the second singlet and that there
is a forbidden $^1A_g^-$ state slightly ($\sim$0.2 eV) below it in energy.
However, our present calculations which are consistent with the
recent work of Birge et al.[37] and of Schulten[38] clearly indicate

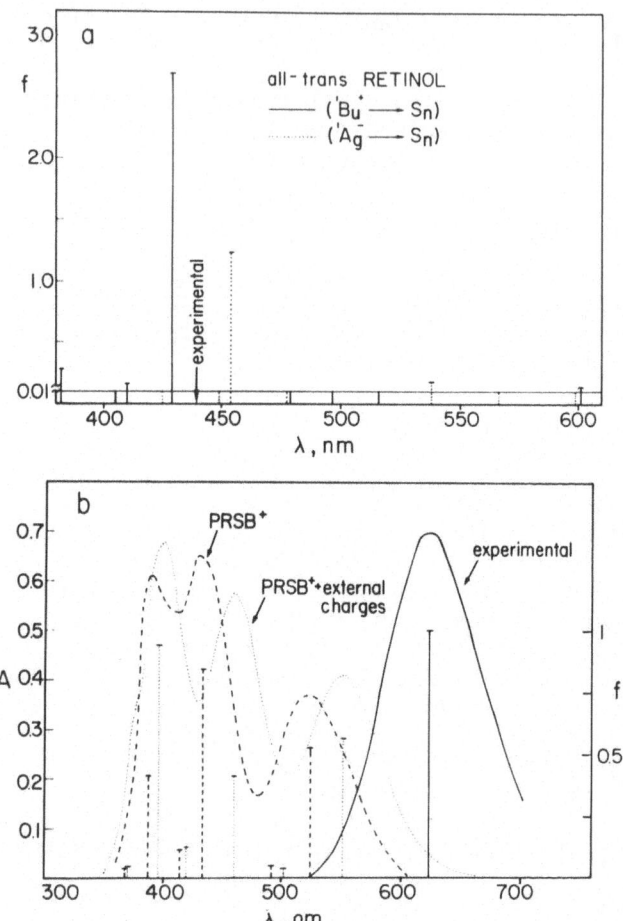

Fig. 2 – Calculated $S_n \leftarrow S_1$ spectra of retinol (<u>a</u>) and (<u>b</u>) of a protonated Schiff base of retinal free of environmental inter-
actions (PRSB$^+$) and of all-trans bacteriorhodopsin (PRSB$^+$
+ external changes). f (vertical lines) is the integrated
area under the corresponding absorption bands relative to
that of the main ground state absorption of the PRSB$^+$ molecule, which is defined as unity. Note that an accurate f
value is not available for the $S_n \leftarrow S_1$ band of retinol reported in ref.[33] (denoted in <u>a</u> as "experimental") but the
results suggest that it is of relatively high intensity.
The curves in <u>b</u> represent the wavelength dependence of A,
a normalized extinction coefficient proportional to the
molar extinction coefficient (see text). The experimental
curve in <u>b</u> is taken from ref. 12.

that in protonated Schiff bases the $^1B_u^+$ state is considerably lower
in energy than the $^1A_g^-$ state. As shown in Fig. 2b this implies that
the lowest excited singlet of BR_{570}^t corresponds to a strongly allowed
transition. Since as discussed above J_{625} could be an all-trans
excited (S_1) state only if it were weakly allowed, the calculations
appear to be in variance with this suggestion.

Next we turn to the results of the calculations for the $S_n \leftarrow S_1$
transition energies and oscillator strengths. It is evident from
Fig. 2a that the agreement between calculations and experiments in
retinol is very satisfactory, independently of the particular ($^1B_u^+$
or $^1A_g^-$) assignment. However, the spectrum of J_{625} is markedly
different from that predicted for the all-trans excited pigment.
These results provide strong independent evidence for the conclusion
that J_{625} cannot be identified with an all-trans excited state of
BR_{570}^t. In fact, the very close similarity between the spectra of
K_{610} and J_{625}, in terms of both transition energy and oscillator
strength, strongly supports the identification of the latter as a
perturbed ground state chromophore.

Fluorescence kinetics

Any discussion of the feasibility of an excited-state assignment for either J_{625} or I should also consider fluorescence emission
from bacteriorhodopsin. In fact, a low fluorescence yield (ϕ_f =
$10^{-4} - 10^{-5}$) from BR_{570}^t samples[39-42] (no emission has been observed
in visual pigments) has induced considerable interest as a potential
tool in elucidating the primary photochemical event in BR_{570}^t. Still
controversial is the question as to whether the emission originates
from excited BR_{570}^t[39-43] or from an excited state of a photoproduct,
different from any of the commonly known intermediates of the photocycle[44,45]. However, even assuming that the emission is due to
BR_{570}^t, it appears that J_{625} cannot be identified with the fluorescent
state. We base this conclusion on the emission lifetime (τ_f) measurements of Alfano et al.[40] and Shapiro et al.[43] (τ_f = 60±15 psec
at 77°K, 40 psec at 90°K) which do not match the rates of the
$J_{625} \rightarrow K_{610}$ process[12] (11 psec at 295°K, 20 psec at 77°K, 36 psec
at 4°K and 52 psec at 1.8°K). The 15 psec lifetime measured for the
fluorescent state at room temperature by Hirsch et al.[46] is not
too far from the 11 psec decay of J_{625}, an observation which could
suggest identity between the two species. However, by cooling to
77°K, the emission quantum efficiency increases by a factor of ~70
[43] while the risetime of K_{610} changes by only a factor of ~3[12].
Since the yields of J_{625} and K_{610} are independent of temperature[12],
it is evident that the two processes exhibit a markedly different
temperature sensitivity, excluding the possibility that the fluorescence originates from J_{625}.

Shapiro et al.[43] suggested, on the basis of extrapolation of
their low temperature data, that a 1.4±0.5 psec emitting state exists

at room temperature, and that it can be attributed to the 1.0±0.4
psec (I) precursor of Ippen et al.[20]. This interpretation, though
appealing, appears to contradict the observation that the 77°K and
90°K lifetimes are much longer than the risetime of J_{625}, established
by Applebury et al.[12] to be faster than 6 psec over the whole
1.8°K-298°K interval. Thus, to the extent that I can be identified
with the photochemically important common excited state, the latter
does not appear to be the one responsible for the fluorescence, in
keeping with the conclusions of Govindgee et al.[42]. The data
therefore suggest that neither I nor J_{625} is fluorescent, and that
emission originates from an excited state produced in a photochemically unimportant side path or, alternatively, in an excited photoproduct of BR_{570}^t.

Conclusions

Several conclusions can be derived on the basis of the above
analysis which impose serious limitations on any model suggested for
the primary event in bacteriorhodopsin (and most probably also
rhodopsin). First, since preceded by the 1.0±0.5 psec formation of
J_{625}, it is evident that the deuteration-sensitive generation of K_{610}
is not the primary dark reaction. Second, we have ruled out the
possibility (Fig. 1A) that J_{625} is the S_1 excited state retaining
the all-transoid configuration of BR_{520}^t(15). It was also shown that
J_{625} cannot be identified with the photochemical important "common
excited state" which determines the cycling yield in rhodopsins.
Such arguments, when combined with the lack of fluorescence from
J_{625} and with the spectral resemblance between J_{625} and K_{610}, make
the conclusion that J_{625} is a ground state chromophore a most feasible
one (Fig. 1B). It thus becomes evident that a structural transformation causing a large bathochromic shift in the ground state, (i.e.,
the generation of J_{625}) precedes the proton transfer reaction. These
conclusions are in variance with models (b, c) which assume that
proton translocation is the first dark reaction, taking place on an
excited surface. They are, however, consistent with model a in which
the generation of J_{625} is assumed to correspond to a trans-cis isomerization of the chromophore while proton transfer (associated with
the $J_{625} \rightarrow K_{610}$ step) is attributed to a ground state relaxation in
the protein (7-9). This model was originally proposed on the basis
of analogies to visual pigments where the evidence for isomerization
at the stage of bathorhodopsin is particularly compelling[7-9].

One point that has been used to argue[14,20] against the isomerization model in bacteriorhodopsin is that in rhodopsin the isomerization is cis→trans while in BR_{570}^t it would have to be trans→cis
(presumably 13-cis). Since both lead to a red shifted photoproduct
and since the thermal trans→13-cis process associated with converting
light adapted (BR_{570}^t) to dark adapted (50% BR^{13-cis}) bacteriorhodopsin
leads to a blue shift, an apparent contradiction is involved. However, we have recently shown[9] that photochemical isomerization,

whether cis→trans or trans→cis, should always lead to a red shifted primary photoproduct. The established point is that isomerization will remove the protonated Schiff base nitrogen from the vicinity of its counter-ion so that a red shift will necessarily follow. Thus, there are no difficulties associated with assuming a photo-isomerization as the primary photoevent in BR_{570}^t. On the contrary, it appears to be the only model consistent with all the available data that is capable of accounting for the universality of the red shift of the primary photoproducts of all pigments. We wish to point out in this respect that the only other explanation[11,13,14] for the red shift in K_{610} that has been proposed attributes it to proton transfer to the Schiff base nitrogen. In addition to being inconsistent with Resonance Raman experiments[26,47,48] (showing that changes in the degree of protonation occur only later in the photocycle) this approach gives no satisfactory explanation for the red shifted J_{625} as a ground state species formed before proton transfer.

A few comments are appropriate concerning the details of the proton translocation reaction. It has been suggested that the generation of K_{610} originates in a higher electronic state[11-16] or in a "hot" ground state[17,18] after trapping in a shallow (<1 Kcal) potential minimum (i.e., J_{625}), from which a proton is transferred by activated barrier crossing or, at low temperatures, by tunnelling. This picture seems quite unlikely since there is no reason to assume that optical excitation which results in a very large excess of vibrational energy will lead to selective thermalization in the shallow trap along the proton coordinate and consequently to the observed temperature dependence[11,12]. Our suggestion is that proton transfer occurs between protein groups which are vibronically detached from the chromophore. Such a process would involve a fully thermalized ground state system and hence accounts for the observed temperature (proton translocation) features, in a mechanism which is free from the difficulties associated with the presence of a large excess of vibrational energy after excitation.

Finally, it should be recalled that most of the conclusions derived in this paper for bacteriorhodopsin are based on experimental data and theoretical calculations concerning the I, J_{625} and K_{610} intermediates of this system. Unfortunately, similar information (e.g., time resolution of the BATHO' generation and its absorption spectrum) is not yet available for rhodopsin. Nevertheless, on the basis of the extensive analogies between the two systems[7-9] we believe that the postulated mechanism for BR_{570}^t also reflects the basic sequence of events in visual excitation.

Acknowledgement: The authors acknowledge a research grant from the Israel Commission for Basic Research. They wish to thank Dr. P.M. Rentzepis for valuable comments and discussions.

REFERENCES

1. Oesterhelt, D. and W. Stoeckenius, 1971. Nature (London) New
   Biol. 233: 149-152.
2. Stoeckenius, W., H.L. Lozier and R.A. Bogomolni. Biochim. Biophys. Acta Bioenergetics Reviews, in press.
3. Honig, B., 1978, Ann. Rev. Phys. Chem., 1978, 29, 31-57.
4. Ottolenghi, M. Advances in Photochemistry, Vol. 13, to be
   published.
5. Oesterhelt, D., and W. Stoeckenius, 1973. Proc. Natl. Acad.
   Sci., U.S.A., 70: 2853-2857.
6. Danon, A. and W. Stoeckenius, 1974. Proc. Natl. Acad. Sci.,
   U.S.A, 71: 1234-1238.
7. Rosenfeld, T., B. Honig, M. Ottolenghi, J. Hurley and T.G. Ebrey,
   1977, Pure and Appl. Chem. 49: 341-351.
8. Hurley, J., T.G. Ebrey, B. Honig and M. Ottolenghi, 1977.
   Nature, 270: 540-542.
9. Honig, B., T. Ebrey, R. Callender, U. Dinur and M. Ottolenghi,
   1979. Proc. Natl. Acad. Sci., U.S.A., 76: 2503-2507.
10. Schulten, K., 1978, in Energetics and Structure of Halophilic
    Microorganisms, ed. S.R. Caplan and M. Ginzburg, Elsevier/North
    Holland, Biochemical Press, New York, pp. 331-335.
11. Peters, K., M.L. Applebury and P.M. Rentzepis, 1977. Proc.
    Natl. Acad. Sci., U.S.A, 74: 3119-3123.
12. Applebury, M.L., K.S. Peters and P.M. Rentzepis, 1978. Biophys.
    J. 23: 375-382.
13. Favrot, J., J.M. Leclercq, R. Roberge, C. Sandorfy and D. Vocelle,
    1978. Chem. Phys. Lett., 53: 433-435.
14. Favrot, J., J.M. Leclercq, R. Roberge, C. Sandorfy and D. Vocelle,
    1979. Photochem. Photobiol. 29: 99-108.
15. Lewis, A. 1978, Proc. Natl. Acad. Sci., U.S.A., 75: 549-553.
16. Lewis, A., M.A. Marcus, B. Ehrenberg and H. Crespi, 1978.
    Proc. Natl. Acad. Sci., U.S.A., 75: 4642-46.
17. Warshel, A. and C. Deakyne, 1978. Chem. Phys. Lett., 55:459-63.
18. Warshel, A., 1978. Proc. Natl. Acad. Sci., U.S.A., 75: 2558-62.
19. Aton, B., R. Callender and B. Honig, 1978. Nature, 273:784-5.
20. Ippen, E.P., A. Shank, A. Lewis and M.A. Marcus, 1978. Science
    200, 1279-1281.
21. Monger, T., A. Doukas, R. Alfano and R. Callender, 1979.
    Byophys. J., 27, 105-116.
22. Lewis, A., R.S. Fager and E.W. Abrahamson, 1973. J. Raman
    Spect., 1: 465-470 (1973).
23. Oseroff, R.A. and R.H. Callender, 1974. Biochemistry, 13:
    4243-48.
24. G. Eyring and R. Mathies, 1979. Proc. Natl. Acad. Sci., U.S.A.
    76: 33-37.
25. B. Aton, A. Doukas, D. Narva, R. Callender, U. Dinur and B. Honig,
    1980. Biophys. J., in press.
26. Lewis, A., J. Spoonhower, R.A. Bogomolni, R. Lozier and W.
    Stoeckenius, 1974. Proc. Natl. Acad. Sci., U.S.A., 71: 4462-66.

27. Mendelshon, R., 1976. Biochim. Biophys. Acta, 427: 295-301.
28. Del Bene, J. and H.H. Jaffe, 1968. J. Chem. Phys., 18, 1807.
29. Schulten, K., I. Ohmine and M. Karplus, 1976. J. Chem. Phys.,
     64, 4422.
30. Dinur, U. and B. Honig, 1979. J. Am. Chem. Soc., 101: 4553.
31. Dewar, M. and T. Morita, 1969. J. Am. Chem. Soc., 91: 796.
32a. Honig, B., A.D. Greenberg, U. Dinur and T.G. Ebrey, 1976.
     Biochemistry, 15: 4593-99.
  b. Balogh-Nair, V., K. Nakanishi and B. Honig, to be published.
33. Rosenfeld, T., T. Alchalal and M. Ottolenghi, 1973. Chem.
     Phys. Lett., 20: 291.
34. Birge, R.R., J.A. Bennett, B.M. Pierce and T.M. Thomas, 1978.
     J. Am. Chem. Soc., 100, 1533.
35. Waddell, W.H., A.M. Schaffer and R.S. Becker, 1973. J. Am.
     Chem. Soc., 95: 8223-8227.
36. Becker, R.S., G. Hug, P.K. Das, A.M. Schaffer, T. Takemura,
     N. Yamamoto and W. Waddell, 1976. J. Phys. Chem. 80: 2265-73.
37. Birge, R.R. and B.M. Pierce, J. Chem. Phys., in press.
38. Shulten, K., private communication.
39. Lewis, A., J. Spoonhower and G. Perrault, 1976. Nature 260:
     675-678.
40. Alfano, R.R., W. Yu, R. Govindjee, B. Becker and T.G. Ebrey,
     1976. Biophys. J. 16: 541-545.
41. Sineshchekov, V.A. and F.F. Litvin, 1977. Biochim. Biophys.
     Acta 462: 450-466.
42. Govindjee, R., B. Becker and T.G. Ebrey, 1978. Biophys. J.
     22: 67-77.
43. Shapiro, S.L., A.J. Campillo, A. Lewis, J. Perreault, J.P.
     Spoonhower, R.K. Clayton and W. Stoeckenius, 1978. Biophys.
     J. 23: 383-393 (1978).
44. Gillbro, T., A. Kriebel and U.P. Wild, 1977. EEBS Lett., 78:
     57-60.
45. Kriebel, A.N., T. Gillbro and U.P. Wild. Submitted for
     publication.
46. Hirsch, M.D., M.A. Marcus, A. Lewis, H. Mahr and N. Frigo,
     1976. Biophys. J. 16: 1399-1409.
47. Marcus, M.A. and A. Lewis, 1977. Science, 195: 1328-30.
48. Campion, A., M.A. El-Sayed and J. Terner, 1977. Biophys. J.
     20: 369-75.

NEW EXPERIMENTAL APPROACHES TO THE STUDY OF THE PHOTORESPONSES OF

FLAGELLATED ALGAE

C. Ascoli, C. Frediani, T. Ristori

Laboratorio per lo studio delle proprietà fisiche di
biomolecole e cellule - C.N.R. - Pisa

INTRODUCTION

Light affects the motion of a wide class of microorganisms.
In flagellated algae phenomena of this type have often been studied
by observing the behaviour of cells. Both individual cell methods
and population methods have been used. A complete review of the
most recent literature on the topic can be found in Nultsch and Häder
(1979); this paper also surveys the older literature, while a review
of experimental methods can be found in Nultsch (1975). The observed
photoresponses have been classified (according to Nultsch and Häder
(1979)) as phototaxis (an oriented motion towards or away from a
light source), photokinesis (dependence of swimming speed on light
intensity), and as photophobic reactions. These last are due to a
temporal change in light intensity; typically when such a change
occurs abruptly, the cell stops and moves off again in a new direction. In the literature it is an open question if these different
photoresponses are related to different mechanisms or to the same
one; even the classification of photoresponses is debatable (Diehn
et al., 1977).

Photoresponses are the result of complex chain reactions which
start from light reception and ultimately give a motor response
through sensory transduction; thus observation of photobehaviour
gives information about all the mechanisms involved. Action spectra
of·the photoresponses have offered the first approach to the identification of the photoreceptor pigment, while photoresponse changes
induced by metabolic drugs can provide information about sensory
transduction and motor activity. More direct information about these
successive steps has appeared to be a precondition for a deeper
understanding of the phenomena involved.

The direct observation of the behaviour of individual cells
(by microscopic observation or by cinematographic tracking) seems
the best way of distinguishing among different effects, but this
approach becomes very difficult and labor-intensive if used to obtain
statistically meaningful data, and it requires subjective selection
of single microorganisms to be tracked. Population methods appear
to yield a better quantification of photobehaviour; they consist in
measuring differences in optical density, induced by a light stimulus
in an initially homogeneous suspension. These measurements only give
information about the final result - photoaccumulation - and do not
provide direct information about how this result is obtained. Phototaxis, phobic reactions and even photokinesis can produce photoaccumulation or photodispersal; so, without carefully selecting the experimental conditions, it is impossible to distinguish among the
different effects and so to measure them. We have therefore developed a new experimental approach, founded on laser Doppler spectroscopy, capable of measuring a component of the velocity and the sign
of this component for a population. So, by applying this technique
to a study of the photoresponses of flagellated algae, we can measure
the number of microorganisms swimming towards and away from the light
source. The application of this technique to Haematococcus pluvialis
and to Euglena gracilis allows the advantages to be discussed with
respect to the usual population methods.

We are also developing another experimental approach for studying the transduction phenomena between photoreception and photoresponse in flagellated algae: intracellular recording of light-
-dependent potential changes.

Interesting information has already been obtained by extracellular recording of a light-dependent potential (Litvin et al.,
1978). The preliminary results obtained in our Laboratory by intracellular recording in Haematococcus pluvialis are reported here,
and the prospects of such research are briefly discussed.

LASER DOPPLER SPECTROSCOPY

Operative principles

Laser Doppler spectroscopy has been used to measure the statistically averaged properties of swimming cell populations. A bibliography on this topic can be found in Ascoli et al. (1978a).

The principle operative in this technique consists in revealing
the Doppler shift of the light scattered from a moving particle.
The optical spectrum of the laser light scattered by a population
of moving point scatterers is the envelope of individual Doppler
shifts and is homologous to the distribution of the component of
the velocity in a given direction, dependent on the geometry of

scattering (Ascoli et al., 1978a). If the particles are moving iso-
tropically, the optical spectrum will be symmetrically distributed
around the laser frequency; but, if there is some preferential direction in the sample, this symmetry disappears. This situation is
illustrated schematically in fig. 1a. The optical spectrum can be
revealed by heterodyne detection (Bergé and Dubois, 1973; Ascoli et
al., 1978a). In this case the photodetector collects the light
scattered by the sample, together with the unperturbed laser beam
at ν_0, and the photocurrent contains the beatings between the Doppler
shifted scattered light and the local oscillator beam (unperturbed
laser light). The photocurrent power spectrum reproduces at low
frequencies the superposition of the right and left sides of the
optical spectrum (fig. 1b). So, if the optical spectrum is symmetrical, the heterodyne spectrum represents the distribution of a component of the velocity. When, as in the case of fig. 1a, the optical

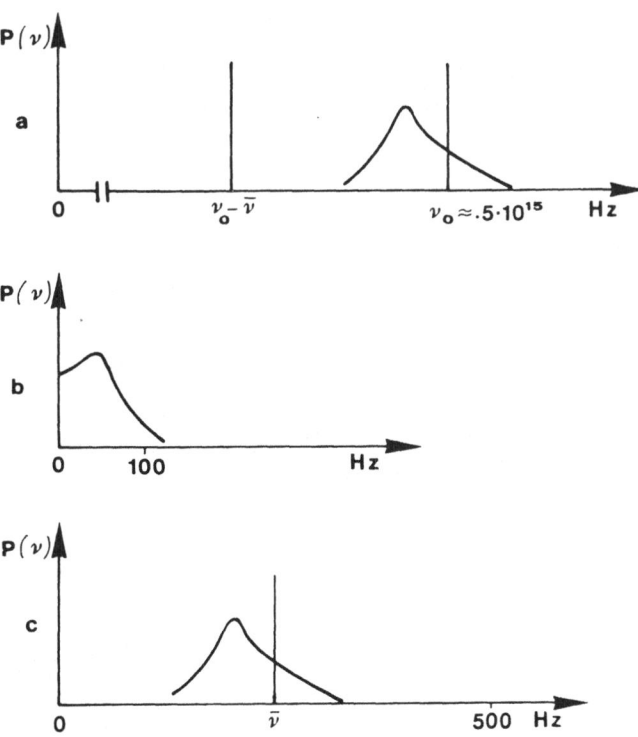

Fig. 1 - Schematic drawing showing: a) optical spectrum of a population of moving particles. The line at $\nu_0-\bar{\nu}$ is the
frequency-shifted laser line used to obtain the spectrum
shown in c; b) heterodyne spectrum; c) frequency-shifted
heterodyne spectrum.

spectrum is not symmetrical around ν_o, any information about this asymmetry is lost in the usual heterodyne spectrum, which is, therefore, an unsuitable tool for the detection of asymmetry in population motion. To obtain a complete reproduction of the optical spectrum, the usual heterodyne spectrometer has been improved (fig. 2). Using an electro-optic phase modulator, the local oscillator has been frequency shifted by a given amount $\bar{\nu}$ with respect to the beam which falls on the sample. The photocurrent will then contain the beatings between the local oscillator, with frequency $\nu_o - \bar{\nu}$, and the light scattered by the sample. As a result, the photocurrent spectrum will completely reproduce, around $\bar{\nu}$, the optical spectrum, without averaging left and right bands (fig. 1c). This frequency shifted heterodyne spectrometer appears to be a useful tool for studying any effect which introduces asymmetries in the motion of particles.

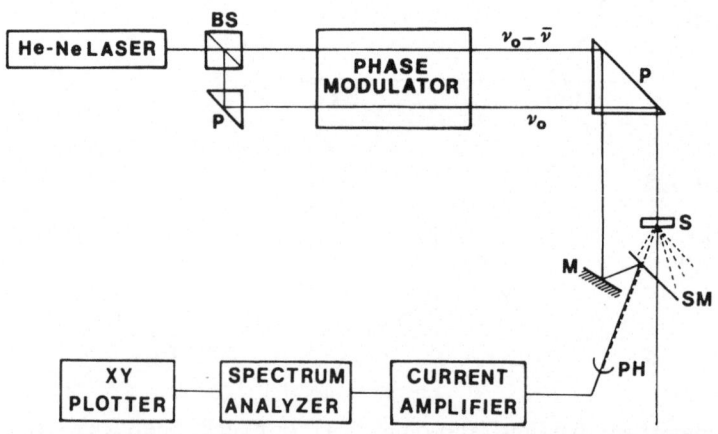

Fig. 2 - Block diagram of the frequency-shifted heterodyne spectrometer. BS, beam splitter; P, prism; S, sample; SM, semireflecting mirror; M, mirror; PH, photodiode.

Application to flagellated microorganisms

We have applied this technique to a study of light responses
in H.p. and E.g. These microorganisms are shown in fig. 3. E.g.
(fig. 3a) is an elongated unicellular microorganism, with dimensions
10x10x60 μm. It shows many structures, notably, chloroplasts, a
photoreceptor - the paraflagellar body - and the stigma, which is
generally believed to play a role in phototaxis. The pigments
present in the paraflagellar body and in the stigma have been studied
using a microspectrophotometric technique (Benedetti et al., 1976;
Benedetti and Checcucci, 1975). H.p. (fig. 3b) is a quasi-spherical
unicellular microorganism, 20 μm in diameter. It too shows photoaccumulation, but its photoreceptor is still unknown. Many experiments have been carried out on photobehaviour, especially on E.g.,
but the nature of photoaccumulation is not yet completely clear for
either of these microorganisms. Phototactic and photophobic components have both been considered in explaining photoaccumulation
(Feinleib and Curry, 1971; and Feinleib, 1978). In E.g. a photokinetic response has been observed, too (Wolken, 1967; Ascoli et al.,
1975). This effect does not seem to be controlled by the photoreceptor which controls photoaccumulation; it depends instead on
the photosynthetic process.

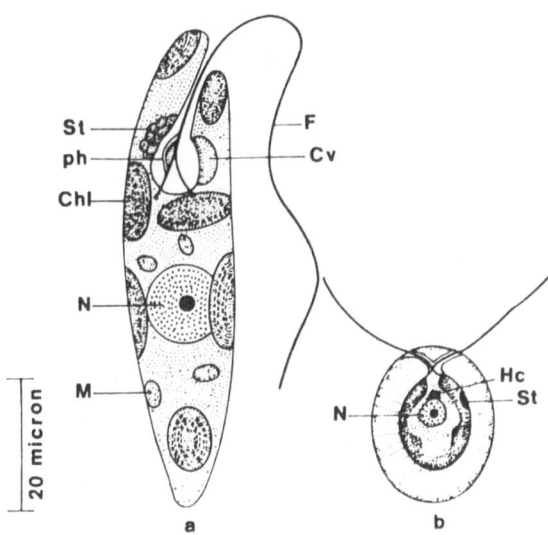

Fig. 3 - Schematic drawing of: a) Euglena gracilis; b) Haematococcus
    pluvialis; N, nucleus; St, stigma; F, flagellum; Chl,
    chloroplast; Hc, haematochrome; Ph, photoreceptor;
    M, mitochondria; Cv, contractile vacuole.

The description given above for the operative principle of
laser Doppler spectroscopy is not adequate for flagellated microorganisms such as E.g. and H.p.; these are not point scatterers and
their motion is not only translatory. Full information about the
dynamics of the motion of these microorganisms can be found in the
photocurrent spectrum, but interpretation of the spectra is complicated. For E.g. the spectrum contains contributions from different
motion parameters (swimming velocity, flagellar beating, body rotation) and the weights of individual contributions depend on each
cell's orientation, as a result of the anisotropic nature of the
light scattering pattern from a non-spherical cell (Ascoli et al.,
1978a).

Quantitative interpretation of spectra is very difficult; the
best approach is that of choosing suitable experimental conditions
in which it is possible to measure the distribution of each of the
motion parameters. By orienting E.g. in a radiofrequency field
(Ascoli et al., 1978b) and by choosing suitable positions of the
photodetector, the mean speed and the frequencies of flagellar beating and cell rotation have been measured, but this experimental
approach can be applied only to microorganisms which (like Euglena)
are elongated and live in a low-conductivity medium. Additional
sources of difficulty are the changes in physico-chemical properties
required to orient the cells and the time required to prepare the
sample. Therefore in the phototaxis experiments we have used unoriented samples; the spectrum obtained for E.g. by the frequency
shifted heterodyne spectrometer is shown in fig. 4. The measurements

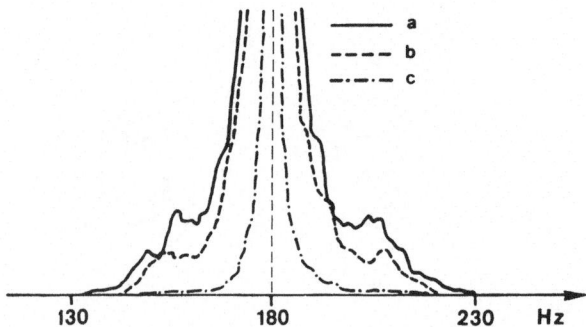

Fig. 4 - Frequency-shifted (180 Hz) heterodyne spectra of a popula-
         tion of E.g. illuminated by the light spot in fig. 5b;
         a) dark-adapted cells; b) 12 minutes after switching on
         the light; c) 16 minutes after switching off the light.

taken on oriented and unoriented samples of <u>Euglena</u> (Ascoli et al.,
1978a; Ascoli and Frediani, 1979) allow us to extract the following
information from the spectrum of fig.4: i) the lateral bands at about
30 Hz from the frequency ν (250 Hz) are the distribution of flagellar
beating frequencies; ii) the large band around ν̄ is not homologous
to the distribution of the effective component of the velocity, since
the different intensities scattered by differently oriented cells,
produce distorsions. These distortions do not introduce asymmetries
in the band; thus we can use the frequency-shifted heterodyne spectrum as an indicator for the asymmetry in the motion of <u>Euglena</u>
cells, although slight asymmetries could be masked.

The scattering pattern and motion feature of <u>H.p.</u> are still
unknown, but its small dimensions and its quasi-spherical shape allow
us to expect, as a first approximation, the light scattering pattern
to be isotropic. Thus the optical spectrum will give the distribution of a component of the velocity for this organism. Fig. 5 shows
the spectrum of a single <u>H.p.</u> cell. We can observe one line narrower
than the bandwidth of the spectrum analyser and two lateral bands
about 40 Hz from that line. We ascribe the lateral bands to the
flagellar beating (Ascoli and Frediani, 1979), while the narrow
central line is due to the Doppler shift only, as for a point scat-
terer. Besides, for small values of the scattering angle the maximum
Doppler shift is less than the distance between the lateral bands
and the Doppler line in the spectrum of each cell; thus the flagellar
beating distribution does not affect the Doppler spectrum. These

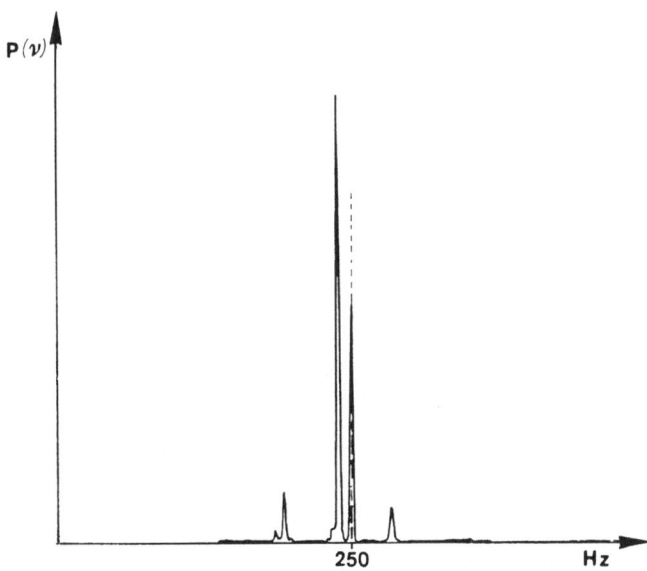

Fig. 5 - Frequency-shifted (250 Hz) heterodyne spectrum of a single
Haematococcus pluvialis.

characteristics of <u>H.p.</u> allow us to use Laser Doppler spectroscopy
not only to observe the asymmetry in the motion of a cell population,
but also to measure the distribution of a selected component of the
velocity in that population.

Phototaxis experiments

To clarify the mechanism of the photoresponses in <u>E.g.</u> and <u>H.p.</u>
we have used two different techniques for illuminating the sample.
In the first, shown in Fig. 6a, the microslide containing the microorganisms ($50 \times 5 \times 0.5$ mm^3) is placed horizontally and a collimated
actinic beam (at 450 nm) forms an angle of $20°$ with the long axis
of the slide; the entire preparation lies within the beam. Since
there are no light intensity gradients, we can reasonably exclude
phobic reactions, and we can ascribe photoaccumulation to phototaxis.

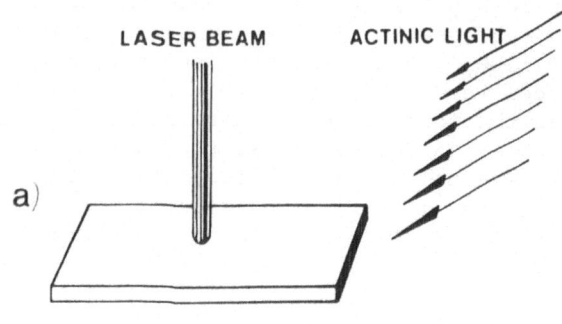

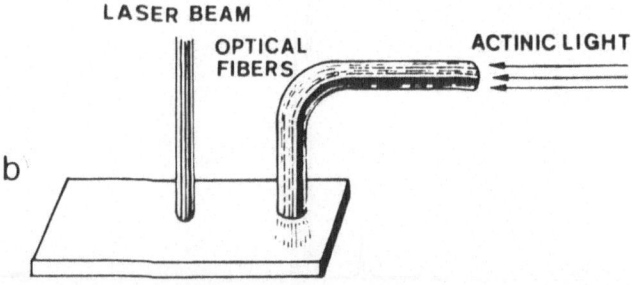

Fig. 6 - Schematic drawing of: <u>a</u>) uniform lateral illumination
($\lambda = 450$ nm, irradiance 10 W/m^2); <u>b</u>) light spot illumination
(white light, 130 W/m^2 irradiance, 2 mm spot diameter).

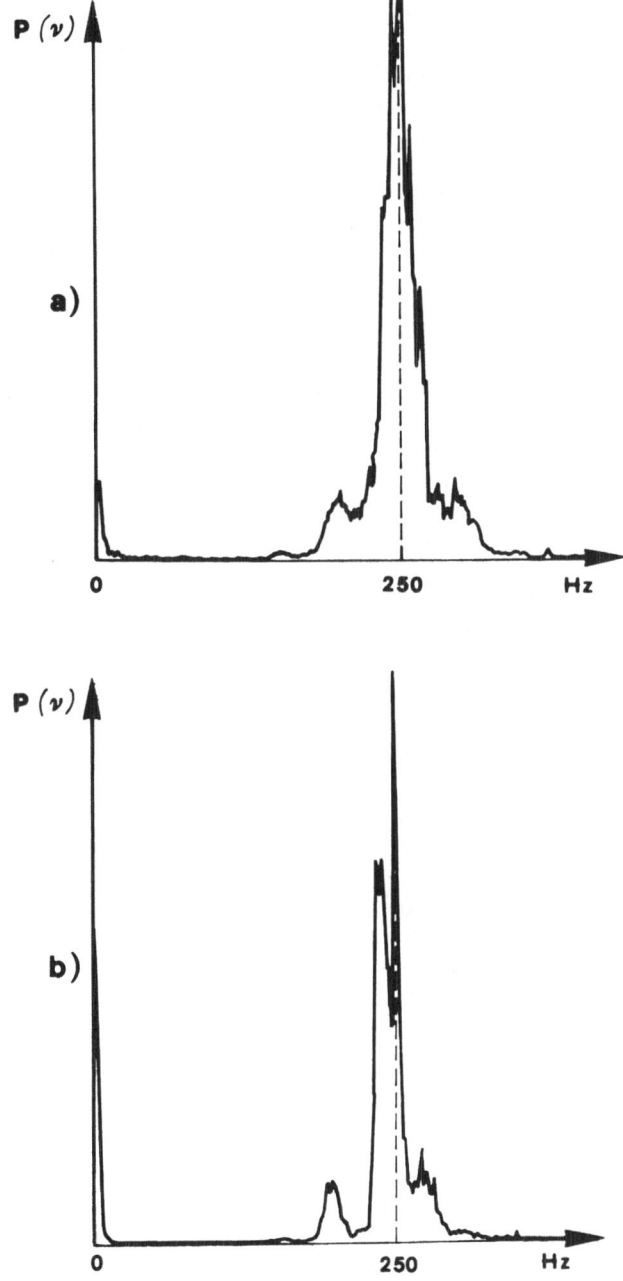

Fig. 7 - Frequency-shifted heterodyne spectra of a population of
Haematococcus pluvialis: a) in dark; b) after 8 minutes in
uniform light. The peak to the left of the frequency ν =
250 Hz indicates that the mean component of the velocity
is directed towards the light source.

Fig. 7 shows the spectra obtained for H.p. in the dark and after
illumination. In the dark (fig. 7a) the spectrum is symmetrical
(the motion of the cell population is isotropic) whereas during illumination (fig. 7b), a marked asymmetry becomes apparent and the
mean component of the velocity is directed towards the light source.
Observation of the microslide at 100x magnification shows an accumulation on the side nearest the light. By contrast, when a sample
of E.g. is illuminated under these conditions, the spectra remain
symmetrical; observation through the microscope confirms that there
is no photoaccumulation. The first result to emerge from this experiment is that H.p. is phototactic under these conditions, whereas
this is not apparent in the case of E.g.

To check this result we have also used another type of illumination, shown in fig. 6b. The geometry of this optical system resembles that used by Lindes et al. (1965), Diehn (1969) and Checcucci
et al. (1976). A bundle of optical fibers is used to direct the
actinic beam perpendicular to the long axis of the microslide; the
light forms a circular spot at distance of 1.5 cm from the laser
beam. Again, the H.p. spectra show a marked asymmetry during illumination, whereas the E.g. spectra (fig. 4) remain symmetrical, but
show a progressive emptying of the zone under the laser beam during
illumination. For both these microorganisms microscopic observation
shows an accumulation under the light spot. We may thus conclude
that H.p. gives a phototactic response (oriented motion toward light
scattered from cells in the illuminated zone), while E.g. does not.
Moreover, the accumulation of Euglena occurs only in the presence
of a light gradient (i.e. in second illumination system) and therefore is probably due to phobic responses.

INTRACELLULAR RECORDING FROM HAEMATOCOCCUS pluvialis

Extracellular recording of the photodependent potential of H.p.
has been performed by Litvin et al. (1978). Graded responses to
light occurred at all intensities and, at high intensity, action
potentials were recorded as well. The investigators suggest that
the active portion of the cell membrane -that producing the electrical
cell activity- is in the anterior end of the cell body.

Intracellular recording from this microorganism has been performed in our laboratory. H.p. (fig. 3b) has a single cup-shaped
chloroplast, a stigma (near the cell periphery and, presumably, inside the chloroplast envelope) as well as a nucleus and a haemato-
chrome. The locus (or the loci) of photoreception is still unknown,
but it is reasonable to assume that it has some relation to the
stigma. The cell wall is covered by a gelatinous coat; this structure
probably plays a role in protecting the cell membrane during the
impalement. Fig. 8 shows the experimental system used for impaling
the cell. A suction electrode is used to hold the microorganism;

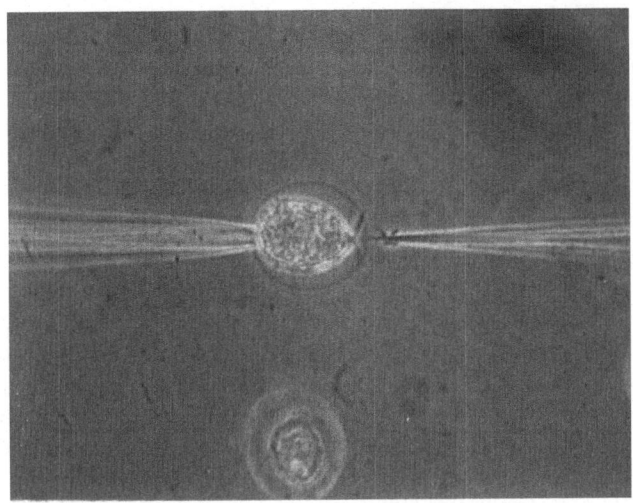

Fig. 8 - The impaling of H.p. Microscope magnification is 800.

the impaling micropipette is pushed forward slowly by a micromanipu-
lator until it touches the cell and then a decisive thrust is given
by a piezoelectric device.

Fig. 9 shows a graded response to light; the resting potential
is about -15mV and the amplitude of the response is about -10 mV.

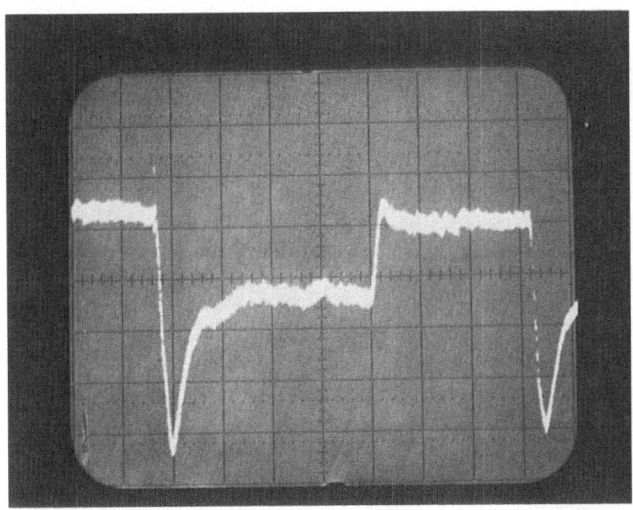

Fig. 9 - Graded response to light; time scale 1 sec/div; amplitude
scale 2 mV/div.

Both hyperpolarizing and depolarizing light responses have been
found. It has been possible to show that the photoresponses have
an equilibrium potential; in fact, by injecting current, the sign
of the light response can be reversed. A sustained discharge of
action potential (regenerative responses), probably due to cell
injury, has occasionally been observed (fig. 10). The polarity
and the time course of such pulses are very different from those of
the spikes in higher animals.

CONCLUDING REMARKS

     The experimental results reported confirm the usefulness of
both types of experiment - laser light scattering and intracellular
recording - to a clarification of the nature of the photoresponses
in flagellated algae and the mechanisms responsible for them.
Laser Doppler spectroscopy has been shown to provide quantitative
information about photoorientation in a population of microorganisms.
We believe that the combination of the frequency-shifted heterodyne
spectrometer with a system for measuring differences in optical
density yields a better technique than is offered by the usual photo-
taxigraphs. In fact our method makes it possible to discriminate
among different types of photoresponses independently of the kind
of light stimulus; the kinetics of the population response can be
followed by obtaining a series of spectra. Another useful feature
of laser Doppler spectroscopy is that it allows quantitative measurements on single microorganisms.

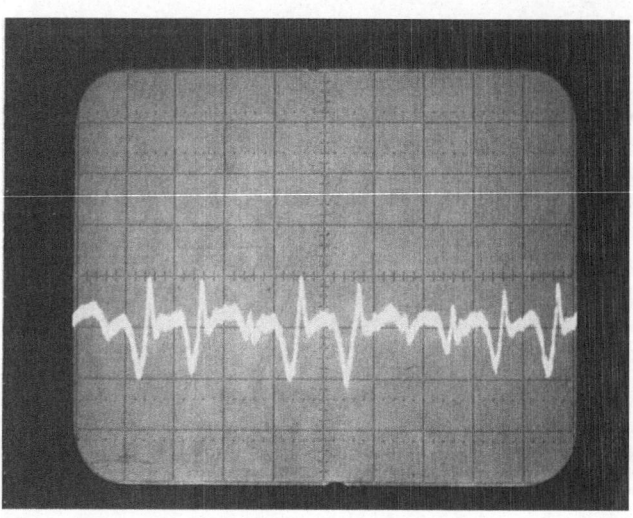

Fig. 10 - Regenerative potentials in damaged cells.

The recording of photodependent electrical potentials is a useful complementary approach. The recording of electrical responses
seems to be the most effective experimental system for studying the
early stages of transduction from photoreception to motor activity.
This technique can also be used to find the site of the photoreceptor
by scanning the cell body with a very small light spot (2 - 3 μm in
diameter) and recording the photopotential. Preliminary experiments
show the feasibility of this experimental system.

REFERENCES

Ascoli,C., Frediani,C., and Nultsch,W., 1975 - A photokinetic effect
    in Euglena gracilis. P116, 5th International Biophysics
    Congress - Copenhagen
Ascoli,C., Barbi,M., Frediani,C., and Muré,A., 1978a - Measurements
    of Euglena motion parameters by laser light scattering.
    Biophys. J., 24:585.
Ascoli,C., Barbi,M., Frediani,C., and Petracchi,D., 1978b - Effects
    of electromagnetic field on the motion of Euglena gracilis.
    Biophys. J., 24:601.
Ascoli,C., and Frediani,C., 1979 - Quasi elastic light scattering
    in the measurement of motion of flagellated algae. In Proceed-
    ings of "Workshop on Quasielastic light scattering studies of
    fluid and macromolecular solutions". June 11-12, Milan. (in
    press)
Benedetti,P.A., and Checcucci,A., 1975 - Paraflagellar body (PFB)
    pigments studied by fluorescence microscopy in Euglena gracilis.
    Plant Sci. Letters, 4:47
Benedetti,P.A., Bianchini,G., Checcucci,A., Ferrara,R., Grassi,S.
    1976 - Spectroscopic properties and related functions of the
    stigma measured in living cells of Euglena gracilis. Arch.
    Microbiol., 111:73
Bergé,P., and Dubois,M., 1973 - Dispositif de mesures optiques adapté
    à l'étude du mouvement de microorganismes vivants., Rev. Phys.
    Appl. 8:89.
Checcucci,A., Colombetti,G., Ferrara,R., Lenci,F., 1976 - Action
    spectra for photoaccumulation of green and colorless Euglena;
    evidence for identification of receptor pigments., Photochem.
    Photobiol., 23:51.
Diehn,B., 1969 - Action spectra of the phototactic responses in
    Euglena. Biochim. Biophys. Acta, 177:136.
Diehn,B., Feinleib,M.E., Haupt,W., Hildebrand,E., Lenci,F., Nultsch,
    W., 1977 - Terminology of behavioral responses in motile micro-
    organisms. Photochem. Photobiol., 26:559.
Feinleib,M.E., and Curry,G., 1971 - The nature of photoreceptor in
    phototaxis. In: Loewenstein,W.R. (Ed.); Handbook of sensory
    physiology. Vol. 1, Springer, Berlin.
Feinleib,M.E., 1978 - Photomovement of microorganisms. Photochem.
    Photobiol., 27:849.

Lindes,D., Diehn,B., Tollin,G., 1965 – Phototaxigraph: recording instrument for determination of rate of responses of phototactic microorganisms to light of controlled intensity and wavelength. Rev. Sci. Instrum., 36:1721.

Litvin,F.F., Sineshchekov,O.A., Sineshchekov,V.A., 1978 – Photoreceptor electric potential in the phototaxis of the alga Haematococcus pluvialis, Nature, 271:476.

Nultsch,W., 1975 – Phototaxis and photokinesis. In "Primitive sensory and communications systems; the taxes and the tropism of microorganisms and cells." Academic Press, London.

Nultsch,W. and Häder,D.P., 1979 – Photomovement of motile microorganisms. Photochem. Photobiol., 29:423.

Wolken,J.J., 1967 – "Euglena", Meredith Publishing Co., Desmoines, Iowa.

PHOTOMOTILE RESPONSES IN Ochromonas danica

M.A. Di Pasquale, M.P. Bizzaro, and M. Ciliberti

Laboratorio Studio Proprietà Fisiche di Biomolecole e
Cellule, C.N.R.
via F. Buonarroti, 9 - 56100 PISA (ITALY)

INTRODUCTION

It is well known that many freely motile unicellular organisms
are able to detect variations in the external environmental conditions, and to react to such variations with alterations in their
motile behavior. In particular, this may happen when light is the
external stimulus: in this case the organisms are said to show photomotile responses.

There are several types of such responses and the interested
reader can find more information on that in the literature (1).
Here we limit ourselves to the so called photophobic reactions, in
other words to those alterations in motion shown by cells upon a
sudden variation in the external illumination. We want to recall
that in case of a motile response to an increase in light intensity,
we speak of a step-up photophobic response (shortly step-up response),
and, in case of a decrease in illumination, of a step-down photophobic response; we will also talk about photoaccumulation, that is
the cell ability to accumulate in an illuminated area.

Photoreceptor structures, dedicated to the detection of the
environmental light stimuli, have been shown to be present in some
known photoreactive eucaryotes, such as for instance Euglena gracilis.
In other cases, such as, for example, Ochromonas danica, no information was until now available on its photomotile responses, but
ultrastructural research has shown the presence of possible photoreceptor structures, as we will see below; therefore, we decided to
investigate whether Ochromonas showed a photomotile behavior and our
preliminary results are reported in what follows.

MATERIALS AND METHODS

Ochromonas danica is a unicellular photosynthetic flagellate
of the class of Crysomonads.

The nucleus and the chlorophyll A-containing chloroplast are
at the anterior end of the cell, posterior to them is the large
leucosin vacuole bounded by a thin rim of cytoplasm. At the posterior end of the cell there is the tail, which often causes the
organisms to attach to the surface of the cover slip. The flagellar
apparatus is at the anterior end of the organism and consists of
two flagella: the long one, bearing two rows of mastigonemes, pulls
forward the cell, in the same direction of wave propagation (2).
The short one, which appears motionless during swimming (3), is
simple and presents close to its base an organelle that could act
as a photoreceptor. Near to this organelle and within the chloroplast is the stigma (4).

For our studies, axenic cultures of Ochromonas danica were
grown in a chemically defined, heterotrophic medium, as described
by Aaronson (5). The green photosynthetic cells were cultured under
continuous illumination (about 5 W/m^2), at 24°C; the dark-bleached
strain was obtained keeping the cells in complete darkness at the
same temperature.

For studying Ochromonas danica photobehavior we performed mass
and single-cell measurements. The former ones were obtained by
means of the phototaxigraph: an instrument that measures changes
in turbidity of a cellular suspension when illuminated by an actinic
light, as described by Checcucci et al. (6). As actinic light source
we used a Tungsten lamp (about 15 W/m^2) shielded by means of a
Calflex anticaloric filter to prevent heating effects. Wavelengths
of actinic light were selected with glass coloured filters which
cut off about 80% of the light intensity. We could not use interference filters because the maximum light intensity available in
this case was not enough to cause a detectable signal in the phototaxigraph. We tested photoaccumulation under red and blue light
using 2-3 days-old cultures of both green and dark-bleached cells.
Cell concentration was about 2 x 10^7 cells/ml. In some experiments
we treated the cells with ethanol solutions of 3-(3-4 dichlorophenyl)
1-1-dimetylurea (DCMU), a drug specifically affecting photosthesis
by inhibiting the electron flow from PSII to PSI. The final DCMU
concentrations varied from 10^{-8} to 10^{-6} M. It was verified on
control samples that this amount of ethanol had no effect on cell
motility. The initial slope of the taxigram was used as index of
photomotile response.

Single-cell photomotile responses were examined using a Leitz
Ortholux II optical microscope, coupled to a Philips B/N LDH 25
videocamera equipped with a IR sensitive SIVICON LDM 4461/00 tube,

connected to a Philips EL 34021 videotape recorder and to a video
monitor. Cells were observed under an inactinic low intensity light
of 730 nm. As actinic light source we used an OSRAM XBO 75 W/L high
pressure Xenon lamp, adjusting the light intensity on the sample by
means of neutral density filters. Following qualitative observations
we decided to use actinic blue light. Cell suspension was placed
on a microscope slide accurately closed by means of a cover slide,
bounded by a thin rim of vaseline. Care was taken to avoid air
bubbles. Concentrations of about 4 x 10^6 cells/ml were used: these
were obtained by using one day-old cultures or by deluting older
ones with their own medium. The swimming paths were traced on
transparent plastic sheet placed over the monitor, by examining
the videotape recordings frame by frame.

RESULTS

First of all, we tested whether Ochromonas danica accumulated
under red and blue light. We found that both green and dark-bleached
cells accumulated under blue light, while red light was only active
for the green cells. In fact, red light was unable to cause any
accumulation in white cells; sometimes, however, a small dispersal
was exhibited by the bleached cells. Usually photoaccumulation takes
place after a delay.

In the following experiments we treated green cells with DCMU:
we found that this drug affects photoresponse in Ochromonas, reducing the accumulation rate under blue light to about 50% of the
control and essentially to zero under red light. These results
were obtained in the drug concentration range of 10^{-7} - 10^{-6} M.
Ethanol was found to have no effect on photoaccumulation. Moreover,
qualitative observations suggest that DCMU does not affect the
response of the dark-bleached cells to blue light.

Our measurements on single cells indicate that Ochromonas shows
both step-up and step-down photophobic responses. The step-up
response in both green and dark-bleached cells is always present
when cells are presented with a proper stimulus, that is a sudden
light intensity increase. Frame by frame analysis of the videotape
recordings shows that, in dark, cells have a linear swimming path,
whereas the light stimulus immediately causes continuous directional
changes in cells trajectories, with occasional rotations on the spot.
The cells resume their normal swimming if the stimulus lasts more
than 10-15 seconds, or, of course, when it is removed.

On the contrary, a sudden light intensity decrease, after a
few seconds of proper illumination, is not enough to cause step-down
photophobic response in Ochromonas. In fact, green cells were found
to show this response only some minutes (about 10) after they were
accurately sealed in the observation chambers, even if sometimes

they do not show it at all. Because of these experimental difficulties, we are still working in order to determine the parameters involved in step-down response. We also observed in phase contrast
the cells attached through their tail to the slide, during the step-
up and step-down responses. It seems that the stimulus causes a
sudden variation of the insertion angle of the long flagellum, so
that it snaps of about 90° toward the cell body and even its beating
frequency appears to change. Even if it was impossible to find out
anything about the possible events concerning the short flagellum
during photophobic responses, we can say that, differently from what
reported by Jahn (3), it seems to beat, both in light and dark.

DISCUSSION

     Our results, both on cell population and on single cells,
demonstrate that Ochromonas danica is able to show motor responses
to light stimuli. To our knowledge it is the first time that this
photobehavior is reported in the literature. Microscopic observations have shown that the elementary motor reaction of Ochromonas
consists of a reorientation of the locomotory flagellum that snaps
down toward the cell body, as already mentioned in the previous
section. This is true both for the step-up and the step-down photophobic responses and is very similar to what happens in Euglena.

     The macroscopic result of this flagellar reorientation is an
increase in the tumbling of cells during their motion, in other
words, during a step-up or step-down response, cells show the so
called klinokinesis. As mentioned before, after a certain time
(of the order of about 10-15 sec) cells adapt to the new illumination
conditions and resume their normal straight swimming, as expected in
a photophobic response. This reaction is shown both be green and
dark-bleached cells and can be classified among the "blue-light
response", even if more work is necessary in order to have a better
knowledge of its wavelength dependence.

     The step-down response can be observed only (but not always)
when the cells have been kept sealed for 6-10 minutes in the observation chamber: this could indicate a possible role for metabolites
or external agents (O_2 perhaps) inducing (or inhiting) this photobehavior. With regard to mass measurements, it should be noted
that accumulation rates are much lower than those observed, under
the same conditions, in other microorganisms (Euglena, for instance).
This probably indicates a smaller efficiency of cell trapping in a
light trap (until now we have not been able to observe any accumulation in a light trap under the microscope), but we do not have
yet definitive ideas about the possible explanations for this.
Accumulation under red light is probably brought about by the photosynthetic system, as indicated by the results with DCMU and the fact
that dark-bleached Ochromonas does not show any red-light induced

accumulation.

    In analogy with the case of Euglena (7)· we could think that this effect of red light is mediated via photosynthetic O_2 evolution (DCMU) selectively blocks it), but more work would be necessary to clarify this point.

REFERENCES

1) Lenci, F., Colombetti, G. (1978) Ann. Rev. Biophys. Bioeng. 7, 341-361.
2) Halwill, M., Peters, P. (1974) J. Cell. Biol. 62, 322-328.
3) Jahn, T., Landman, M., Fonseca, J.R. (1964) J. Protozool. 11, 291-296.
4) Schuster, F., Hershenov, B., Aaronson, S. (1968) J. Protozool. 15, 345-346
5) Aaronson, S., Scher, S. (1960) J. Protozool. 7, 156-158.
6) Checcucci, A., Favati, L., Grassi, S., Piaggesi, T. (1975) Monitore Zool. Ital. 9, 83-98.
7) Checcucci, A., Colombetti, G., Del Carratore, G., Ferrara, R., Lenci, F. (1974) Photochem. Photobiol. 19, 223-226.

PHOTOSENSORY TRANSDUCTION IN Euglena gracilis:

INEFFECTIVENESS OF SOME PROTONOPHORES

A.Castiello, E.Mikolajczyk[+], V.Passarelli

Lab. Biomolecole e Cellule, Via Buonarroti,9
56100 PISA (Italy)
+ Nencki Institute of Experimental Biology
Pasteura 3, 02-093 WARSAW (Poland)

INTRODUCTION

The different steps involved in the photosensory transduction chain of microorganisms have been investigated in recent years by several research groups[1], but many problems are still not clarified. It is generally accepted that the components of a stimulus-response system can be schematized as follows:

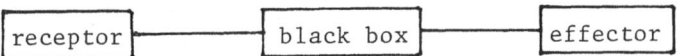

In particular in Euglena gracilis the receptor and the effector have been reasonably well identified[1], whereas not much information is available on the molecular mechanisms that constitute the so called black box. Some hypotheses have been put forward and they concern mainly:
1) a possible role of metabolic pathways, such as photosynthesis and oxidative phosphorylation;
2) membrane potentials;
3) ion gradients and fluxes, in particular light-induced ΔpH variations.

Previous studies in our research group have excluded[2] a correlation between photosynthesis or oxidative phosphorylation and the photobic step-down response in Euglena. Moreover, also a possible role of a light-induced ΔpH was shown to be unlikely. In order to further investigate points 2) and 3) we have performed a series of experiments using substances such as Gramicidin D and Carbonycyanide-chlorophenylhydrazone (CCCP), that are known to affect specifically proton transport across biological membranes[3,4,5].

MATERIALS AND METHODS

Axenic cultures of Euglena gracilis, strain Z, were grown in an automatic continous culture apparatus at 24°C, under constant illumination[6]. The photobehaviour of single cells was examined by means of a microscope (Leitz Ortholux) coupled to a video-recording system described elsewhere[7]. The recordings are played back and analyzed at lower speed, in order to determine parameters such as the percentage of motile cells, their speed and the percentage of cells showing step-down and step-up responses. Oxygen uptake in the dark was measured by means of an oxygen meter HACH. The drugs from Sigma were used throughout our experiments without further purification. They were dissolved in ethanol and control samples were prepared adding to the cell-suspension the same amount of ethanol used in the experiments with the highest drug concentration.

RESULTS AND DISCUSSION

The experimental results are summarized in following table:

DRUG	% Motile Cells test / % Motile Cells con.	Swim Rate Test / Swim Rate Contr.	% Step-up Test / % Step-up Contr.	% Step-down test / % Step-down Contr.	O_2 Uptake in Dark Test/Contr.
CCCP (5μM)	0.2±0.03	1.0±0.1	1.0±0.2	1.0±0.2	0.45
Gram. D (10μM)	0.6±0.1	1.0±0.1	1.0±0.2	1.0±0.2	1.0

The effect of CCCP on the photoresponses of Euglena was studied in the range 1 μM – 5 μM. The percentage of swimming cells decreases from 100% to 20% increasing concentration, whereas the speed of the motile cells is not significantly affected even at 5μM. A further increase of CCCP concentration up to 10μM does however cause a drastic decrease of swimming speed, but only a very few cells were still motile and it was therefore not possible to obtain a statistically meaningful determination of the cell speed.

Oxygen uptake inthe dark was clearly inhibited only at 5 μM decreasing down to about 45% of untreated cells. Both step-up and step-down photophobic responses were not affected by CCCP up to the highest concentration used.

Similar results were obtained with Gramicidin D in the concentration range 0.1 – 10 μM. The latter value is probably overestimated since the drug is not so soluble in ethanol. A possible time-dependent effect of the drugs was studied, but not significant effect could be detected within 6 hours from beginnig of the experiments.

In conclusion, CCCP and Gramicidin D, drugs that are able to alter selectively proton conductance across biomembranes seem to affect Euglena cells only in an unspecified way at high concentrations, as previously observed with DNP[2].

In no case it was possible to observe any effect of these drugs on the step-up or step-down photophobic responses.

Even cells that are almost immobilized but still able to shake the flagellum, show both photophobic reactions when given a light stimulus with flagellar erection and sometimes also a full rotation on the spot together with body contractions.

These results seem therefore to indicate that agents affecting proton conductance do not alter the sensory transduction chain of Euglena gracilis.

REFERENCES

1. Lenci, F., and G. Colombetti (1978) Ann.Rev.Biophys.Bioeng. 7, 341-361.
2. Barghigiani,C., G.Colombetti, F.Lenci, R.Banchetti, M.P.Bizzarro (1979) Arch.Microbiol. 120, 239-245.
3. Kessler, R.J., C.A.Tyson, D.E.Green (1976) Proc.Natl.Acad.Sci. U.S.A. 73, 3141-3145.
4. Manson, M.D., P.Tedesco, H.C.Berg, F.M.Harold, C.Van der Drift (1977) Proc.Natl.Acad.Sci.U.S.A. 74, 3060-3064.
5. Pressman, B.C. (1976) Ann.Rev.Biochem. 45, 501-530.
6. Ferrara, R., S.Grassi, G.Del Carratore (1975) Biotechnol.Bioeng. 17, 985-996.
7. Barghigiani, C., G.Colombetti, B.Franchini, F.Lenci (1979) Photo-chem.Photobiol. 29, 1015-1019.

BIOMACROMOLECULES: PROTEIN STRUCTURE AND FUNCTION

MAGNETIC PROPERTIES AND STRUCTURE OF OXYHEMOGLOBIN

AND CARBONMONOXYHEMOGLOBIN

M. Cerdonio, S. Morante and S. Vitale

Facoltà di Scienze, Università di Trento

38050 Povo, Trento - Italy

Abstract

The case is presented about the magnetic state of HbO_2 and HbCO to show, on the basis of available experimental data, that the controversy on the possible presence of unpaired spins in the iron-ligand complex is not yet resolved. We point out the need for the identification of a fully diamagnetic state of the protein to be taken as reference and outline a possible method. A few preliminary results on packed red cells are presented.

1 - The low-lying magnetic states of HbCO and HbO_2: present status

The classic experimental work of Pauling and collaborators on the magnetic properties of liganded hemoglobin describes the iron--ligand complex as diamagnetic both for HbCO and for HbO_2 (1-3). Subsequent experimental studies had proposed a correction to the Pauling's value of the diamagnetism of these derivatives of hemoglobin, but failed to discern even the possibility of the presence of unpaired electrons (4). On the grounds of new experimental data (5,6) we are forced to question this description of the electronic structure of HbCO and HbO_2.

Many theoretical models have been developed in recent years for the electron distribution in the iron-ligand bond and in the ferrous heme (7-13), which may be appropriate for hemoglobins, and attempts have been made towards a detailed comparison to spectroscopic properties (14-16). However the magnetic properties are a direct probe of the spin and charge distribution in the ground state of the complexes and so they are aften taken as crucial experimental datum either for

input or for discussion of the theory. The importance of the magnetic datum may also be recognized when one considers that it is
also needed in conjunction with the interpretation of experimental
data from other methods like Mössbauer, Raman, infrared and magnetic
circular dichroism spectroscopy. As none of these methods can un-
equivocally assign the spin structure of the ground states, they
rather must be supplemented with a priori assumptions about the mag-
netism of these states in order to attempt interpretation. An example of this state of affairs is presented by the interpretation
given in recent years of the temperature dependence of the Mössbauer
quadrupole splitting of iron in oxyhemoglobin discovered by Lang and
Marshall (17). Although such a behaviour is strongly suggestive of
thermal unpairing of spins in the iron-oxygen bond, as at that time
the iron-oxygen complex was described as fully diamagnetic it was
necessary to explore the possibility of thermally activated rotations
of the oxygen molecule to explain simultaneously magnetic and Mössbauer data (18). Such a rotation is indeed allowed in model compounds (18), but is forbidden by steric hindrance in hemoglobin (19)
and myoglobin (20). On the other hand, if one now accepts the view
of a paramagnetic Fe-O_2 center in HbO_2, for which we have given experimental evidence (5,6), it is easy to reconcile Mössbauer and
magnetic data within a plausible electronic structure assuming a
fixed position for the O_2 molecule (21).

From the experimental point of view the main difficulty concerns
how to identify a fully diamagnetic state of the protein solution to
be taken as reference when evaluating paramagnetic contributions.
In the past, after the work of ref (1-3), the carbonmonoxy derivatives of hemoglobins were taken as such a system in the belief that
the FeCO would be fully diamagnetic irrespective of the species from
which the Hb was obtained and of the solution condition regarding
pH, ionic strength, etc. However the experimental evidence for this
is unsatisfactory. The conclusion that Hb CO is diamagnetic in ref
(1) is indirect as it is founded on a comparison made at a single
temperature with the properties of hemochromogens, which were determined in separate work (22). We have found human carbonmonoxy
Hb A fully diamagnetic in the frozen solution state, giving convincing evidence of that through the temperature dependence of the susceptibility of the solution (5). Results at room temperature indicated that, for hemolysate and stripped Hb solutions at pH 7, the
same solutions would preserve a fully diamagnetic state also in the
room temperature liquid state (6). On the other hand a remarkable
effect has been recently discovered for solutions of HbCO from carp:
addition of inositol exaphosphate to carp HbCO solutions at the low
pH appropriate for the R → T conformational switch results in an
increase of the diamagnetism of the solution (23). The only interpretation we could offer was that the FeCO complex shows a variability in the magnetic moment reverting, upon addition of IHP, from
a paramagnetic state with $\mu_C \approx 1.4$ Bohr magneton to a diamagnetic state.
Thus the HbCO system may not invariably display full diamagnetism

and care should be taken in using it as reference state.

We believe that it is now wanted an extensive experimental study of the magnetic properties of HbCO and HbO$_2$ of different species and in different solution conditions. On one hand this will resolve controversies among existing experimental results and be a firmer guidance for theoretical work concerning the nature of the low lying electronic states of the heme. On the other hand it will be of great interest to use the variability of magnetic moment of the iron-ligand complex as a probe to study relations between electronic structure and biological function. We have started such a program, with the aid of a high resolution superconducting magnetometer developed in our laboratory (24,25).

2 - Magnetic state of carbonmonoxy hemoglobin in human red blood
    cells

The first aim of our program is to clearly define the magnetic properties of the iron-ligand system in conditions as close as possible to physiological ones. Whole blood is hard to work out due to density gradients formation and so packed herytrocytes seem to be the best system to start with. We note that in order to avoid to break the cells and to maintain physiological conditions, freezing must be avoided and measurements must be performed in the liquid state. In such a restricted temperature range (0-40°C) any small paramagnetic component induces a very weak temperature dependence on the susceptibility of the solution and thus it can be resolved more easily by a direct comparison with a well established fully diamagnetic reference system. We try to establish this fully dia-magnetic reference state in two independent ways:

i) the temperature dependent paramagnetic contribution of the sus-ceptibility of deoxygenated red cells can be measured with enough resolution in the liquid state because the magnetic moment of the iron is large: if a true Curie-law behaviour is obtained, then the $1/T = 0$ extrapolation of the solution susceptibility represents the diamagnetic part of the susceptibility itself.
ii) The magnetic moment of CN$^-$ methemoglobin is well known from the temperature dependence of magnetic susceptibility (26,27). Values obtained in different conditions by different researchers agree with a small error and so the diamagnetic part of the susceptibility of CN metHb containing herytrocytes can be calculated just subtracting the known value of the paramagnetism of the FeCN complex.

In the following sections we present preliminary results of this method in connection with an attempt to study the susceptibility of carbonmonoxide-equilibrated packed red cells.

3 - Materials and Methods

Red cells from human blood of a local bank were washed and packed by a standard procedure: whole blood was centrifuged for 10 min at 1500 g and the white cells and plasma were discarded; the red cells were resuspended in four volumes of 0.1 M NaCl solution and then centrifuged for 10 min at 1500 g and the supernatant discarded; washing was repeated five times. The hemoglobin in packed cells was converted to the CN met form by addition of 5 mole KCN and, soon after, of 1.2 mole sodium nitrite per mole heme. The carbonmonoxy derivative was obtained by flushing and equilibrating the oxygenated sample with pure CO gas in a tonometer for about 1 hr; during the equilibration time the sample was kept in the cold at 4°C and gently shaken to increase the exposed surface to the CO. The deoxy-Hb form was prepared as follows: the oxygenated sample was placed in a test tube connected to a water pump and to a pure nitrogen tank; evacuation and refilling with nitrogen was repeated several times for 15 ÷ 30 min to obtain complete deoxygenation; then a small amount of sodium dithionite was added and the solution was transferred anaerobically in the quartz tube to perform magnetic measurements: the tube was sealed with an elastomer cork and again purged from air, by means of a syringe needle connected to a water pump and a nitrogen tank.

Concentrations, expressed per heme, were determined by the Drabkin method and were all close to 18 mM. Concentrations of CO--heme were also determined breaking the cells in distilled water and using known extinction coefficient of CO-hemoglobin; results of both methods were invariably found to be equal in the experimental error. Magnetic volume susceptibility measurements were performed with our superconducting susceptometer (25) as described elsewhere (23). The overall accuracy of the method for protein aqueous solutions referred to calibrations with pure deoxygenated water is 7×10^{-10} cgs units, or 1/1000 of the volume susceptibility of water.

4 - Results and discussion

a) search for a reference diamagnetic state

The volume magnetic susceptibility of a suspension of packed deoxygenated herytrocytes is given in figure 1 as a function of the inverse of temperature. As it can be seen a Curie-law behavior is exhibited and the best fit line of the figure can be written as

$$10^6 \, \chi_{cc}^{susp} = - (0.752 \pm 0.008) + \frac{(68.3 \pm 0.2)}{T} \qquad (1)$$

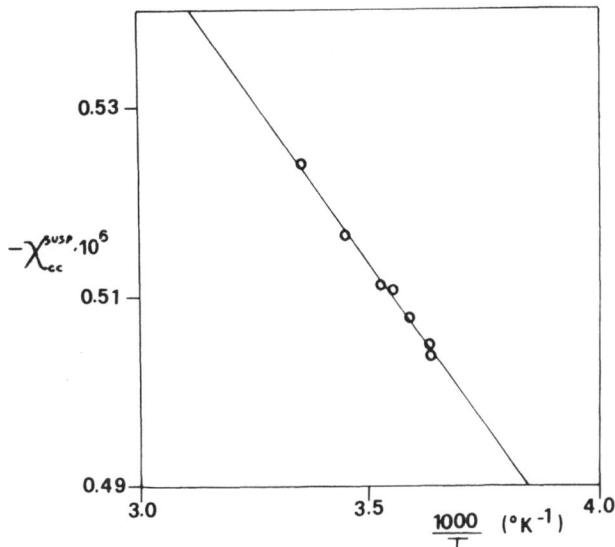

Fig. 1 - Volume magnetic susceptibility of a suspension of packed
human red cells, deoxy derivative, as a function of inverse
temperature in the 0°C to 27°C interval, at a heme concentration of 19.8 mM; (O) experimental data, (———) best fit
line (see text); error bars are within the size of the dots.

the first figure on the r.h.s. of eq 1) represents χ_{cc}^{dia}, the diamagnetic part of the susceptibility, while the second, in the hypothesis
that only the heme iron contributes to the overall paramagnetism,
corresponds to a magnetic moment of (5.3 ± 0.1) B.M. for the iron.
This value is in good agreement with the measured magnetic moment
of deoxyhemoglobin in phosphates containing solutions (28).

In Table I we give the magnetic susceptibility of a suspension
of red cells containing hemoglobin fully converted to the CN_{met} form.
From this value, assuming a magnetic moment of (2.44 ± 0.18) B.M.
for the FeCN complex, which is the average of the values given by
Coryell et al. (29), Iizuka et al. (26) and Messana et al. (27), the
diamagnetism of the suspension can be calculated. This value is
found to be $10^6 \chi_{cc}^{dia} = -$ (0.745 ± 0.008) where the error bar comes
from the relatively wide spreading of the published values for the
moment of the FeCN complex.

TABLE I

Volume magnetic susceptibility χ_{cc}^{susp} of packed human red cells suspensions at room temperature.

derivative	heme conc.(mM)	$10^6 \chi_{cc}^{susp}$ (cgs) experimental	$10^6 \chi_{cc}^{dia}$ (cgs) calculated	
deoxy	19.8 ± 0.1	(see fig. 1)	-0.752 ± 0.008	a)
CNmet	17.7 ± 0.1	-0.6973 ± 0.0007	-0.745 ± 0.008	b)
CO + 9mM KCN	17.7 ± 0.1	-0.7288 ± 0.0007		
CO	18.3 ± 0.1	-0.7312 ± 0.0007		
CO + 9mM KCN	18.5 ± 0.1	-0.7314 ± 0.0007		

a) obtained from an extrapolation to 1/T = 0 of the data of fig. 1 (see text)

b) obtained by subtracting from the experimental value the paramagnetic contribution due to μ_e = 2.44 B.M. (see text).

As it can be seen this value for the diamagnetism of the suspension is in good agreement, within the experimental errors, with
that extracted from the deoxy-derivative. Thus we can take the
average of the two values and we get $10^6 \cdot \chi_{cc}^{dia} = -(0.749 \pm 0.006)$ at
an average heme concentration of 18.8 mM.

b) <u>magnetic susceptibility of packed, CO-equilibrated, human
red blood cells</u>

In Table I the magnetic susceptibilities at 293°K of CO-equilibrated suspensions of packed red cells are reported. Different
samples come from distinct blood samples and so have slightly different heme concentrations. The average value of the susceptibility
is $10^6 \chi_{cc} = -(0.7310 \pm 0.0004)$ at an average heme concentration
of 18.2 mM. If we assign the difference

$$10^6 (\chi_{cc}^{HbCO} - \chi_{cc}^{dia}) = +(0.019 \pm 0.006)$$

to a paramagnetic contribution χ^{para} due to the heme CO complex we
find $\chi^{para} = +(1000 \pm 350) \times 10^{-6}$ cgs/mole at 293°K corresponding
to a magnetic moment μ_{eff} of $\mu_{eff} = (1.5 \pm 0.2)$ B.M. This assignment is supported by the following considerations. In two samples
KCN was added in 0.5 mole/mole of heme ratio to revert a possible
aquomet fraction to the cyano form, which is low spin; to produce
a paramagnetism of the order of that quoted above, one needs around
7-8 mM of CNmetHb, i.e. 40% of total heme concentration, which would
be largely evident in optical spectra. Full vis-uv absorption
spectra of the samples were performed before and after the magnetic
measurements breaking the cells in distilled water and a detailed
inspection of the features of the spectra did not reveal the presence
of any heme-iron derivative other than HbCO. Heme concentrations
were determined using the known values for the HbCO extinction
coefficients ε_{mM} (539 nm) = 14.36 ε_{mM} (568.5 nm) = 14.31 and after
that the samples were fully converted to CN meth form by Drabkin
method and heme concentration was again determined with the extinction coefficient of the cyano form ε_{mM} (540 nm) = 11.0. The concentrations determined by both methods were invariably found to be equal
in the experimental errors, giving another strong argument to claim
that no other form of heme derivative was present in the samples.
As shown in a previous work (6) our methods do not produce detectable
amounts of non heme-bound metal ions and so we are forced to conclude
that the relatively low paramagnetism shown by the CO-equilibrated
cells must be attributed to the CO-iron complex.

We must note that the deviation we found of the susceptibility
of CO-herytrocytes from the diamagnetic reference state is equal,
in the experimental errors, to the difference between the values
quoted for the magnetic susceptibilities of HbCO aqueous solutions

at 18 mM heme concentration, respectively by us (6), $10^6 \chi_{cc}^{HbCO}$ = −0.744 ± 0.004, and by Pauling and coworkers (1), $10^6 \chi_{cc}^{HbCO}$ = 0.727 ± 0.001; moreover this deviation coincides as well with the change in the susceptibility of solutions of Carp-CO-hemoglobin, at equal concentration, induced by the addition of IHP at low pH (23). All this suggests that the heme-CO complex in hemoglobin is able to take at least two magnetic states; at present we have no clear indication either about the specific conditions that can induce such a change of state in human CO-hemoglobin, either about the structural meaning of this finding. However we must emphasize that our results clearly show that the CO-hemoglobin cannot be considered as an always fully diamagnetic compound and that its susceptibility cannot be taken as an invariable reference value for the diamagnetism of the protein.

Acknowledgements

    We are indebted with dr. Anna Calì for giving us the human blood we used for this work and with dr. Ernesto Di Iorio and dr. Giorgio Giacometti for helpful suggestions.

REFERENCES

1) L. Pauling, C.D. Coryell, Proc. Natl. Acad. Sci. USA 22, 210 (1936).
2) D.S. Taylor, C.D. Coryell, J. Am. Chem. Soc. 60, 1177 (1938).
3) C.D. Coryell, L. Pauling, R.W. Dodson, J. Phys. Chem. 43, 825 (1939).
4) R. Havemann, W. Haberditzl, G. Rabe, Z. Phys. Chem. 218, 417 (1961).
5) M. Cerdonio, A. Congiu-Castellano, F. Mogno, B. Pispisa, G.L. Romani, S. Vitale, Proc. Natl. Acad. Sci. USA 74, 398 (1977).
6) M. Cerdonio, A. Congiu-Castellano, L. Calabrese, S. Morante, B. Pispisa, S. Vitale, Proc. Natl. Acad. Sci. USA 75, 4916 (1978).
7) L. Pauling, Nature 203, 182 (1964).
8) L. Pauling, Proc. Natl. Acad. Sci. USA 74, 2612 (1977).
9) J.J. Weiss, Nature 203, 183 (1964).
10) A. Dedieu, M.M. Rohmer, A. Veillard, in "Metal-Ligand Inter-actions in Organic Chemistry and Biochemistry", eds. B. Pullman and N. Goldblum. part. 2, 101 (1977) by Reidel Publishing Co. Dordrecht-Holland.
11) B.D. Olafson, W.A. Goddard III, Proc. Natl. Acad. Sci. USA 74, 1315 (1977).
12) B.H. Huynh, D.A. Case, M. Karplus, J. Am. Chem. Soc. 99, 6103 (1977).
13) R.F. Kirchner, G.H. Loew, J. Am. Chem. Soc. 99, 4639 (1977)
14) M.W. Makinen, A.K. Churg, H.A. Glick, Proc. Natl. Acad. Sci. USA 75, 2291 (1978).

15) B.I. Greene, R.M. Hochstrasser, R.B. Weisman, W.A. Eaton, Proc. Natl. Acad. Sci. USA 75, 5255 (1978).
16) T. Kitagawa, Y. Kyogoku, T. Iizuka, M.I. Saito, J. Am. Chem. Soc. 98, 5169 (1978).
17) G. Lang, W. Marshall, J. Mol. Biol. 18, 385 (1966).
18) K. Spartalian, G. Lang, J. Chem. Phys. 63, 5375 (1975).
19) E.J. Heidner, R.C. Ladner, M.F. Perutz, J. Mol. Biol. 104, 707 (1978).
20) S.E.V. Phillips, Nature 273, 247 (1978).
21) M. Bacci, M. Cerdonio, S. Vitale, Biophys. Chem. 10, 113 (1979).
22) L. Pauling, C.D. Coryell, Proc. Nat. Acad. Sci. USA 22, 159 (1936).
23) M. Cerdonio, S. Morante, S. Vitale, A. de Young, R.W. Noble (submitted for publication).
24) M. Cerdonio, C. Cosmelli, G.L. Romani, C. Messana, C. Gramaccioni Rev. Sci. Instr. 47, 1 (1976).
25) M. Cerdonio, S. Morante, S. Vitale, Proc. XV Int. Congress Refr. (in press).
26) T. Iizuka, M. Kotani, Biochim. Biophys. Acta 194, 351 (1969).
27) C. Messana, M. Cerdonio, P. Shenkin, R.W. Noble, G. Fermi, R.B. Perutz, M.F. Perutz, Biochemistry 17, 3652 (1978).
28) Y. Alpert, R. Banerjee, Biochim. Biophys. Acta 405, 144 (1975).
29) C.D. Coryell, F. Stitt, L. Pauling, J. Am. Chem. Soc. 59, 633 (1937).

EFFECTS OF ORGANIC SOLVENTS ON THE CONFORMATIONAL AND FUNCTIONAL STABILITY OF SOME BIOLOGICAL MACROMOLECULES

L. Cordone, A. Cupane, M.A. Dolce, S.L. Fornili, V. Izzo,
P.L. San Biagio, G. Sgroi and E. Vitrano

Istituto di Fisica dell'Università and
Gruppo Nazionale Struttura della Materia (CNR)

via Archirafi, 36 - 90123 Palermo (Italy)

INTRODUCTION

As it is well known, solvent perturbation due to the presence of organic solvent in the solution medium affects the conformational equilibria of biological macromolecules[1,2,3]. Since the biological functions of macromolecules are strongly correlated with their conformation, an effect on functional properties is also expected.

In the present report we summarize the results of experiments aimed at ascertaining the effect of some organic solvents (mainly monohydric alcohols) on the functional stability of two proteins (human hemoglobin and calf liver β-galactosidase) and on the helix--coil transition of calf thymus DNA. Aim of this report is to give a comprehensive picture of the work we have been performing on the role played by the solvent on conformational and functional stability of biological macromolecules.

CALF LIVER β-GALACTOSIDASE

The effects of methanol, ethanol, iso-propanol and n-propanol on the kinetics of hydrolysis of o-nitrophenyl-β-D-galactosidase (ONPG) by calf liver βgalactosidase have been studied[4]. We shall report a short summary of the main features of this work.

The rate of reaction was slowed down by the presence of alcohols, the effect increasing with alcohol concentration and alkyl group size. The analysis of data allowed to discard an interpretation in terms of enzyme inhibition due to the binding of a single alcohol molecule to the protein. Data have been interpreted in terms of general

solvent effects. The logarithm of the fraction of functional β-
-galactosidase enzyme molecules was found to depend in an almost
identical way on the number of methylenes per alcohol chain as:
1) the lowering of melting temperature of ribonuclease[5]; 2) the
non bulk-electrostatic contributions to the free energy difference
between R and T states of hemoglobin[6], and 3) the logarithm of the
fraction of non denatured DNA molecules in the low alcohol concentration region[7].

This fact supports the suggestion that the effects measured
are to be related to the varied interactions (in the presence of
alcohols) among hydrophobic groups on the enzyme surface and the
solvent. In particular it was suggested that β-galactosidase enzymes
can exist under functional or non functional conformations (differing
by the extent of hydrophobic surface exposed to the solvent), the
quilibrium between the two forms being function of alcohol concentration and alkyl group size.

HUMAN HEMOGLOBIN

The effects of solvent perturbation on the functional properties
of hemoglobin have also been studied. The study was started by
measuring the effects of methanol, ethanol, iso-propanol and n-
-propanol on the oxygen affinity of hemoglobin[6].

Following the two state model of Monod, Wyman and Changeux[8]
(MWC) one has:

$$P_{50} = L^{1/4}K_R; \quad \text{when } L \gg 1 \text{ and } K_R \ll K_T \tag{1}$$

In eq.(1), P_{50} is the measured oxygen pressure at half hemoglobin saturation, L is the allosteric parameter and K_R and K_T are
the oxygen dissociation constants for the oxy (R) state and for the
deoxy (T) state respectively. If K_R is not affected by the perturbant, effects on the oxygen affinity can be put in relation with
effects on the free energy difference between the oxy (R) and deoxy
(T) conformations of the hemoglobin molecule, according to the
following equation:

$$4RT \ln \frac{P_{50}(C_p)}{P_{50}(0)} = RT \ln \frac{L(C_p)}{L(0)} = \Delta\Delta G \tag{2}$$

where C_p is the concentration of perturbant. Differential spectroscopy in the visible region of the oxy-hemoglobin absorption spectrum
indicated K_R not to be affected by the presence of alcohols[6]; this
fact allowed the use of eq.(2).

As it is well known the hemoglobin T conformation is stabilized
by salt bridges[9]. An effect of the bulk dielectric constant on the

T $\rightleftarrows$ R conformational equilibrium therefore might be relevant. This[10,6]
eventual contribution to $\Delta\Delta G$ can be written in the following form:

$$\Delta\Delta G_{es} = - ne^2 N_A (\frac{1}{r_R} - \frac{1}{r_T}) (\frac{1}{\varepsilon(C_p)} - \frac{1}{\varepsilon(0)}) \qquad (3)$$

In eq.(3), n reflects the effective number of couples of unitary
electronic charges (opposite in sign) which move from an initial
effective separation r_T (in T state) to a final effective separation
r_R (in R state). In deriving eq.(3) the electrostatic screening due
to the presence of counterions in solution has been neglected. This
approximation is justified by the fact that in our conditions the
Debye length is much greater than r_T. Eq.(3) implies that a straight
line for all the alcohols must be obtained in a $\Delta\Delta G$ vs. $1/\varepsilon(C_p)$ plot,
if the effects are to be attributed only to variations of bulk
dielectric constant. Such a plot is reported in Fig. 1. As can
be seen, bulk-electrostatic effects (that act to stabilize the hemoglobin T conformation) are relevant at low alcohol concentration
(i.e. for values of $\varepsilon(C_p) \simeq \varepsilon(0)$). In fact for $1/\varepsilon(C_p) \to 1/\varepsilon(0)$
points relative to all alcohols fall on a single straight line.
If one assumes that this straight line represents the bulk-electrostatic contributions at all alcohol concentrations, the quantity
$\Delta\Delta G_{nes} = \Delta\Delta G - \Delta\Delta G_{es}$ gives the sum of all other contributions, non
related to the variations of the bulk dielectric constant. Fig. 1
shows that these non bulk-electrostatic effects (that act to stabilize the hemoglobin R conformation), opposite in sign to bulk-electrostatic ones, become relevant by increasing alcohol concentration.

A check of the above result has been obtained by studying the
effect of formamide and acetamide[11]. These perturbants are known
to increase the value of bulk dielectric constant of the solutions[12].

Results are shown in Fig. 1. As can be seen:
1) In agreement with the above results, bulk-electrostatic effects,
opposite in sign to those obtained in the case of alcohols, are again
dominant at low perturbant concentration.
2) Non bulk-electrostatic effects become again relevant at high
perturbant concentrations and are of the same sign as those obtained
in the case of alcohols.

From Fig. 1 it can be seen that in the limit $1/\varepsilon(C_p) \to 1/\varepsilon(0)$
curves relative to experimental $\Delta\Delta G$ values approach two different
straight lines for alcohols and amides. This fact is not to be
expected on the basis of eq.(3). This discrepancy might be attributed e.g. to slight alterations (in the presence of amides) of the
pK of ionizable groups on the hemoglobin surface, that could cause
a decrease of the parameter n in eq.(3). Work is in progress at our
laboratory to understand the reason of this discrepancy. In addition,
it should be mentioned that, on the basis of spectroscopic evidence,
very slight effects of amides on K_R cannot be excluded.

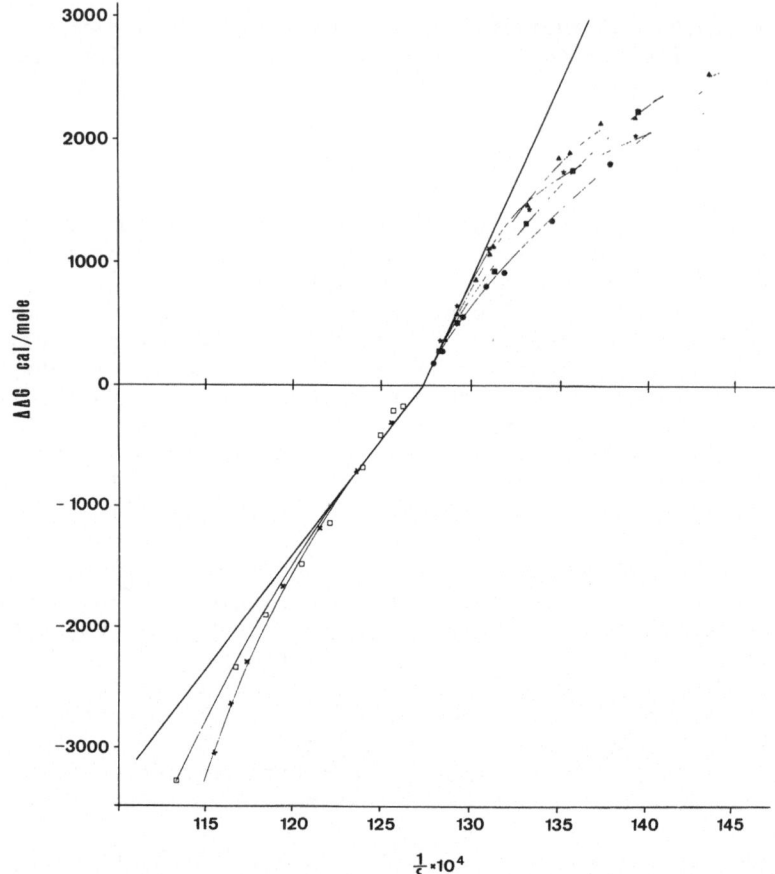

Fig. 1 - ΔΔG, calculated from experimental values of log P_{50} following eq.(2) in the text, as a function of the inverse bulk die-lectric constant of the solvent. ● Methanol; ▲ ethanol; ■ iso-propanol; ★ n-propanol; □ Formamide; ✕ Acetamide. The straight lines, tangent to experimental ΔΔG curves in the limit $1/\varepsilon(C_p) \rightarrow 1/\varepsilon(o)$, represents the bulk-electro-static contributions, according to eq.(3) in the text.

Values of dielectric constants at 25°C have been obtained by interpolating values reported by Åkerlöf[13] (in the case of alcohols) and by Rohdewald and Moldner[12] (in the case of amides).

Values of non bulk-electrostatic contributions to hemoglobin T → R transition are in good agreement with the hydrophobic contri-bution estimated by correlating data previously reported in the liter-ature[6,14,15]. This fact supports the suggestion that non bulk--electrostatic contributions are mainly related to protein solvent

TABLE I

ALCOHOL	c_A (molar units)	$\Delta\Delta G_{nes}$ T=20°C (Kcal/mole)	$\Delta\Delta H_{nes}$ (Kcal/mole)	$\Delta\Delta S_{nes}$ (e.u./mole)
Methanol	2.0	− .4	9±2	33±7
Ethanol	2.1	− .9	12±1	45±1
Iso-propanol	1.6	−1.5	10±1	39±4
N-propanol	1.6	−1.6	19±1	69±3

Values of thermodynamic parameters for the non bulk-electrostatic contribution. $\Delta\Delta G_{nes}$ values were determined as described in the text. $\Delta\Delta H_{nes}$ and $\Delta\Delta S_{nes}$ values were obtained from Vant'Hoff plots of $\Delta\Delta G_{nes}$ by least squares fitting of experimental points with straight lines. Standard deviations of $\Delta\Delta H_{nes}$ and $\Delta\Delta S_{nes}$ values are obtained directly from the least squares analysis.

hydrophobic interactions arising from the lowered (in the presence of perturbants) free energy needed to expose hydrophobic surface to the solvent, following hemoglobin T → R transition[16]. Moreover, as already noted, these non bulk-electrostatic contributions exhibited the same dependence on the number of methylenes per alcohol chain as the logarithm of the fraction of functional β-galactosidase enzyme molecules[4], the lowering of melting temperature of ribonuclease[5] and the logarithm of the fraction of non denatured DNA molecules in the low alcohol concentration region[7].

A further check was obtained through the study of the effects of alcohols on the oxygen affinity of hemoglobin at various temperatures[11]. This allowed the determination of non bulk-electrostatic contributions, due to the presence of alcohols, to the standard enthalpy and entropy differences between T and R states of hemoglobin.

Typical results of this study are shown in Table 1. Both $\Delta\Delta H_{nes}$ and $\Delta\Delta S_{nes}$ were found to be positive quantities for all alcohols at all concentrations studied, whereas $\Delta\Delta G_{nes}$ values were found to be always negative. This fact implies that, at the investigated temperatures, the non bulk-electrostatic contributions reflect entropy driven processes. Moreover $\Delta\Delta G_{nes}$ values are always much smaller than the corresponding $\Delta\Delta H_{nes}$ and $T\Delta\Delta S_{nes}$ values; i.e. they arise from the composition of large enthalpic and entropic contributions that partly cancel each other. According to current literature these features are characteristic of hydrophobic interactions [17,18].

CALF THYMUS DNA

Although the effects of organic solvents on DNA helix to coil transition have been extensively studied[2], some further experiments have been performed in our laboratory in order to see if the effects of alcohols on DNA conformational stability have features common to those reported above on the functional properties of proteins.

To detect the relevance of bulk-electrostatic effects, the same procedure as for hemoglobin has been followed. In Fig. 2 is reported the fraction of non denatured DNA molecule vs. the inverse bulk dielectric constant of the solution. If effects of bulk-electrostatic nature would be dominant in some concentration range, all curves in Fig. 2 should overlap in that particular range. An inspection of Fig. 2 indicates that effects of bulk-electrostatic nature do not play a dominant role in DNA denaturation induced by alcohols. The fact that, contrary to what happens for hemoglobin, effects related to variations of the bulk dielectric constant of the solvent do not play a fundamental role could seem surprising. Indeed, the repulsion among charged phosphate groups plays a fundamental role in the stabilization of DNA double helix. However, it can be shown, following standard Debye-Hückel approach, that (due

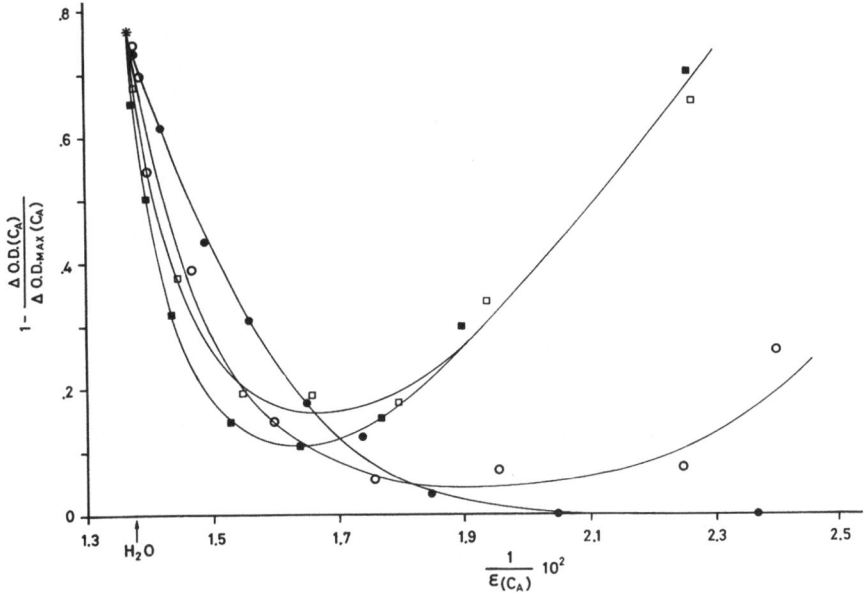

Fig. 2 – Fraction of non denatured DNA molecules at T = 67.2°C as a
function of the inverse bulk dielectric constant of the
solvent. ● Methanol; ○ Ethanol; □ iso-propanol;
■ n-propanol; ✳ water.
Under our salt conditions (10^{-2}M phosphate buffer pH 7.4)
the denaturation at 67.2°C in the absence of alcohol is
about 20%. DNA concentration ∿ 25 mg/ml.

to the different distances among interacting charges) the derivative
of the bulk-electrostatic free energy with respect to the bulk dielectric constant of the solvent is in the case of DNA much smaller
than in the case of hemoglobin.

From Fig. 2 it can be seen that at low concentration alcohols
play a destabilizing role on DNA helix conformation: this effect increases with alcohol concentration and alkyl group size. At high
concentration, on the contrary, the fraction of denatured DNA, in
the case of methanol, reaches saturation whereas for the other alcohols it reaches a maximum and then decreases. The rate of decrease
is higher for alcohols with longer alkyl chains.

In the low alcohol concentration region the fraction of non
denatured DNA was found to depend on the number of methylenes per
alcohol chain in the same way as for β-galactosidase enzymes and for
hemoglobin. This fact allowed to suggest that the destabilizing
effect of alcohols on DNA structure can be related to the varied
hydrophobic interactions between DNA bases and solvent.

In order to explain the behavior observed in the high concentration region it is in order to recall that in non polar solvents purine and pyrimidine bases associate by hydrogen bonding, whereas in water association between bases by hydrogen bonding does not occur (see e.g. ref. 2 and references therein cited). Therefore the decreased extent of DNA denaturation at high alcohol concentration can be ascribed to the lack of possibilities for DNA bases to form hydrogen bonds with solvent. This effect becomes relevant at quite high concentrations of long chain alcohols i.e. under conditions where it was not possible to perform experiments with β-galactosidase or hemoglobin because of protein denaturation.

CONCLUSIONS

In Fig. 3 are reported, as a function of the "effective" numbers of methylenes per alcohol chain:

a) the logarithm of the fraction of functional calf liver β-galactosidase enzymes

b) the non bulk-electrostatic contributions to the free energy difference between R and T states of hemoglobin

c) the logarithm of the fraction of non denatured DNA molecules in the low alcohol concentration region

d) the lowering of melting temperature of ribonuclease. Data in Fig. 2,d have been taken from ref. 5.

All plots in Fig. 3 have been obtained by first drawing a curve joining the points relative to the values of the various quantities at $1\underline{M}$ concentration of methanol, ethanol and $\underline{n}$-propanol. The values corresponding to $1\underline{M}$ concentration of $\underline{iso}$-propanol have been then put on the curves. The values of the relative abscissae have been taken as the effective number of methylenes per $\underline{iso}$-propanol chain.

It is interesting to note that:
1) all curves in Fig. 3 have the same shape
2) the effective numbers of methylenes per $\underline{iso}$-propanol chain obtained in the different cases fall close to each other.

As mentioned above, in the case of hemoglobin values of $\Delta\Delta G_{nes}$ are in good agreement with the hydrophobic contribution estimated by correlating different data previously reported in the literature; moreover, values and signs of $\Delta\Delta G_{nes}$, $\Delta\Delta H_{nes}$ and $\Delta\Delta S_{nes}$ indicated that non electrostatic contributions reflect entropy driven processes. These features support the suggestion that alcohols affect the oxygen affinity of hemoglobin by altering protein-solvent hydrophobic interactions.

The strong similarity of curves plotted in Fig. 3 indicates

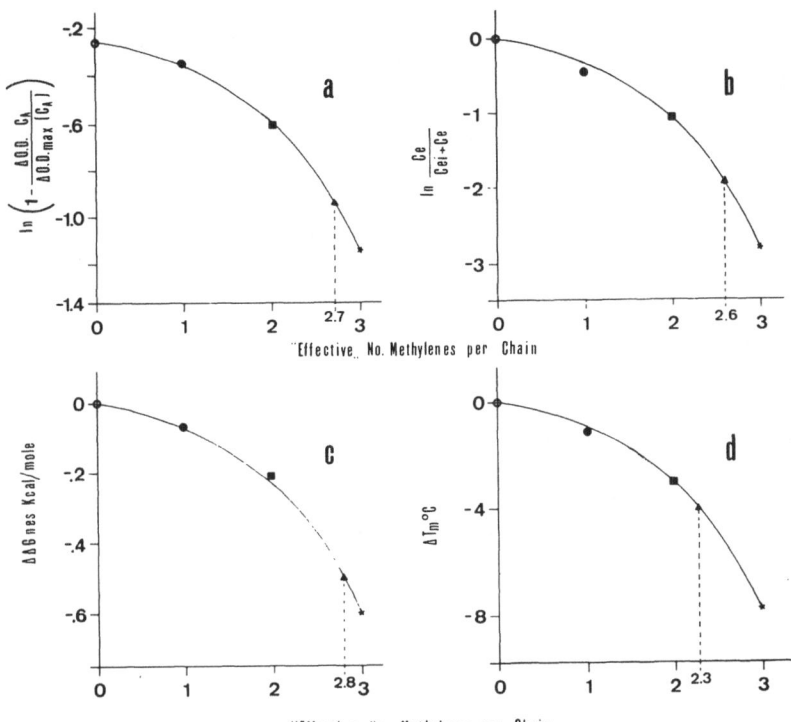

Fig. 3 - a) Logarithm of the fraction of non functional calf liver
β-galactosidase enzymes, b) non bulk-electrostatic contributions to the free energy difference between R and T states
of hemoglobin, c) logarithm of the fraction of non denatur-
ated DNA molecules in the low alcohol concentration region,
d) lowering of melting temperature of ribonuclease, as a
function of the "effective concentration" of methylenes in
1M alcohol solutions. ● Methanol; ■ Ethanol; ▲ iso-
-propanol; ★ n-propanol. Scales were normalized by making
coincident values in the absence of alcohols and in the
presence of 1M n-propanol. Note that curves in Figs. 3a),
b), c) and d) have identical shapes.

that they represent processes based on the same molecular mechanism.
A comprehensive look at the reported results suggests that the alteration introduced by solvent perturbants on macromolecule-solvent
hydrophobic interactions play a relevant role in this mechanism.

Among the authors V. I. is on the staff of the Istituto Nazionale di Fisica Nucleare (I.N.F.N.) and P.L.S.B. is a research fellow
of Comitato Regionale Ricerche Nucleari e Struttura della Materia
(C.R.R.N.S.M.).

REFERENCES

1. Eagland, D. (1975) in "Water-A Comprehensive Treatise", vol. IV, Franks, F. Ed., Plenum Press, New York, pp. 305-518.
2. Edelhoch, H. and Osborne, J.C.Jr. (1976) Adv. Protein Chem. 30, 183-249.
3. Herskovits, T.T. (1962) Arch. Biochem. Biophys. 97, 474-484.
4. Cordone, L., Izzo, V., Sgroi, G. and Fornili, S.L. (1979) Biopolymers 18, 1965-1974.
5. Von Hippel, P. and Wong, U.Y. (1965) J. Biol. Chem. 240, 3909-3923.
6. Cordone, L., Cupane, A., San Biagio, P.L. and Vitrano, E. (1979) Biopolymers 18, 1975-1988.
7. Cordone, L., Dolce, M.A., Fornili, S.L., Izzo, V. and Sgroi, G.: manucript in preparation.
8. Monod, J., Wyman, J. and Changeux, J.P. (1965) J. Mol. Biol. 12, 88-118.
9. Perutz, M.F. (1970) Nature 228, 726-739.
10. Laidler, K.J. and Bunting, P.S. (1973) 'The Chemical Kinetics of Enzyme Action', Oxford University Press, New York, pp. 216-220.
11. Cordone, L., Cupane, A., San Biagio, P.L. and Vitrano, E.: manuscript in preparation.
12. Rohdewald, P. and Moldner, M. (1973) J. Phys. Chem. 77, 373-377.
13. Åkerlöf, G. (1932) J. Am. Chem. Soc. 54, 4125-4139.
14. Chothia, C. (1974) Nature 248, 338-339.
15. Nozaki, Y. and Tanford, C. (1971) J. Biol. Chem. 246, 2211-2217.
16. Chothia, C., Wodak, S. and Janin, J. (1976) Proc. Natl. Acad. Sci., USA 73, 3793-3797.
17. Ben-Naim, A. (1978) J. Phys. Chem. 82, 874-885.
18. Tanford, C. (1978) Science 200, 1012-1018.

ISOTOPIC EFFECTS ON THE KINETICS OF THERMAL DENATURATION OF

MET-HEMOGLOBIN

Antonio Cupane^, Daniela Giacomazza, Francesco Madonia,
Pier Luigi San Biagio and Eugenio Vitrano^

Istituto di Fisica dell'Università
^and Gruppo Nazionale Struttura della Materia (CNR)
via Archirafi 36, 90123 - Palermo, Italy

ABSTRACT

We studied the effects of total and partial deuteration on the
kinetics of thermal denaturation of met-hemoglobin. The kinetics
were shown to be first order with respect to protein concentration:
this was true both in H_2O and in D_2O within the entire range of
temperatures examined. Deuterium oxide increased the stability of
the native conformation of met-hemoglobin: this effect increased
progressively by increasing the amount of D_2O in the solution.
Extension of the experiments to the amplest possible temperature
range (50-63°C) allowed the determination of the isotopic effect on
the activation enthalpy and entropy of the denaturation reaction;
the isotopic effect resulted to be mainly entropic.

Our results do not support an interpretation in terms of purely
steric effects due to the deuteration of the protein; rather, they
suggest that solvent effects, possibly through protein-solvent hydrophobic interactions, are relevant.

INTRODUCTION

Substitution of water by deuterium oxide is known to affect both
the functional properties and the conformational stability of bio-
molecules (1). The interpretation of such kind of effects is difficult, in view of the variety of phenomena involved. These are essentially of two types: i) effects due to the replacement of hydrogen by deuterium inside the protein (in polypeptides, the deuterium
bond appears to be stronger and longer than hydrogen bond (2,3)),
ii) solvent effects (e.g., the dielectric constant of D_2O is lower
than that of H_2O, hydrophobic interactions appear to be stronger in

D_2O than in H_2O (5,6)).

Effects of deuteration on the conformational stability and on the functional properties of hemoglobin have been already reported (7) These effects have been attributed by the authors to asymmetric partial deuteration of the protein (7); such interpretation has been questioned by other authors (8).

In the present work we have studied the effect of total and partial deuteration on the kinetics of thermal denaturation of met-hemoglobin. Extension of the experiments to the amplest possible temperature range allowed the determination of the effects of D_2O on the activation enthalpy and entropy of the process. Results show that the effect of isotopic substitution is mainly entropic.

Our findings do not support an interpretation in terms of deuteration of the protein, rather, they suggest that bulk solvent effects, possibly through protein-solvent hydrophobic interactions, are relevant.

MATERIALS AND METHODS

Hemoglobin was prepared from freshly drawn heparinized human blood, as already described (9). Met-hemoglobin was prepared following the procedure described by M.F. Perutz (10). Deuterium oxide (99.8%) was obtained from BOC Ltd. (Prochem), London. All other chemicals were reagent grades. All experiments were performed with samples in 0.1 M phosphate buffer ($KH_2 PO_4$ + K_2HPO_4). The pH of the solutions was measured with a Radiumeter PHM72 digital pH-meter and a type E 5021 microelectrode. Unless otherwise stated, pH-meter reading was 7.0 for all samples, including those containing D_2O.

The kinetics of thermal denaturation were measured following the "irreversible assay"; the experimental technique was essentially identical to that used by Tomita and Riggs (7). Aliquots of met-hemoglobin solution ($\simeq$ 5 ml) were put in glass tubes accurately sealed to prevent sample evaporation and incubated in the water bath of a cryothermostat Colora WK5, thermostated ($\pm.1°C$) at the desired temperature. After the desired time, the solutions were rapidly cooled in an ice bath and centrifuged at 27,000xg for 15 min. at 4°C. The absorbance of the supernatant solution was measured at 20°C and at $\lambda=500$ nm with a spectrophotometer Cary 118C. In experiments with low met-hemoglobin concentration (0.02%), the absorbance was measured at $\lambda=405$ nm, in the Soret band region of the spectrum. Temperatures were measured with a Leeds-Northrup precision bridge and a platinum thermometer calibrated against a similar one having a NBS certificate. The fraction of native met-hemoglobin was calculated with the following expression:

$$\frac{N(t)}{N(o)} = \frac{O.D(t)}{O.D(o)} \cdot$$

Analysis of data was performed with a HP2100-A computer.

RESULTS AND DISCUSSION

A) <u>Analysis of kinetic data at various temperatures.</u>

Fig. 1 shows the kinetics of thermal denaturation of met-hemo-globin in normal water at various temperatures.

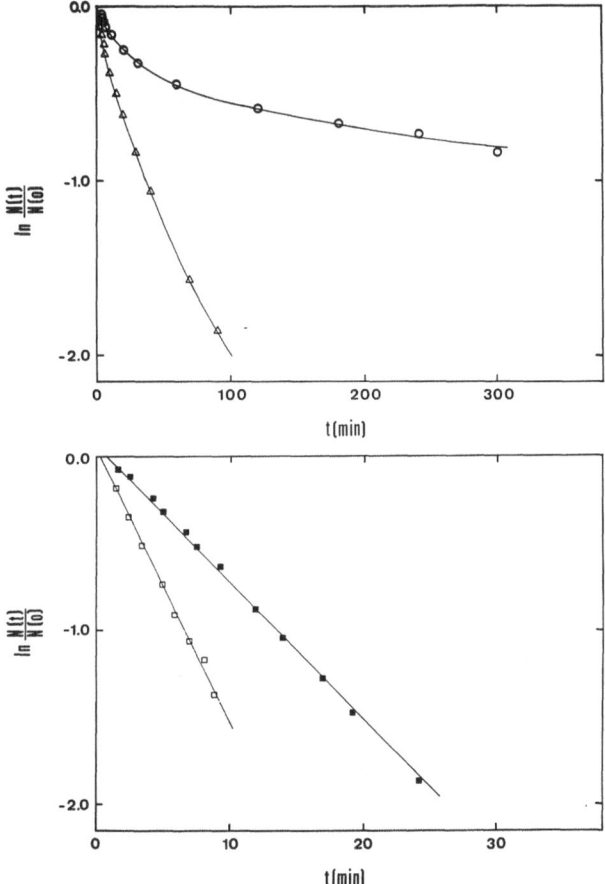

Fig. 1 - Kinetics of thermal denaturation of met-hemoglobin in normal
water. Met-hemoglobin concentration: $3x10^{-5}M$ (tetramer);
0.1M phosphate buffer; pH=7.0.
□ : T=61.5°C; ■ : T=60°C; △ : T=57°C; ○ :T=54°C
Note the small delay time (≈40 sec) that is evident in the
figure referring to the experiments at 61.5 and 60°C. This
is the time needed to attain thermal equilibrium in the water
bath of the thermostat. This delay time is not evident in
the figure referring to the experiments at 57 or 54°C, due
to the different scale in the abscissae.

At temperatures higher than 57°C the experimental points are well fitted by a single straight line and the denaturation reaction is first order with respect to time. At temperatures lower than 57°C deviations from the linear behavior are evident and the order of the reaction with respect to time is appreciably greater than unity.

Fig. 2 shows Log $(dN/dt)_{t=0}$ vs Log $N(0)$ at T = 54°C, where $N(0)$ is the initial concentration of met-hemoglobin and $(dN/dt)_{t=0}$ is the initial slope of the denaturation kinetics. Fig. 2 clearly shows that the reaction is first order with respect to protein concentration and thus that the denaturation process can be considered, in our concentration range, as a unimolecolar reaction. This is true both in H_2O and D_2O, within the entire range of temperatures examined (52–63°C).

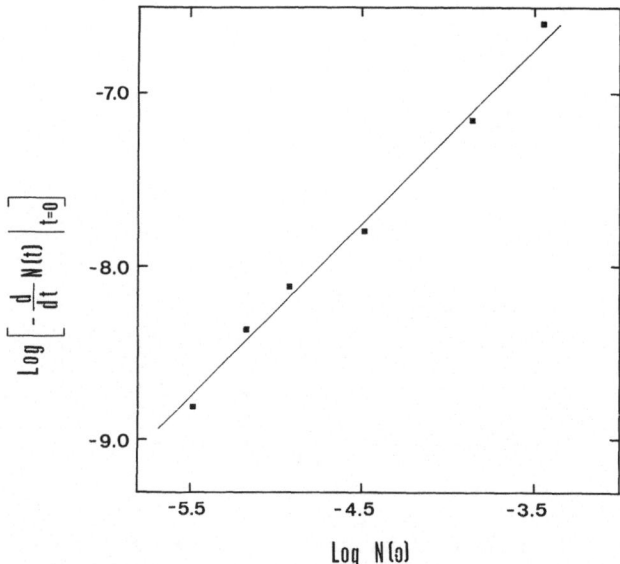

Fig. 2 – Initial slope of the denaturation kinetics in normal water as a function of the initial concentration of met-hemoglobin. Concentrations are expressed in molar units; initial slopes in mole sec^{-1}; 0.1M phosphate buffer pH=7.0, T=54°C. The straight line of slope 1.0 is a computer best fit to the experimental points.

We interpret the data shown in Fig.1 in the following way: if we write the reaction as

$$N \underset{K_2}{\overset{K_1}{\rightleftarrows}} D \overset{K_3}{\rightarrow} A$$

where N is the concentration of native hemoglobin, D is the concentration of denatured hemoglobin, and A is the concentration of denatured hemoglobin which aggregates and precipitates, then at $T > 57°C$ it is $K_2 \simeq 0$ and $dN/dt = -K_1 N$ i.e. the order with respect to time is unity; at $T < 57°C$ it is $K_2 \bar{0} \neq 0$ and the expression of N as a function of time is complicated, simulating a reaction of order higher than 1. It is clear however that, independently of any particular interpretation, at all temperatures the quantity $(1/N(0))(dN/dt)_{t=0}$ is a measure of the rate constant of the denaturation process, K_1.

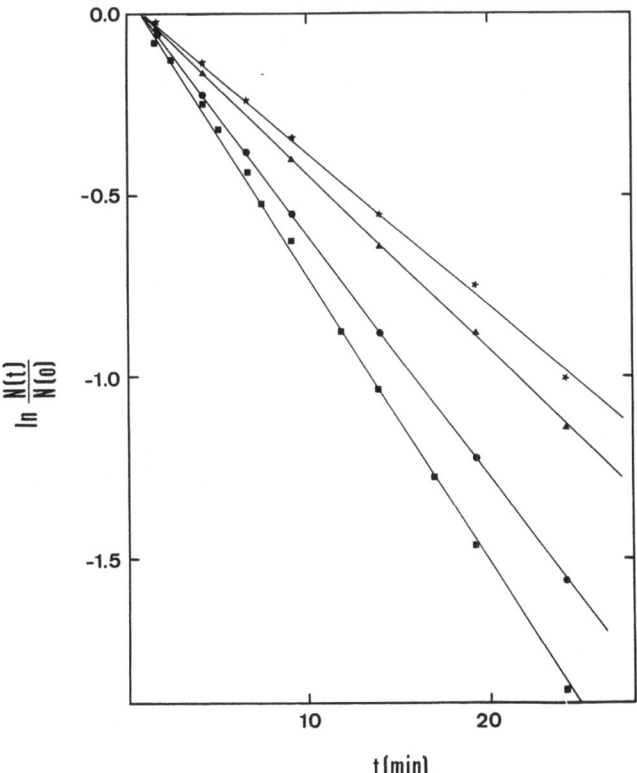

Fig. 3 – Denaturation kinetics of met-hemoglobin. Met-hemoglobin concentration: $3 \times 10^{-5}M$ (tetramer); 0.1M phosphate buffer $pH_M = 7.0$; $T = 60°C$.
■ : H_2O; ● : 25% D_2O in H_2O; ▲ : 75% D_2O in H_2O; ★ : D_2O

B) Effect of D_2O on K_1

Fig. 3 shows denaturation kinetics of met-hemoglobin at $T=60°C$ in normal water and in solutions containing various proportions of D_2O in H_2O. Actual K_1 values are shown in Table I. Deuterium oxide stabilizes the native conformation of met-hemoglobin; this effect increases progressively by increasing the amount of D_2O in the solution; moreover it does not depend upon the time of incubation of the sample, for incubation times greater than $\simeq 4$ hrs.

Data in Table I were taken at constant reading of the pH-meter; however, possible differences by up to 0.4 pH units between pD values and actual meter readings (11,12,13) were considered as a possible origin of artifacts. For this reason we measured the dependence of K_1 upon pH near neutrality. Results are shown in Table II. From Table II we see that if one assumes, following Lumry, Smith and Glanz (11) and Glasoe and Long (12), $pD = pH_{meter} + 0.4$ and compares data in H_2O and D_2O taken at pH=pD, then the isotopic effect on K_1 would be increased. That is, differences between pH and pD values cannot be invoked as the source of the isotopic effect that we have measured. In what follows we will compare, conservatively, data taken at constant pH-meter reading.

The temperature dependence of the isotopic effect on the kinetic constant K_1 was also studied. According to the transition state theory it is

$$\ln \frac{K_{1H}}{K_{1D}} = \frac{\Delta\Delta G_A}{RT}$$

where K_{1H} and K_{1D} stand for the kinetic constants in H_2O and D_2O respectively and $\Delta\Delta G_A = \Delta G_A (D_2O) - \Delta G_A (H_2O)$ is the isotopic effect on the activation free energy of the denaturation reaction. Fig. 4 shows a plot of $\ln K_{1H}/K_{1D}$ vs $1/T$. From such plot, quantitative informations on the isotopic effect on the activation enthalpy (ΔH_A) and entropy (ΔS_A) of met-hemoglobin thermal denaturation can be obtained. The experimental points in Fig. 4 have been fitted with a straight line by linear least squares regression; the results of the fitting are:

$$\Delta\Delta H_A = \Delta H_A(D_2O) - \Delta H_A(H_2O) \simeq 0$$

$$\Delta\Delta S_A = \Delta S_W(D_2O) - \Delta S_A(H_2O) \simeq -1 \text{ e.u./mole.}$$

that is, the isotopic effect on the thermal denaturation of met-hemoglobin is essentially entropic.

In a previous work (7) Tomita and Riggs reported on the effects

TABLE   I

% D_2O in H_2O	$K_1 \times 10^3$ (sec^{-1})
0	1.3
25	1.1
75	0.8
100	0.7

Dependence of the kinetic constant K_1 of the met-hemoglobin denaturation reaction upon isotopic composition of the solvent. Met-hemoglobin concentration 3×10^{-5} M (tetramer); 0.1 M phosphate buffer $pH_M = 7.0$; T=60°C.

TABLE   II

PH_M	$K_1 \times 10^3$ (sec^{-1})
6.6	1.0
7.0	1.3
7.4	1.9

Dependence of the kinetic constant K_1 of met-hemoglobin denaturation reaction upon pH near neutrality. Met-hemoglobin concentration: 3×10^{-5} (tetramer); 0.1M phosphate buffer; T=60°C.

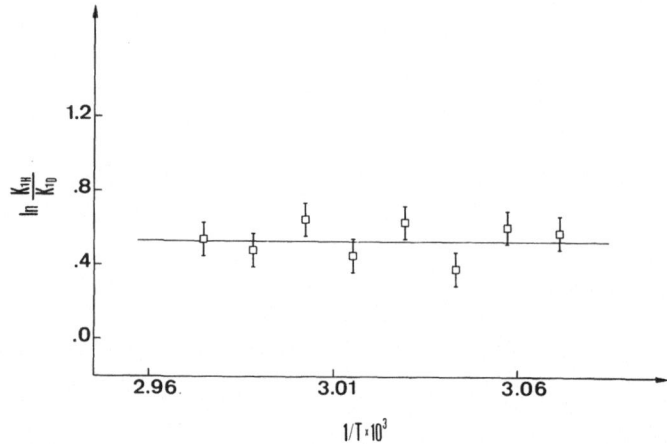

Fig. 4 – Arrhenius plot for the isotopic effect on the kinetic
constant of the thermal denaturation of met-hemoglobin.
Other experimental conditions are as in Fig. 3. The
continuous line is the computer best fit to the experimental data.

of D_2O on the conformational stability and on the functional properties of hemoglobin. Denaturation was found by these authors to
stabilize the native conformation of met-hemoglobin (in agreement
with the present data), as well as to alter the oxygen dissociation
curve of oxyhemoglobin; this effect depended upon the time of incubation with D_2O and on the proportion of D_2O in H_2O. They interpreted
their results in terms of distortions caused in the hemoglobin molecule by asymmetric deuteration of the globin; these distortions were
expected to reach a maximum at about 70-80% D_2O in H_2O. This interpretation is not supported by the present results; indeed:
1) The stabilizing effect on the native conformation of met-hemoglobin
increases monotonically by increasing the proportion of D_2O in H_2O
(see Fig. 3). Nothing "strange" happens at 70-80% deuteration.
2) The activation enthalpy of the denaturation process seems not to
be altered by isotopic substitution of the solvent (see Fig. 4).
If distortions caused by the greater strength and length of deuterium
bonds inside the protein were relevant, a sizeable enthalpy variation
should be expected.
3) The activation entropy of the denaturation process is lowered by
the isotopic substitution of the solvent ($\Delta S_A(D_2O) < \Delta S_A(H_2O)$). It
is difficult to see how this could be done other than through solvent
effects, since the protein configurational entropy can hardly be
affected by replacement of hydrogen by deuterium inside the macro-
molecule.

In conclusion, purely steric effects due to the replacement of hydrogen by deuterium inside the protein seem not to be relevant to the isotopic effect on the thermal denaturation of met-hemoglobin. An interpretation of our data in terms of solvent effects seems more likely. In particular, the finding that the isotopic effect is "entropy driven" may suggest the relevance of the strengthening of protein-solvent hydrophobic interactions upon substitution of H_2O with D_2O. According to this interpretation, the negative $\Delta\Delta S_A$ value found in the present work would be indicative of a higher degree of local order imposed on D_2O molecules by non polar groups of the protein moiety that, upon denaturation, become exposed to the solvent.

ACKNOWLEDGEMENTS

The authors thank Proff. M.U. Palma and L. Cordone for discussion and criticism. The technical help of Mr. G. Izzo and Ms. A. La Franca is also acknowledged. Support to this paper was provided by the Comitato Regionale Ricerche Nucleari e Struttura della Materia (CRRN-SM). One of authors (P.L.S.B.) is research fellow of CRRN-SM.

REFERENCES

1. Katz,J.J. and Crespi,H.L. (1972) in "Isotope Effects in Chemical Reactions", Collins,C.J. and Bowmann, N.S., Eds., V.R.R. Comp. New York, pp. 286-363.
2. Némethy,G. and Scheraga,H.A. (1964) J.Chem.Phys. 41, 680-689.
3. Tomita,K.J., Rich,A., De Loze,C. and Blont,E.R. (1962) J.Mol. Biol. 4, 83-92.
4. Eisenberg,D. and Kauzmann,W. (1969) "The Structure and Properties of Water", Oxford at the Clarendon Press.
5. Kresheck,G.C., Schnaider,H. and Scheraga,H.A. (1965) J.Phys.Chem. 69, 3132-3140.
6. Kresheck,G.C. (1975) in "Water-A Comprehensive Treatise" vol. IV Franks, F. Ed., Plenum Press, New York and London, pp. 119-120.
7. Tomita,S. and Riggs,A. (1970) J. Biol. Chem. 245, 3104-3109.
8. Cupane,A., Palma,M.U. and Vitrano,E. (1974) J.Mol.Biol. 82, 185-192.
9. Cordone,L., Cupane,A., San Biagio,P.L. and Vitrano,E. (1979) Biopolymers 18, 1975-1988.
10. Perutz,M.F. (1972), Nature 237, 495-499.
11. Lumry,R., Smith,E.L. and Glantz,R.R. (1951) J.Amer.Chem.Soc. 73, 4330-4333.
12. Glasoe,P.K. and Long,F.A. (1960) J.Phys.Chem. 64, 188-190.
13. Bates,R.G. (1968) Anal. Chem. 40, 28A-38A.

THE EFFECT OF HYDROGEN PEROXIDE ON SPIN-LABELED ALCOHOL DEHYDRO-

GENASE FROM EQUINE LIVER

P.R.Crippa[+], L.Donelli, R.Favilla[+] and A.Vecli[+]

[+]Unità di Biofisica Molecolare, GNCB-CNR

Istituto di Fisica, Università di Parma

SUMMARY

Hydrogen peroxide has been shown to act both as a substrate
and a strong modifying agent on the enzyme alcohol dehydrogenase
from equine liver (LADH). The effects of H_2O_2 on spin labeled LADH
have been investigated by ESR technique. The obtained results can
be summarized as follows:

1. Spin label alkylation of Cys-46 enzyme residue induces a parallel
decrease of both alcohol dehydrogenatic and peroxidatic activities.
2. H_2O_2 induces a fast quenching of ESR signals of spin labeled
LADH, without altering their shape.
3. A slow quencing process, partially inhibited by the presence of
coenzymes, takes place subsequently, with a decrease of amplitude
and change of shape of the ESR spectrum.
4. Cupric ions accelerate the slow quenching process.
5. The ESR spectrum of the enzyme denatured with urea is remarkably
different from that obtained after prolonged treatment of the spin
labeled enzyme with H_2O_2, although both reagents unfold the protein
architecture.

INTRODUCTION

In a previous paper from our laboratory (1), it was pointed
out that hydrogen peroxide can act as a substrate for equine liver
alcohol dehydrogenase (LADH) (EC 1.1.1.1), giving rise to a catalytic
process that has been extensively studied (2-4).

On the other hand, hydrogen peroxide is well known to be a
strong oxidizing agent for several protein residues, such as cysteine,

methionine and others (5).   These oxidizing reactions result in
drastic changements of some chemico-physical parameters, of the
proteins, such as loss of biological function, partial unfolding
and so on.   In particular the inactivation of LADH proceeds more
and more rapidly as the concentration of hydrogen peroxide is increased, whereas the presence of coenzymes result in a protective
effect.

Measurements of optical rotatory power, ORD and fluorescence
on solutions of LADH in the presence of H_2O_2 suggest that the enzyme
undergoes a progressive unfolding process, occurring in parallel with
its chemical modification (6).

The results obtained by ESR-spin labeling technique reported
in this paper allowed us to classify the occurring phenomena in two
main groups: fast quenching effects preceding inactivation and slow
quenching effects in parallel to optical rotatory power and fluorescence temporal variations.

EXPERIMENTALS

Horse liver alcohol dehydrogenase was purchased from Boehringer
Mannheim as a crystalline suspension in 0.02 M phosphate buffer at
pH 6.5 containing 10% ethanol.   It was normally dialysed 4x12 hours
against 0.1 M phosphate buffer at pH 7.5 at 4°C, and then filtered
through a Millipore membrane (pore size 0.45 μm) in order to eliminate denatured enzyme.   The enzyme concentration was determined
spectrophotometrically at 280 nm ($\varepsilon = 3.54 \times 10^4$ $M^{-1}cm^{-1}$).   The alcohol
dehydrogenase activity was assayed according to Dalziel (7) and the
peroxidase activity according to Favilla and Cavatorta (1).

Iodoacetamide spin label (ISL-101: 4-(2-iodoacetamido)-2,2,6,6-
-tetramethylpiperidinoxyl) was purchased from Syva Company, Palo Alto.
Other chemicals used without further purifications were: 30% H_2O_2
solution (Peridrol, Merck); β-NAD$^+$ and β-NADH disodium salt (both
grade I, Boehringer); isobutyramide (IBA, Eastman); pyrazol (Schu-
chardt); $CuSO_4$ (C.Erba, Milan); sodium phosphate salts (C.Erba) and
urea (BDH).   Quartz bidistilled water was used throughout.

The iodoacetamide spin label (ISL-101) was allowed to react
selectively with Cys-46 residue, which is bound to the active site
zinc ion of LADH, as described elsewhere (8).   All ESR measurements
were performed on a JEOL JES-ME-3X apparatus operating in X band
and equipped with a flat quartz cell for aqueous samples at constant
instrumental conditions: microwave power 2×10^{-3} watts, modulation
width 4 Oe, gain 1200, time constant 1 second, sweep time 5 minutes.

According to preliminary results obtained at variable spin label
concentration, a linear relationship between first moment and nitro-
xide free radical concentration was assumed.   The first moment

calculation was made following the procedure suggested by Wyard (9). The errors in this kind of measurements have been evaluated to be about 2%. Comparisons between different spectra were done taking into account another source of error, arising from cell positioning into the spectrometer cavity, experimentally evaluated to be about 5%.

RESULTS AND DISCUSSION

Relationship between activities during LADH spin labeling

Alcohol dehydrogenase activity (ADH) and peroxidase activity (POD) measurements performed during the process of LADH spin labeling, confirm the previously found evidence that both types of activities utilize the same enzyme active site (1). Actually, as shown in Fig. 1, the two catalytic activities decay at the same rate, suggesting a very similar effect of the spin label sensitive Cys-46 residue alkylation on both of them.

ESR spectra of free and LADH bound spin label ISL-101

The ESR spectrum of ISL-101 in aqueous solution exhibits the well known characteristics in shape, peak to peak separation and relative height. Addition of H_2O_2 at final concentration below 0.2 M does not affect the spectrum over a time interval comparable to that of our experiments.

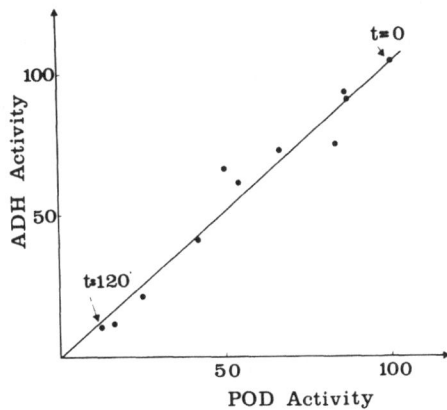

Fig. 1 - Time decay of alcohol dehydrogenase (ADH) and peroxidase-(POD) activities during the first two hours of the label attach to LADH.

The ESR spectra of spin labeled LADH in aqueous solution we
obtained are practically identical to those reported by Spallholz
and Piette (8). These authors suggest the presence of two superimposed components of different relative mobility. Their hypothesis
was indirectly confirmed by Sloan and Mildvan (10) who, working on
yeast alcohol dehydrogenase, were able to separate by gel filtration
the spin labeled enzyme (strongly immobilized component) from the
aggregated fraction with twice molecular weight (less immobilized
component).

This molecular situation complicates the analysis of the ESR
data and makes impossible to treat them in terms of rotational
correlation times.

Effect of H_2O_2 on spin labeled LADH ESR spectrum

Addition of H_2O_2 to a spin labeled enzyme solution gives rise
to a very rapid quenching of ESR spectral intensity, followed by a
slower quenching process which has been investigated as a function
of time. The magnitude of the rapid quenching process depends on
H_2O_2 concentration in an exponential fashion, as shown in Fig. 2.

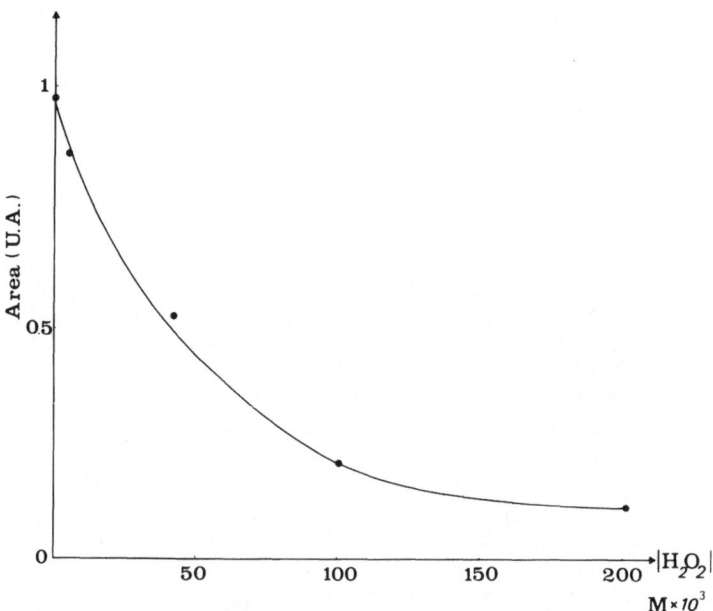

Fig. 2 - Dependence of the "instantaneous" quenching of the ESR
        spectrum of SL-LADH on the H_2O_2 concentration. The effect
        is observed only after the setting time of the spectrometer
        (about 1 min).

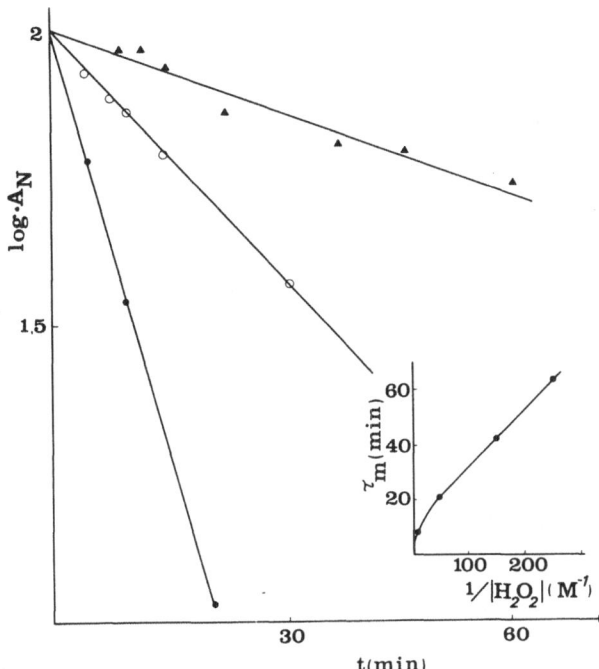

Fig. 3 - Time course of the slower quenching of the ESR spectrum
area of SL-LADH. The concentrations of H_2O_2 are:
▲ 4 mM; o 20 mM; ● 200 mM.   Insert: halftimes of the
decay of the ESR signal as a function of the H_2O_2 concentration.

The time course of the slower quenching process is reported in
Fig. 3 as the logarithm of the residual area of the ESR spectrum,
normalized by extrapolation at zero time, versus time. For the
large H_2O_2 concentration range below 0.2 M, the experimental points
lie, within the experimental errors, on straight lines, indicating
that the slow quenching is a pseudo first order process. The insert
in Fig. 3 shows the halftimes, deduced from the slope of lines
corresponding to different H_2O_2 concentrations, against the reciprocal H_2O_2 concentrations. Only at high H_2O_2 concentrations (> 0.2 M)
linearity is lost, suggesting more complicated reaction patterns.
Concurrently with the low quenching process, the ESR spectrum exhibits shape modifications: the second line, corresponding to the
less immobilized component, decreases more rapidly than the first

one, while the other lines become broader.

Effect of H_2O_2 on spin labeled LADH ESR spectrum in the presence of coenzymes

As found by Spallholz and Piette (8), addition of NADH alters the distribution of the two components in the ESR spectrum towards a situation of reduced mobility of the enzyme bound spin label. Addition of isobutiramide (IBA) does not further alter the shape and the intensity of the ESR spectrum. Addition of H_2O_2 to this ternary complex gives rise to a slow quenching of the spectrum, whose rate is even lower than that observed with spin labeled LADH alone (Fig. 4).

The presence of NAD^+ in the spin labeled enzyme solution does not modify appreciably the ESR spectrum. Addition of H_2O_2 induces a slow quenching as well, whose rate appears rather insensitive to H_2O_2 concentration.

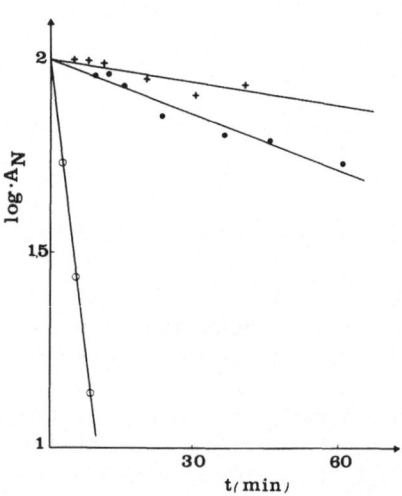

Fig. 4 - Time course of the slow quenching of the area of the ESR spectrum of SL-LADH in the presence of 4 mM H_2O_2:
o SL-LADH+H_2O_2 (control); + SL-LADH+NADH.1mM)+IBA(1mM)+ H_2O_2; o SL-LADH+Cu^{++} (50 μM) + H_2O_2.

Effect of Cu^{++} ions on spin labeled LADH ESR spectrum

Addition of paramagnetic metal ions, such as Cu^{++} ions, to a spin labeled LADH solution containing H_2O_2, accelerates appreciably the slow quenching of ESR signals (figure 4), but does not alter the shape of the ESR spectrum when compared with that observed for a spin labeled enzyme solution with H_2O_2.

The ESR signal peculiar of Cu^{++} ions is off range and not observable at the concentration used (50 μM). The presence of NADH strongly protects the ESR intensity from being quenched by H_2O_2 plus Cu^{++}.

Effect of urea on spin labeled LADH ESR spectrum

When increasing concentrations of urea from 0 to 8 M are added to a spin labeled LADH solution, the ESR signals change their shape towards that more peculiar of a less immobilized spin label (Fig. 5). The third line maintains a reduced amplitude suggesting that the enzyme covalently bound spin label moiety is not completely free to rotate in spite of the progressive unfolding of the enzyme. The isosbestic points in Fig. 5 suggest that the action of urea can be interpreted as a variation of the relative populations of the two components present in the ESR spectrum. Elimination of urea by dialysis allows the ESR spectrum to reassume a shape very similar to that observed before the denaturation process.

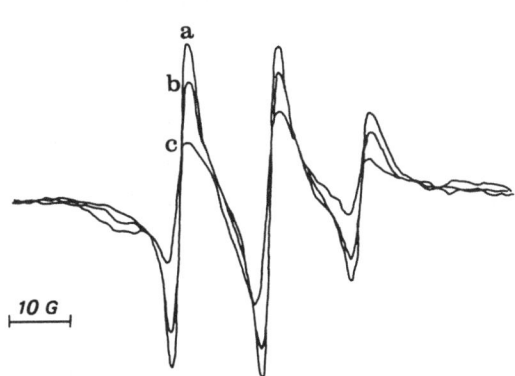

Fig. 5 – Effect of urea on the ESR spectrum of SL-LADH:
(a) SL-LADH+urea 6 M; (b) SL-LADH+urea 3 M; (c) SL-LADH
(control).

CONCLUSIONS

     The use of ESR spectroscopy as a tool to investigate the effect
of H_2O_2 and urea on spin labeled LADH in aqueous solution at pH 7
has provided informations concerning conformational processes the
enzyme undergoes in different conditions. The dependence of the
fast quenching of ESR signals on H_2O_2 concentration and the small
protective effect observed when binary or ternary complexes with
coenzymes are formed, are consistent with a mechanism of quenching
which involves the formation of a complex between SL-LADH and some
radical species coming from H_2O_2.

     Concerning the slow quenching process, two distinct effects
are observed: the first one is a diminution of amplitude, the second
one, which occurs concomitantly with the preceding one, concerns a
change of shape of ESR spectrum. The chemical modifications brought
about by H_2O_2 are responsible of a progressive unfolding of the
enzyme molecule, as detected with other techniques (ORD and fluorescence (3)). The decrease of the EPR signal and its change of shape
are consistent with that unfolding process, the rate of which depends
upon the H_2O_2 concentration in a Michaelis-Menten fashion (see Fig.
4, insert) indicating an at least two step process, the slowest one
reflecting a conformational change which brings the spin label moiety
into a microenvironment susceptible of attack by adventitious radical
species always present in solution. At the same time the conformational changes associated with the progressive unfolding of the
enzyme changes the relative mobility of the residual enzyme bound
spin labels, which is reflected in a change of ESR spectral shape,
towards an unexpected more immobilized state.

     The slow quenching process, which reflects the unfolding process,
is slowed down by the presence of coenzymes. The accelerating effect
on the slow process produced by cupric ions is consistent with the
metal ion assisted attack of the enzyme structural integrity by
H_2O_2 derived radical species. The paramagnetic nature of Cu^{++} could
hence be responsible of this accelerated unfolding process. We
think that the quenching effect due to a direct magnetic interaction
between Cu^{++} ions and the spin label free radical (10) can be excluded on the basis of our experimental observations that ESR spectra
of systems composed by free spin label or spin labeled LADH plus
Cu^{++} but in absence of H_2O_2 do not show any quenching. The effect
of magnetic quenching was observed in several other cases (11) (12)
and requests a stable binding of the ion to the labeled enzymes.

     On account of our results, we interpret the slow quenching
phenomenon as due to the attach of the $OH^{\cdot}$ radical produced by
spontaneous homolytic dissociation of H_2O_2 (2)(13) to the unfolding
spin labeled LADH, which progressively brings the spin label moiety
in an accessible microenvironment.

For what concerns the effect of urea, figure 5 clearly shows that the unfolding process brought about by this reagent is of a different nature with respect to that provoked by H_2O_2, let apart the chemical modifications introduced by the peroxide. It does not change the area but only the shape of the ESR spectrum which tends towards a less immobilized situation, as expected for a complete unfolding of the protein. The effect can be reversed by elimination of urea, which suggests an at least partial refolding of the structure of the enzyme.

REFERENCES

1. R.Favilla and P.Cavatorta, FEBS Lett. 50, 324 (1975).
2. R.Favilla, A.Mazzini, A.Fava and P.Cavatorta, Eur. J. Biochem. (in press).
3. R.Favilla, P.Cavatorta, A.Mazzini and A.Fava, Eur. J. Biochem. (in press).
4. A.Mazzini, E.Dradi, R.Favilla, A.Fava, P.Cavatorta and M. Abdallah, Eur. J. Biochem. (in press).
5. G.E.Means and R.E.Feeney, Chemical modifications of proteins, Holden-Day Inc., San Francisco (1971).
6. P.Cavatorta, R.Favilla and A.Mazzini (in preparation).
7. K.Dalziel, Acta Chem. Scand. 11, 397 (1957).
8. J.E.Spallholz and L.H.Piette, Arch. Biochem. Biophys. 148, 596 (1972).
9. S.J.Wyard, J. Sci. Instr. 42, 769 (1965).
10. D.L.Sloan and A.S.Mildvan, Biochemistry 13, 1711 (1974).
11. J.S.Leigh Jr., J. Chem. Phys. 52, 2608 (1970).
12. I.D.Campbell, R.A.Dwek, N.C.Price, G.K.Radda, Eur. J. Biochem. 30, 339 (1972).
13. R.A.Dwek, H.R.Levy, G.K.Radda, P.J.Seeley, Biochem. Biophys. Acta 377, 26 (1975).
14. W.C.Schumb, C.N.Sutterfield, R.L.Wentworth, Hydrogen peroxide, chapter 7, Reinhold Pub.Corp. (1955).

POLYFUNCTIONALITY OF HORSE LIVER ALCOHOL DEHYDROGENASE

Roberto Favilla and Alberto Mazzini

Unità di Biofisica Molecolare, GNCB-CNR

University of Parma, Italy

Horse liver alcohol dehydrogenase (LADH; EC 1.1.1.1) is today one of the most thoroughly investigated enzymes. Since its crystallization in 1948 (1) many authors have studied its properties in great details: the Swedish school, first with Theorell and then with Bränden and their co-workers, has extensively reviewed our knowledges on this and related dehydrogenases (2,3). The active form of LADH is known to be dimeric and its three-dimensional structure has been solved at 2.4 Å resolution (4).

Despite these very precise structural informations, its main physiological role has long remained a puzzling dilemma. Its presence in large amounts in the horse liver (more than two grams of purified enzyme can be recovered from a single liver) as well as its important role in the ethanol metabolism have greatly stimulated many researchers' interest. In the last years the observed very large substrate specificity of LADH has been taken as an emerging evidence of enzymatic polyfunctionality, whose efficiency could be dependent on many factors, among which substrates and cofactors availability in different intracellular compartments could play a main role.

LADH known catalytic activities are summarized here below:

1. Alcohol dehydrogenating activity

This activity refers to the oxidation of primary alcohols to aldehydes and of secondary alcohols to ketones, as well as to the corresponding reverse reductions. Besides simple aliphatic mono and polyalcohols, such as ethanol and ethylene glycol, some steroids, such as 3-keto and 3-beta-hydroxysteroids have also been found to

be substrates of an isoenzymatic form of LADH (5), whereas an intermediate compound along the biosynthetic pathway leading to the biliar
salts, namely 5-beta-colestan-3-alpha,7-alpha,12-alpha,26-tetrol,
as well as omega-hydroxylated fatty acids, intermediates in the omega
degradation of fatty acids, have been discovered to be substrates of
both ethanol-active and steroid-active LADH isoenzymes (6,7). Although it is now certain that LADH plays a main role in the metabolism of ingested ethanol to acetaldehyde, other systems, among which
catalase and the ethanol oxidizing microsomal system (MEOS) have been
claimed to play some role (8). In man a steroid active form has also
been found (9) among the various existing isoenzymes and recently a
pyrazol insensitive activity has been suggested to be a determinant
of alcoholism (10).

2. Aldehyde dehydrogenating activity

This second kind of activity, concerned with the oxidation of
aldehydes to the corresponding carboxylic acids can act upon aliphatic and steroid aldehydes (11). A combination of both alcohol
and aldehyde dehydrogenating activities leads to dismutation reactions, in which aldehydes are converted to the corresponding alcohols and acids in the presence of the coenzyme NAD^+ (12). Very
recently Dutler has communicated that LADH is able to catalyze the
oxidation of histidinol to histidine, through the intermediate
histidinal (13), suggesting its possible involvement in the terminal
steps of the histidine biosynthesis, since no other histidinol dehydrogenase activity has been isolated from mammals and the two
steps oxidation from histidinol to histidine, catalyzed by a single
enzyme, is under the control of pyridoxal and pyridoxal phosphate,
two cofactors involved in the regulation of amino acids biosynthesis,
which can bind to LADH.

3. Oxidative decarboxilating activity

Together with the above mentioned activity towards histidinol,
Dutler has communicated (13) that coproporphyrinogen is converted
by LADH to protoporphyrinogen, through a reaction of oxidative de-
carboxylation, whose pattern appears to be rather complex and has
not yet been clarified. Even in this case LADH has been suggested
to play a physiological role, because it shows many similarities
with the known coproporphyrinogen oxidases from both bacterial and
mammalian sources (14).

4. Esterolytic activity

During the last year Tsai was able to demonstrate the estero-
lytic activity (15) of LADH using octanoate esters as substrates
(the ability of this enzyme to catalyse a given reaction generally
increases with the chain length of the substrate to reach a maximum
at eight carbon atoms). The esterase activity, assayed following

the rate of either p-nitrophenol release at 400 nm or acid liberation by means of a pH-stat, did not require NAD^+ as essential coenzyme, and was found to be associated with all the isoenzymatic forms of LADH. Furthermore from inhibition studies, the esters are probably hydrolyzed at the same domain where the dehydrogenase activity is also accomplished. No suggestion is done about a possible physiological meaning of this activity.

5. Peroxidatic activity

This activity catalyses the oxidation of the coenzyme NAD^+ by hydrogen peroxide. It was first observed some years ago in our laboratory (16) and then investigated more and more extensively (17-23). The whole reaction is biphasic, a first step consisting of the enzymatic conversion of NAD^+ to a product, named Compound I, which spontaneously converts to NADX, the final stable product, studied with NMR spectroscopy for its structure. It is interesting to note that NADH is a strong inhibitor of this reaction. The main feature of this activity will be here briefly reviewed, together with some recent results.

The time course of the reaction can be easily followed by measuring the absorbance increase at 300 nm. The catalytic events take place within the same active site pocket where also alcohols are oxidized, as demonstrated by a series of observations (inhibition by pyrazol, NADH and ethanol; parallel decrease of both alcohol dehydrogenating and peroxidatic activities caused by hydrogen peroxide itself (17), by spin labelling (24) and by EDTA (21)). Steady state kinetic investigations at pH 7 indicate an ordered sequential mechanism, with NAD^+ first entering the active site; the rate limiting step has not yet been individuated because product inhibition studies have not been performed and transient studies are still in progress. All the factors involved in the alcohol dehydrogenating activity have been found to be also important in the peroxidatic activity, e.g. the active site zinc ion is essential for catalysis, as revealed by the decrease of activity observed during enzyme pre-incubation with EDTA. Also the two sulfidryl groups belonging to Cys 46 and Cys 174 are essential for the catalysis with H_2O_2, as shown by a linear relationship among decrease of peroxidatic activity and increase of oxidised cysteine residues observed when LADH is preincubated with H_2O_2 in the absence of coenzyme (22). Another analogy with the alcohol dehydrogenating activity comes from the amount of protons released during the reaction, which is equivalent to that observed with alcohol substrates (1 mole of protons released per mole of NAD^+ modified). Stoichiometric determinations lead to a final balance of one mole of NAD^+ converted to Compound I per mole of H_2O_2 reacted. The presence of metal ions at low concentrations does not influence the initial rate of the enzymatic step thus allowing us to state that the reaction can take place without metal ion assistance, as instead found for many other oxidation reactions

involving hydrogen peroxide (25). With the aid of chromatographic
techniques we could separate the products of the reaction, but not
the intermediate compound I because too labile. Depending on the
reaction conditions two main different products are detected: when
H_2O_2 is eliminated soon after the fast enzymatic step NADX is recovered by ion exchange column chromatography (its properties are
discussed below); whereas a different compound, named Y, elutes
through the column when H_2O_2 elimination is delayed. This behaviour,
together with its time course, can be rationalized according to the
following minimal scheme:

$$NAD^+ + H_2O_2 \xrightarrow{\text{LADH}} \text{Compound I} \begin{array}{c} \xrightarrow{H_2O} NADX \\ \searrow_{H_2O_2} Y \end{array}$$

The chemical structure of NADX, as derived from ^1H and ^{13}C NMR
measurements, appears to be consistent with the following formula:
$ADPR-NH-CH=C(CHO)-CONH_2$. As one can see, only the nicotinamide
moiety has been modified by H_2O_2, the remaining part (adenosinedi-
phosphoribosyl moiety) being still intact.

As already mentioned above, hydrogen peroxide behaves both as
a substrate and an inactivating agent of LADH. From the values of
the kinetic parameters obtained from initial rates of catalysis and
inactivation a global mechanistic scheme has been deduced, which
quantitatively accounts for both effects at pH 7 (22). It is finally
interesting to note that this peroxidatic reaction is specific not
only for LADH since it is catalyzed also by the homologous enzyme
from yeast (YADH), thus conferring it a more general validity.

REFERENCES

1. R.K.Bonnichsen and A.M.Wassén, ABB 18, 361 (1948).
2. H.Sund and H.Theorell, "The Enzymes", 2nd ed., Vol. 7, p. 25
    (1963).
3. C.-I.Brändén, H.Jörnvall, H.Eklund and B.Furugren, "The Enzymes",
    3rd ed., Vol. 11 Part A, p. 103 (1975).
4. H.Eklund, B.Nordström, E.Zeppezauer, G.Söderlund, I.Ohlsson,
    T.Boiwe and C.-I.Bränden, FEBS Lett. 44, 200 (1974).
5. R.Pietruszko, A.Clark, J.M.H.Graves and H.J.Ringold, BBRC 23,
    526 (1966).
6. G.Waller, H.Theorell and J.Sjovall, ABB 111, 671 (1965).
7. I.Björkhem, H.Jörnvall and Å.Åkeson, BBRC 57, 870 (1974).
8. R.D.Hawkins and H.Kalant, Pharmacol. Rev. 24, 67 (1972).
9. R.Pietruszko, H.Theorell and C.de Zalenski, ABB 153, 279 (1972).
10. T.K.Li, W.Bosron, N.Dafeldecker, L.Lange and B.Vallee, PNAS 74,
    4378 (1977).
11. L.Kendal and A.Ramanathan, Biochem.J. 52, 430 (1952).
12. J.A.Hilson and R.A.Neal, J.Biol.Chem. 247, 7106 (1972).

13. H.Dutler, communication held at the "FEBS Special Meeting on Enzymes", Dubrovnik (1979).
14. R.Poulson and W.J.Poulglass, J.Biol.Chem. 249, 6367 (1974).
15. C.S.Tsai, BBRC 86, 808 (1979).
16. P.Cavatorta, R.Favilla and A.Vecli, in "Atti della Prima Riunione Scientifica Plenaria della Società Italiana di Biofisica Pura ed Applicata, p. 127, Camogli (1973).
17. R.Favilla and P.Cavatorta, FEBS Lett. 50, 324 (1975).
18. R.Favilla and P.Cavatorta, in "Atti III Congresso Nazionale della Società Italiana di Biochimica" p.B44, Siena (1977).
19. A.Mazzini and R.Favilla, in "Atti del III Congresso Nazionale della Società Italiana di Biochimica" p.B45, Siena (1977).
20. R.Favilla and A.Mazzini, in "Atti V Congresso Nazionale della Società Italiana di Biochimica", p. 234, Il Ciocco (1979).
21. R.Favilla, A.Mazzini, A.Fava and P.Cavatorta, Eur.J.Biochem. (in press).
22. R.Favilla, P.Cavatorta, A.Mazzini and A.Fava, Eur.J.Biochem. (in press).
23. A.Mazzini, E.Dradi, R.Favilla, A.Fava, P.Cavatorta and M.Abdallah, Eur.J.Biochem. (in press).
24. P.R.Crippa, L.Donelli, R.Favilla and A.Vecli (in this volume)
25. G.E.Means and R.E.Feeney, "Chemical Modification of Proteins", p. 163, Holden-Day Inc., San Francisco.

BIOMACROMOLECULES: NUCLEIC ACIDS

APPLICATION OF SMALL ANGLE NEUTRON SCATTERING TO RNA-PROTEIN COMPLEXES

Giuseppe Zaccaï

Institut Laue-Langevin

156X Centre de Tri, 38042 Grenoble Cédex

A small angle scattering experiment on particles in solution provides a simultaneous and accurate determination of two parameters: the forward scattered intensity which depends on the excess scattering mass of the particles and its radius of gyration (1). Excess scattering mass is written:

$$\Sigma b - \rho_s V$$

where Σb is the sum of scattering amplitudes of all atoms in the particle, V is its partial volume and ρ_s is the scattering mass per unit volume of solvent. The expression is completely analogous to Archimedes' principle. The radius of gyration R_G is also analogous to its inertial counter-part;

$$R_G^2 = \frac{\Sigma \ (b_i - \rho_s v_i) \ r_i^2}{\Sigma \ (b_i - \rho_s v_i)}$$

where v_i is the volume of atom i, r_i is its distance from the centre of scattering mass of the particle.

When there are different particles j in the solution, the expression for the forward scattered intensity is written:

$$I(0) = \text{constant} \ \times \ \Sigma_j c_j \ (\Sigma b_j - \rho_s V_j)^2$$

where c_j is the molar concentration of particle j. Because $I(0)$ is proportional to the square of excess scattering mass, it is very sensitive to complex formation; a dimer will scatter four times as

strongly as a monomer, for example. The equivalent expression for
R_G is

$$R_G^2 = \frac{\sum\limits_{j} c_j \, (\Sigma b_j - \rho_s V_j)^2 \, R_{Gj}^2}{\sum\limits_{j} c_j \, (\Sigma b_j - \rho_s V_j)^2}$$

The measured R_G will also be dominated by the larger particles.

   To obtain I(0) and R_G it is usually sufficient to measure in
an angular range such that $0.3 \lesssim R_G Q \lesssim 2.0$, where $Q = 4\pi \frac{\sin\theta}{\lambda}$, θ
is half the scattering angle and λ is the wavelength of the radiation.
By extending the range to $R_G Q \gtrsim 5$, the scattering curve I(Q) is
obtained. There is not a one to one correspondance between I(Q)
and the structure of the particle (1); on the other hand, it gives
indications of the shape of the particle, and is useful in comparisons with curves calculated from models. In the special case of
spherical particles, I(Q) contains a lot of useful information and
extends to quite large angles. Studies on spherical viruses are
reviewed in (2).

   An important advantage of neutron radiation is that the scattering amplitudes of ^1H and ^2H (D) are of different sign. By varying
the $[H_2O]$: $[D_2O]$ ratio in the solvent, ρ_s can be varied over a wide
range, which encompasses the mean scattering densities of protein
and RNA. These are quite different; that of protein is the same as
$\sim 40\%$ D_2O, while that of RNA corresponds to $\sim 70\%$ D_2O. Measuring the
same complex in buffers of different $[H_2O]$: $[D_2O]$ composition,
therefore, allows the small angle scattering parameters to be determined separately for the RNA and protein components of the complex.
Small angle neutron scattering has been applied to ribosomes (3),
RNA viruses (2) and the interaction of aminoacyl-tRNA synthetase
with tRNA (4,5). Reference (6) is a review of the method as applied
to biological structures. Before we describe a study in some detail,
let us discuss the usefulness of this very low resolution structural
information in understanding the molecular biology of the interaction.

   Considering their importance in the biochemistry of the cell,
RNA-protein interactions are still very poorly understood. A good
example of specific RNA-protein recognition is the interaction
between amino-acyl tRNA synthetase and its cognate tRNA (7). Since,
on the ribosome, the tRNA molecule acts only as an adapter between
the amino-acid and the messenger RNA, it is vital for the efficiency
of protein synthesis that the correct amino-acid be bonded to the
correct tRNA. The bonding reaction is catalyzed by a specific amino-
acyl tRNA synthetase for each amino-acid and requires ATP and magnesium ions. The ribosome is an example of another type of protein-
RNA interaction. Whereas the amino-acyl-tRNA synthetase system is
dynamic with an enzymatic turnover (enzyme-tRNA complexes being

formed and dissociated when the tRNA is amino-acylated) a ribosome
is an apparently stable structure of protein and RNA. Presumably,
its precise structure is necessary for it to perform its function
efficiently. RNA viruses are intermediate in this scheme, being
very stable protein-RNA structures which, nevertheless, must undergo
drastic changes during the infection cycle. Details of the molecular
mechanisms of such interactions remain unknown, except for their
classification under broad headings: electrostatic (between basic
protein residues and phosphates in RNA), hydrophobic (base and
aromatic residue stacking), hydrogen bonding. It appears likely
that RNA molecules have tertiary structure which would be defined
by their sequence (in a somewhat analogous way to proteins). One
could then think of the interaction in terms of macromolecular
structure. The folding within the tertiary structure of each component would play an important role in making specific interaction
sites accessible. All the concepts of allostery would also come
into play. The crystal structure of tRNA has of course been solved
and its L shape with the amino-acid and anticodon sites at opposite
extremities related by a pseudo two fold axis fits very neatly into
this approach (8). Another example is ribosomal RNA. There is
evidence now to justify considering ribosomal RNA as a frame on
which the proteins are placed (3). It appears to maintain its
structure relatively well even after removal of all proteins.
The obvious step is to study a protein-RNA crystal at high resolution
and to observe directly which bonds are made. It is proving to be
a very difficult step to take. Despite great efforts by several
groups only RNA virus crystals have been studied successfully. $_\circ$
The structure of Tomato Bushy Stunt Virus has been solved to 2.9 A
resolution (9). No RNA appears in the electron density map, however,
probably due to its lack of long range order in the crystal lattice.
The part of the protein which interacts with the RNA is also missing
from the map. Flexible or disordered sequences also occur in the
2.7 A structure of tyrosyl-tRNA synthetase from Bacillus Stearo-
thermophilus (10). Two examples are not a great deal to speculate
on, yet it could be that flexible or unfolded protein sequences are
a feature of the interaction with RNA. If so, the hope that high
resolution structural information will be obtained from a crystal-
lized complex (amino-acyl tRNA synthetase with tRNA, for example)
could well be frustrated. In this context, small angle neutron
scattering experiments are very useful. They define the stoichiometry and gross topology of a complex with the spatial relationship
of RNA and protein. Structures of components bound within the complex can be compared to their free states and the dependence of
complex formation and structure on external parameters such as pH
or ionic conditions can be observed simply by adjusting the buffer
composition. An interesting example of how small angle information
can complement crystallographic information is the structure of
Tomato Bushy Stunt virus. Neutron experiments in solution have
described the radial distribution of all the protein and RNA in the
particle whereas only the surface protein appears in the crystal
structure (2).

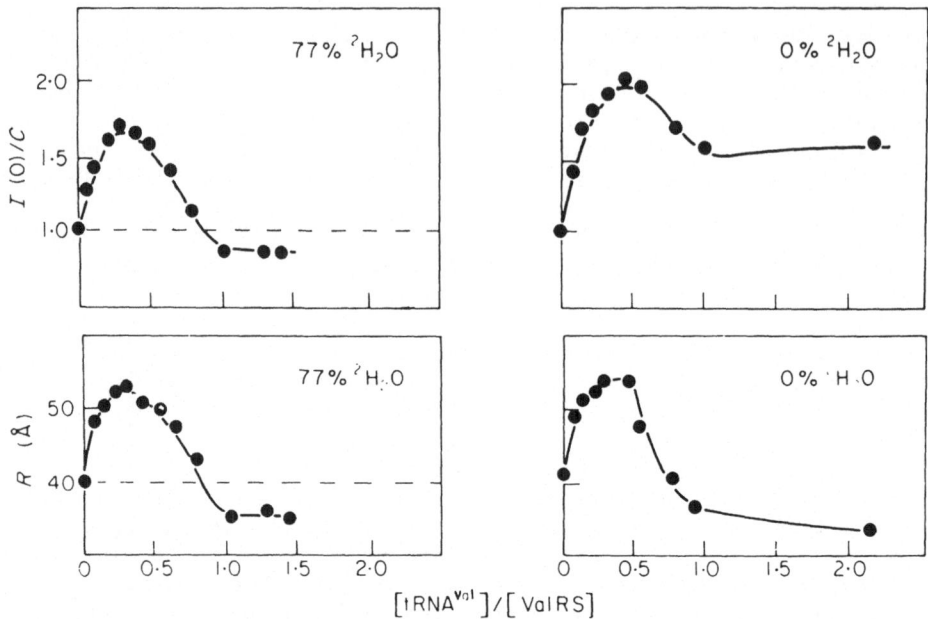

Fig. 1 – Values of I(0)/C (C is the concentration in mg/ml) normal-
        ised to 1.00 for free valyl-tRNA synthetase at each
        $[D_2O]:[H_2O]$ ratio, and R_G plotted as functions of the
        molar ratio of tRNA to enzyme. Concentrations of protein
        were maintained constant at 7 mg/ml. The curves were
        calculated from an equilibrium model given in the text.
        The buffer was 50 mM phosphate (pH 6.3). The figure is
        from reference (5).

        Figure 1 shows part of the data obtained in a study of the
interaction of yeast valyl-tRNA synthetase with different RNA's (5).
It is important to bear in mind that the curves in the H_2O frames
and in the 77% D_2O frames are <u>different views of the same reaction</u>.
In the first, both tRNA and protein contribute to the scattering,
each according to its excess scattering mass; in the second, the
excess scattering mass of tRNA is close to zero so that its contribution is negligible both when it is in the complex and free in
solution. In 77% D_2O, only the protein is "seen". The initial rise
in scattered intensity and radius of gyration must be interpreted in
terms of protein aggregates. As [tRNA] increases, these dissociate
until at a [tRNA] : [protein] ratio of 1, there are only monomers
in the solution. But why are the values of I(0)/C and R_G different
now from the initial values? They simply show the parameters of the
protein within the complex to be different from when it is free.
The smaller R_G indicates a more compact conformation. Considering
that the initial value is fairly high for a protein of that molecular
weight, it is not unlikely that the protein structure "closes in"

around the tRNA in the complex. The drop in scattered intensity
indicates a drop in excess scattering mass. It is difficult to see
how Σb could diminish (the protein has only one polypeptide chain
so that dissociation is excluded); the data show, therefore, that
the partial volume of the complex is smaller than the sum of its
components so that $(\Sigma_b - \rho_s V)^2$ is smaller (in 77% D_2O, $\Sigma b - \rho_s V$ is
negative). This change in V can be estimated quantitatively to be
$\sim 1\%$. A negative specific volume change had been seen in a pressure
gradient study of another amino-acyl tRNA synthetase-tRNA complex
(11). A quantitative value could not be obtained, however. The
rise in the curves in the H_2O frames reflects both the fixation of
tRNA in the complex and the formation of protein aggregates. At
the ratio of 1 the intensity corresponds exactly to that from a
one-to-one complex. It is easy to show that because of the relative
magnitudes and signs of Σb and $\rho_s V$, data in H_2O are not at all sensitive to changes in V. The increase in intensity beyond the ratio
of one is due to the free tRNA being added to the solution. The
curves shown in the figure are calculated from the following reaction
scheme, using the equations for I(0) and R_G given at the beginning
of the paper:

$$M + X \overset{K_1}{\Longleftrightarrow} MX$$
$$M'X + M \overset{K_2}{\Longleftrightarrow} DX$$
$$DX + M \overset{K_3}{\Longleftrightarrow} TX$$

where M is the enzyme monomer and X is tRNA. M'X, DX, TX are 1
enzyme-1 tRNA, 2 enzyme-1 tRNA and 3 enzyme-1 tRNA complexes,
respectively. A good fit was obtained with $K_1 = 400$ M^{-1} and
$K_2 K_3 = 0.001$ M^{-2}.

The experiment explored the structure of several complexes and
showed that both the aggregation of the protein and the conformational changes were modulated by ionic conditions. It is interesting to note, however, that these phenomena exist in conditions which
are quite similar to those in the cell so that one could speculate
on their functional relevance.

The ribosome has been a favorite nucleo-protein for study by
neutron scattering (References in (3)). We shall mention briefly
a recent experiment which made use of the scattering curve I(Q)
and $[H_2O]$: $[D_2O]$ variation in the solvent (3). Electron micrographs
of 16S RNA had been obtained and showed it to have a Y shaped structure. Is this structure maintained within the 30 S ribosome particle?
The scattering curve of the model was calculated and compared with
neutron data. In 42% D_2O, the excess scattering mass of protein is
negligible, so that data collected from the 30 S in this buffer
should only reflect the RNA structure within the particle. The

observed and calculated curves agreed very well, providing further
evidence for the RNA framework model of the ribosome (3).

ACKNOWLEDGEMENTS

I am grateful to B. Jacrot for discussions and advice in compiling
this short review.

REFERENCES

1. A. Guinier and G. Fournet, Small Angle Scattering of X-rays,
   Wiley, New York (1955).
2. B. Jacrot, Structural Studies of Viruses with X-rays and Neutrons,
   Comprehensive Virology, eds.: H. Frankel-Conrat, R.R. Wagner
   (1980).
3. I.N. Serdyuk, A.K. Grenader and G. Zaccaï, Ribosome Structure:
   Neutron X-ray Scattering, J. Mol. Biol., 135:691 (1979), and
   references therein.
4. P. Dessen, S. Blanquet, G. Zaccaï and B. Jacrot, Neutron Scattering of tRNA binding, J. Mol. Biol., 126:293 (1978).
5. G. Zaccaï, P. Morin, B. Jacrot, D. Moras, J.C. Thierry and R.
   Giegé, Valyl-tRNA synthetase-tRNA interactions, J. Mol. Biol.,
   129:483 (1979).
6. B. Jacrot, Neutron Scattering from Biological Structures, Rep.
   Progr. Phys., 39:911 (1976).
7. P.R. Schimmel and D. Söll, Amino-acyl tRNA Synthetases, Ann.
   Rev. Biochem., 48:601 (1979).
8. S.H. Kim, Symmetry Recognition hypothesis model for tRNA binding
   to amino-acyl tRNA synthetase, Nature, 256:679 (1975).
9. S.C. Harrison, A.J. Olson, C.E. Schutt, F.W. Winkler and G.
   Bricogne, Structure of Tomato Bushy Stunt Virus to 2.9 A
   resolution, Nature, 276:368 (1978).
10. M.J. Irwin, J. Nyborg, B.R. Reid and D.M. Blov, 2.7 Å structure
    of Tyrosyl-tRNA synthetase from Bacillus Stearothermophilus,
    J. Mol. Biol., 105:577 (1976).
11. J.R. Knowles, J.R. Katze, W. Konigsberg and D. Söll, Study of
    Leu- and Ser-tRNA synthetase from E.coli, J. Biol. Chem., 245:
    (1970).

A SPECTROPHOTOMETRIC STUDY OF THE STRUCTURE OF rRNA
IN THE RIBOSOME AND IN THE FREE STATE

A. Araco[1], M. Belli[2], F. Mazzei[2], G. Onori[3]

[1]Laboratorio di Biologia Cellulare ed Immunologia,
Istituto Superiore di Sanità, Roma

[2]Laboratorio delle Radiazioni, Istituto Superiore di
Sanità, Roma

[3]Istituto di Fisica, Università di Perugia

INTRODUCTION

The ribosome is a complex structure in which several components
are involved. It is known that Mg^{2+} plays an important role in determining this structure; moreover, experiments on RNA-proteins
interactions have suggested that these two species of molecules can
exert a strong mutual influence on their respective conformation[1].

To understand how the rRNA molecules are fitted into the ribosome, it is important to be acquainted with their secondary and
tertiary structure and to know the extent to which this structure
is affected by the interactions with the other ribosome components,
i.e. proteins and cations.

This work is aimed to clarify some aspects of the role played
by Mg^{2+} ions in the E.coli ribosomes, i.e. to which extent the rRNA
structure, both inside and outside the ribosome, is determined by
the interaction with the Mg^{2+} ions.

In the rRNA molecules there are single and double strand domains
with a large amount of stacking interactions between bases. The
strong absorption band in the ultraviolet region, with a maximum
at 260 nm, is connected mainly with $\pi \to \pi^*$ transitions and therefore
it is affected by alterations in the base-stacking interactions.
This band is a suitable probe to detect modifications in the rRNA
secondary structure (or, more correctly, in those structures which

303

are stabilized by the base-stacking) due to interactions with Mg^{2+} ions and ribosomal proteins. This probe is selective for RNA structure as in the range 240-320 nm the ribosomal proteins do not appreciably contribute to the absorption.

We have found that in intraribosomal RNA the base-stacking interactions are less than in the free RNA, and that the presence of Mg^{2+} alone, does not allow RNA in solution to assume the same conformation as it has in the ribosome, and that proteins play an important role in determining this conformation.

RESULTS AND DISCUSSION

a) Denaturation spectra of E.coli ribosomal subunits and of their isolated RNAs

It si known that temperature largely affects the conformation of nucleic acids in solution. Heating can break the hydrogen bonds between complementary base pairs (thermal denaturation) leading the molecules to a less ordered conformation. At the same time, an increase in the u.v. absorption spectrum occurs, this due mainly to the decrease in the interaction between neighbouring bases. The analysis of the difference spectrum upon denaturation (denaturation spectrum) may give information on the amount of the ordered and stacking-stabilized structure initially present in the RNA moiety[2].

In order to evaluate the amount of this structure in intra-ribosomal and in free RNAs, a number of denaturation spectra for E.coli 50 S and 30 S ribosomal subunits and for their isolated RNAs were collected in the range 230 - 300 nm. Typical denaturation spectra for subunits and their RNAs are reported in Fig. 1. Each denaturation spectrum was recorded by heating the sample from 20°C up to 95°C. The samples were dissolved in a buffer solution at low ionic strength (Tris-HCl 1 mM, pH 7.4; magnesium-acetate 0.08 mM) in order to prevent particle aggregation on heating[2].

The spectra for subunits are lower than that for their RNAs. They can all be obtained as a superposition of the denaturation spectra for A-U and G-C base pairs in double helical polyribonucleotides ("reference" spectra), according to the equation:

$$\Delta A(\lambda) = a \cdot \Delta \varepsilon_{AU}(\lambda) + b \cdot \Delta \varepsilon_{GC}(\lambda) \tag{1}$$

where $\Delta A(\lambda)$ is the observed denaturation spectrum, $\Delta \varepsilon_{AU}(\lambda)$ and $\Delta \varepsilon_{GC}(\lambda)$ are the reference spectra on a molar nucleotide basis[3]. The coefficients a and b of the linear combination are linked to the fraction f_{AU} and f_{GC} of A-U and G-C base pairs in the samples. We have used a best fit procedure[2] in order to get the best estimate of the parameters a and b, and hence the fractions f_{AU} and f_{GC}

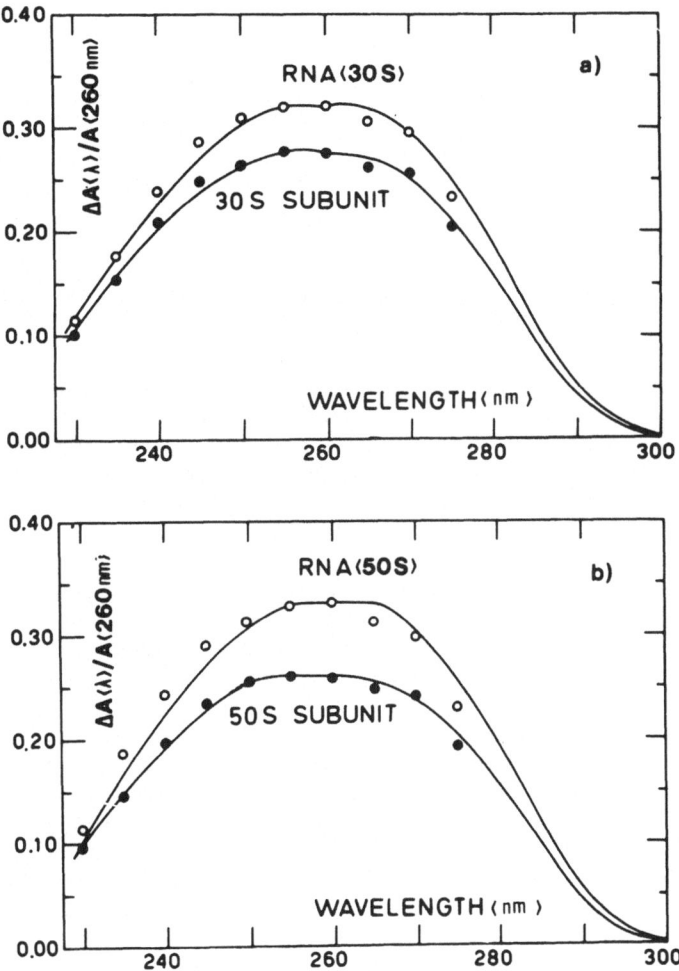

Fig. 1 - Denaturation spectra for a) 30S subunits and their RNA;
b) 50S subunits and their RNA. Solid line represents the
experimental spectra; circles correspond to the values obtained by fitting the experimental spectra with a superposition of the "reference spectra" for denaturation of
A-U and G-C base pairs.

in our sample. This analysis gives only approximate values, as the contribution coming from the stacking in the structured single stranded regions is neglected.

In any case, the results show that the amount of the stacking interaction is less in the intraribosomal RNA than in the isolated RNA molecules. If we assume that the difference is essentially due to a difference in the helical content, our analysis based on Eq.1 indicates that the RNA inside the ribosome has a lower amount of A-U and G-C base pairs than the isolated RNA, the average decrease in f_{AU} and f_{GC} being of 7±2% and 6±2%, respectively. This result suggests that proteins may play a role in determining the secondary structure of RNA inside the ribosome.

b) Influence of Mg^{2+} on the rRNA structure, as revealed by thermal denaturation experiments

It should be noted that the results reported above have been obtained for the specified ionic environment. It is known that this environment can affect the ribosome structure; for example, removal of magnesium bound to E.coli ribosomes causes a conformational change (unfolding) of the two subunits with a modification of their sedimentation coefficients and radii of gyration[4-7].

We have found that magnesium removal also involves changes in the secondary structure of the RNA, both in the ribosome[10] and in free state, as shown by a comparison of their denaturation spectra. As far as the subunits are concerned, our findings indicate that ribosome unfolding is related to a slight but significant increase in the ordered and stacking-stabilized structure of the rRNA. Magnesium removal from isolated rRNAs also causes slight conformational changes, which correspond to a loss in the ordered structure. All these results, taken together, indicate that: the RNA structures in the ribosome and in the free state are different in the presence of Mg^{2+}; the Mg^{2+} removal affects the RNA structure in a different way for the two cases; after Mg^{2+} removal the two structures appear very similar. This last point is well evidentiated by a comparison of the melting profiles for free RNA and subunits with and without Mg^{2+}. The results obtained by these experiments are summarized in Tab. 1, where the value of the maximum hyperchromicity H at 260 nm, the melting temperature T_m and the melting range ΔT (measured between 0.1 H and 0.9 H) are reported for the large subunit and its RNA; similar results have been obtained for the small subunit.

Significant differences may be observed, in denaturation behavior between subunits and isolated RNAs in the presence of Mg^{2+}: the melting profiles for the subunits are sharper (i.e. they melt more cooperatively) and the maximum hyperchromicity at 260 nm is significantly lower than that of their RNAs. On the contrary, in the absence of Mg^{2+}, both subunits and RNAs show a very similar behavior in the

TABLE I

Melting parameters for 50S subunits and RNA (50S) with and without Mg^{2+}

	H(%)	T_m(°C)	ΔT(°C)
Large subunits in 10 mM Tris-HCl, 0.1 mM Mg^{2+} (pH 7.4)	27±1	67±0.5	21±2
RNA (50S) in 10 mM Tris-HCl, 0.1 mM Mg^{2+} (pH 7.4)	33±1	62±0.5	32±3
Large subunits in 10 mM Tris-HCl, 1 mM Mg^{2+}, 1.5 mM EDTA (pH 7.4)	29±1	45±1	48±3
RNA (50S) in 10 mM Tris-HCl, 1 mM Mg^{2+}, 1.5 mM EDTA (pH 7.4)	29±1	43±1	46±3

sense that the hyperchromicity becomes practically the same and the melting transition occurs over the same wide temperature range.

In the absence of Mg^{2+} the RNA structure in the subunits thus appears to be very similar to that of the free RNA.

Incidentally, it should be noted that Mg^{2+} removal decreases the thermal stability of subunits and RNAs, and this results in a shift of the melting temperature of about 20°C towards lower temperatures and in a broadening of the melting profiles measured at 260 nm; this fact implies that the RNA structure in the absence of Mg^{2+} is stable only below 8-10°C (see also ref. 9). Accordingly, we performed all the experiments which involve Mg^{2+} removal at temperature of 4°C in order to avoid the thermal denaturation effect which is likely to occur at room temperature. It may be noted that many of the experiments reported in the literature on the ribosome unfolding have been performed at room temperature and this may explain some discrepancies found among the results[9].

c) Unfolding spectra

The conclusions drawn from the preceding results are confirmed by direct measurements of the absorption changes induced (at 4°C) by the removal of Mg^{2+} from ribosomes and RNAs.

Fig. 2 shows the changes observed in the absorption spectra of the subunits and their RNAs in the range 240-320 nm after complete removal of the bound magnesium following EDTA addition. These "unfolding spectra" clearly indicate that both subunits and RNA are affected by Mg^{2+} removal, but not in the same way. As far as the RNAs are concerned, an increase in the absorbance is observed practically over the entire spectral range. This indicates a decrease in the secondary structure of the RNA or, more precisely, a loss in its stacking-stabilized structure. On the contrary, removal of Mg^{2+} from the subunits leads to a significant decrease in the absorbance over the entire wavelength range.

The unfolding spectra for the subunits are very similar to the difference observed between the denaturation spectrum of the native and that of the unfolded subunits (see ref. 10). This is what one should expect if the ribosome unfolding is related to a modification in the stacking-stabilized ordered structure of the RNA. The experimental results indicate that for the subunits, unlike for free RNAs, magnesium removal causes a net increase in the base-stacking interactions. In principle, this change can be related to a number of different contributions, such as: 1) modifications in the double helical content, 2) changes in the single strand base-stacking or 3) alteration in the tertiary structure; the analysis of the observed spectral changes does not enable us to distinguish among these possibilities.

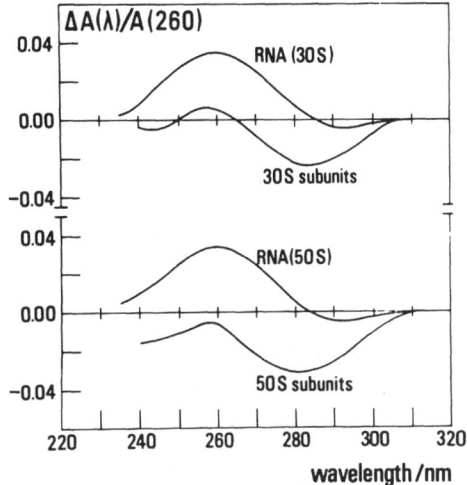

Fig. 2 - Unfolding spectra for the subunits and their RNAs. They
represent the changes in the absorption spectra on Mg^{2+}
removal. The solvent was 10 mM Tris-HCl (pH 7.4) with
0.08 mM Mg-acetate; magnesium removal was accomplished by
EDTA addition to a concentration of 0.2 mM.

A quantitative interpretation of the "unfolding spectra" can
be suggested if the first cause is the predominant one. In this case,
a description should be possible in terms of superposition of the
reference spectra for denaturation of A-U and G-C base pairs in double
helical polyribonucleotides according to the procedure already mentioned for the denaturation spectra, based on Eq.1. This showed that
the spectrum changes on ribosome unfolding are consistent with a
modification in the double helical content of the RNA, that corresponds to an average increase of 6±1% in the fraction f_{GC} of the
total number of the nucleotides in G-C base pairs, besides an average
decrease of 2±1% in the corresponding fraction f_{AU} for A-U pairs[8].
The same analysis, carried out for the changes observed on Mg^{2+} removal from the isolated RNAs, shows that they are consistent with
an average decrease of 7±1% in the f_{AU} fraction.

According to what suggested by the melting profiles, one can assume that, in the absence of Mg^{2+}, the molar absorption coefficient for the RNA inside the ribosome is the same as that for the RNA in the free state:

$$\varepsilon_{RNA}(\text{in the presence of } Mg^{2+}) + \Delta\varepsilon_{RNA}(\text{on } Mg^{2+} \text{ removal}) =$$
$$= \varepsilon_{SUB}(\text{in the presence of } Mg^{2+}) + \Delta\varepsilon_{SUB}(\text{on } Mg^{2+} \text{ removal}) \tag{2}$$

Hence, the difference in the unfolding spectra ($\varepsilon_{RNA} - \varepsilon_{SUB}$) is representative of the difference in the stacking-stabilized structure of the RNA in the ribosome with respect to that of the RNA in the free state (in the presence of Mg^{2+}). As $\Delta\varepsilon_{RNA} - \Delta\varepsilon_{SUB}$ is positive in the wavelength range we have examined (see Fig. 3), the RNA in the ribosome should have a greater absorption coefficient (and therefore a lower amount of stacking-stabilized structure) than in the free state.

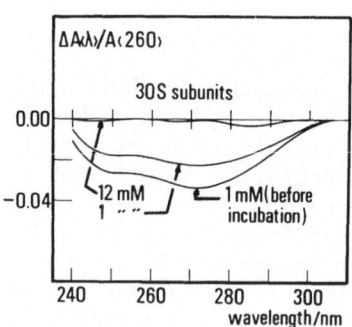

Fig. 3 - Changes in the absorption spectrum of the small subunits after removal and subsequent readdition of various amounts of magnesium. The absorbance values were measured at 4°C after samples incubation at 40°C for 30 min. The concentrations of magnesium added to the samples are shown in the fig.

If one ascribes these differences to a different helical content, the preceding analysis carried out on the unfolding spectra gives that the free RNA has a greater amount of A-U and G-C base pairs, namely Δf_{AU} = 5±2% and Δf_{GC} = 6±1%. This is in agreement, within the experimental errors, with the results obtained from the denaturation spectra previously reported.

d) Refolding of subunits

We made an attempt to refold the subunits and their RNAs at low temperature (about 4°C) by readdition of Mg^{2+} to the unfolded samples. This causes a decrease in the absorption spectrum which was practically the same for the subunits and their RNAs.

As far as the RNAs are concerned, the absorption decrease equals the increase observed on Mg^{2+} removal, so that there are practically no overall spectrum modification within the examined wavelength range. Readdition of Mg^{2+} to the subunits, causes an absorption decrease which adds to that we observed on Mg^{2+} removal, so that no reversibility is obtained in this case. This situation is clearly shown in Fig. 3 (curve a). This implies that readdition of Mg^{2+} at 4°C to the isolated RNAs is capable of restoring their initial stacking--stabilized structure, while this does not hold true for RNAs in the subunits, which show a further increase in their stacking interactions.

The attempt to completely refold the subunits was not successful when the Mg^{2+} readdition was performed at low temperature even if the Mg^{2+} concentration is raised to about 10 mM. On the other hand, the reconstitution experiments of Traub and Nomura[11] showed that incubation at a relatively high temperature (about 40°C) was necessary to obtain a functional particle from the separated RNA and proteins. The same authors were also able to obtain functionally active 30S subunits from the unfolded particles after incubation at 40°C in a proper ionic environment. Following these indications, we attempt to completely refold the subunits by incubating the unfolded samples at 40°C for 30 min. after readdition of various amounts of Mg^{2+}. Fig. 3 shows the changes in the absorption spectrum for the small subunit resulting from this experiment.

As it can be seen, a complete restoration of the initial absorbance is observed after incubation at 40°C for 30 min. in the presence of a relatively high concentration of Mg^{2+} (12 mM). After this treatment the particles are also indistinguishable from the native subunits as far as the sedimentation pattern and the melting profile are concerned.

Similar experiments aimed to refold the large subunits, showed that the absorption spectrum does not completely return to the initial value after incubation, the reversibility being of the order of 50%. Besides ultracentrifuge measurements showed that, under these con-

ditions, only about half of the total number of particles have re-
turned to the initial sedimentation coefficient, while the other
part sediments as a slower band. This behavior could be connected
to the inhomogeneity of the 50S sample, as it was already observed
that a fraction of the 50S subunit population could have nucleolytic
breaks in its RNA chain probably due to the purification procedures
(7).

We have also found that samples of isolated RNA (30S) which
underwent magnesium removal and readdition to a concentration of
12 mM, showed a slight increase in the absorption spectrum when
temperature was raised from 4 to 40°C; these modifications, which
indicate a weakening in the base stacking interactions due to thermal
denaturation, were completely reversed as the sample was brought
back to 4°C.

This result suggests that heating the subunits at 40°C in the
presence of a relatively high Mg^{2+} concentration leads to a more
mobile RNA structure which facilitates the reconstitution of the
initial ribosome conformation. It clearly appears that the protein
component is required to fold the RNA in the conformation it has in
30S subunits, and that incubation at 40°C causes a reorganization
of the particles with the ensuing restoration of the correct
RNA-RNA and RNA-proteins interactions.

REFERENCES

1. Brimacombe,R., Stoffler,G., and Wittmann,H.G. (1978) Ann. Rev.
    Biochem., 47:217-249
2. Araco,A., Belli,M., Giorgi,C., Onori,G. (1975) Nucl. Acids Res.,
    2:373-382
3. Cox,R.A. (1966) Biochem. J., 98:841-857
4. Gavrilova,L.P., Spirin,A.S. (1966) J.Mol.Biol., 16:473-489
5. Gesteland,R.F. (1966) J.Mol.Biol., 18:356-371
6. Miall,S.H., Walker,J.O. (1969) Biochem. Biophys. Acta, 174:
    551-560
7. Reale Scafati,A., Araco,A., Belli,M., Falbo,V., Giorgi,C. and
    Maggini,M. (1976) Nucl. Acids Res., 3:2171-2182
8. Belli,M., Onori,G., Araco,A., Giorgi,C. (1976) Biopolymers,
    15:1229-1232
9. Onori,G., Araco,A., Belli,M., Giorgi,C. (1977) Studia Biophy-
    sica, 2:155-160
10. Araco,A., Belli,M., Onori,G. (1979) Studia Biophysica, 75:
    25-30
11. Traub,P., Nomura,M. (1969) J.Mol.Biol., 40:391-413

POLYPEPTIDE SEQUENCES CODED BY COMPLEMENTARY DNA STRANDS:

A COMPUTER INVESTIGATION OF GENE POLYNUCLEOTIDE SEQUENCES

J. Quartieri,[°] V. Scarlato,[+] G. Mastrocinque,[°]
M. Cipollaro,[x] and A. Cascino[+]

[°]Istituto Elettrotecnico, Facoltà di Ingegneria, University
of Naples
[+]International Institute of Genetics and Biophysics, CNR,
via Marconi 10, Naples
[x]III° Servizio di Analisi, 2nd Medical School, University
of Naples

ABSTRACT

We have analysed both the coding and the complementary strands
of many structural genes of known polynucleotide sequence, looking
for open reading frames initiated with the AUG codon as a necessary
condition for the existence of virtual genes. While in the coding
strand we found, as significative virtual gene, only a codon sequence
corresponding to the real gene, the analysis of the complementary
strand shows, in some cases, virtual gene even longer than the real
protein.

The case of rabbit β-globin is discussed along with other features that make "a priori" indistinguishable, on the basis of actual
knowledge of gene structure, the coding from the complementary
strands.

On the basis of this analysis, we hypothesized the existence
of proteins (c.i.p.) coded by the complementary strand of some genes.

I. INTRODUCTION

Complete nucleotide sequences are nowadays established for a
number of genes. In order to investigate on this basis the primary
structure of polynucleotides, computer techniques and theoretical
information approaches are rapidly developing[1]. Search is generally
pursued outside the coding region of a given gene looking for spe-

313

cific sequences and recurrence relationships to find palindromes or stem loops to be correlated to specific regulatory sites along the DNA molecules.

In this paper, we report a computer analysis to investigate gene sequences themselves. Gene identification is tentatively based on some very simple assumptions, to be reported in the sequel. This analysis has been extended to the complementary strand of the investigated sequences. Computer obtained plots of virtual gene distributions on both the coding strand, where real gene sequences are located, and the complementary strands are presented. Similar gene distributions are found on both strands: in some cases, sequences are found on the complementary strand which might code for a protein even longer than the corresponding product of the coding strand. The hypothetical translation product of the sequence found on the complementary strand is called complementary inverted protein, or c.i.p..

II. MATERIALS AND METHODS

In order to analyse polynucleotide sequences in their three different reading frames, a Fortran computer algorhythm has been developed which is able to display the virtual genes existing in a sequence. A virtual gene is defined as a sequence which, according to a simple criterion to be discussed, is recognized as containing virtual genetic information for a protein. The computer code analyses the sequence and recognizes initiation I (AUG) and termination T_1, T_2, T_3 (UAA, UAG, UGA) codons. The criterion which has been chosen in order to identify the virtual genes is defined by either one of the following two statements:

1) The sequence is included between an initiation and a termination codon in phase.
2) The sequence follows an initiation codon and no termination codon is found until the entire sequence ends. This case is positively considered because it can be supposed that somewhere beyond the examined sequence a termination codon must be found.

When a gene is found, its codon length is computed by the Fortran code. In case 2) the length is computed from the initiation codon to the sequence end. It is also assumed that the valid AUG to initiate the gene sequence is always the farthest from the termination codon or the sequence end. The absolute frequency of virtual genes per unit length is plotted as a function of the length itself for each investigated frame and DNA strand. The procedure results in hystograms such as are represented by Figs. 1 and 2, to be discussed in the next section. For each frame and sequence, all the initiation and termination codons are displayed by the computer algorhythm, yielding plots such as are represented by Fig. 3, where the case of the rabbit β-globin sequence, read only in one frame, is displayed.

III. RESULTS AND DISCUSSION

The investigated sequences are listed in the caption to Fig.1, which shows the virtual gene codon length distribution as found by the computer code in the case of the coding strands. All the expected real genes have been identified by the computer, thus providing a positive check on the numerical code operation. A number of virtual genes have also been found.

The overall gene length distribution can be subdivided into two regions.

The first region contains several virtual genes shorter than 60 codons, which are not supposed to code for a protein. In the second region, which ranges from 90 up to 544 codons, some other virtual genes are less compactly distributed. All these are indeed the sequences that are translated into proteins. The peak in this distribution between 140 and 160 codons corresponds to the various globin genes that, in each and also within different organisms, have similar amino acid length.

In between these two principal regions we found, in the range from 70 to 90 codons, two virtual genes, along with a known gene, and also a sequence contained in the 16S ribosomal RNA molecule[10]. This is the overlapping region between the two previously described distributions.

If we consider only virtual genes longer than 90 codons, then the relative gene frequency per sequence is $F_s = 12/12 = 1$.

Quite similar considerations can be made in the case of the complementary strand, represented in Fig. 2, except that in region II the relative frequency of genes per investigated sequence is rather smaller (i.e. $F_{ns} = 7/12 = 0.58$) and the gene length distribution does not present a peak in the globins' codon length range.

From all these considerations we conclude that the two distributions of virtual genes in the coding and complementary strands are not distinguishable.

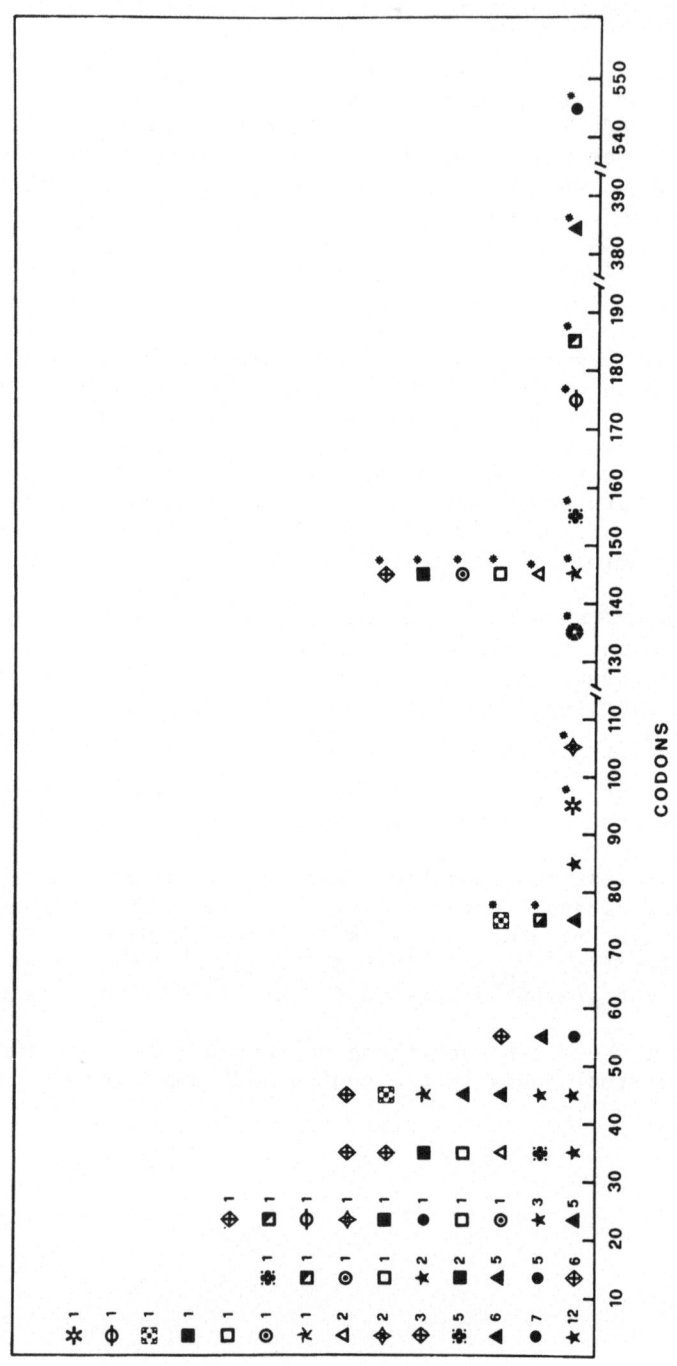

Fig. 1 - Virtual gene length distribution found by the computer in
the coding strands. The absolute frequency of virtual genes
is plotted as a function of the codon length for all three
reading frames. Each sequence is identified by a given
symbol. Coding strand means the DNA strand which, tran-
scribed into a messenger RNA molecule, is translated into
a protein of a known amino-acid sequence. The only exception is the 16S ribosomal RNA sequence[10].
The investigated sequences are the following:

- ◉ Human α-globin[2]
- ◻ Human β-globin[3]
- ⊕ Human γ-globin[2]   ?
- ✶ Human ε-globin[4]
- △ Rabbit α-globin[5]
- ◼ Rabbit β-globin[3]
- ▲ Chicken ovalbumin[6]
- ● MS2 Replicase gene[7]
- ⦂ Embryonic mouse V$_\lambda$ gene[8]   ?
- ✳ Rat preproinsulin I[9]   ?
- ★ E.coli 16S ribosomal RNA gene[10]
- ◧ Sea urchin histone H4[11]   ?
- ⬥ Sea urchin histone H2B[11]   ?
- ✳ Sea urchin histone H3[11]
- ◪ Sea urchin histone H2A[11]
- ⦵ Sea urchin histone H1[11]   ?
- ▨ Mitochondrial structural gene for subunit 9 of yeast ATPase
  complex[12]

The termination codons in yeast mitochondria DNA are UAA and
UAC, since UGA is used as tryptophan codon[13]. Sequences which are
not completely known are flanked by a question mark. These sequences
have been treated in the same way as the others, but the corresponding results are clearly temporary. Real genes are distinguished by
an asterisk in the hystogram. The number flanking each symbol indicates how many virtual genes of a given length have been found on
the same sequence. When only one virtual gene is found, the number
is omitted. The total number of bases contained in the analysed
sequence is about 13 Kb.

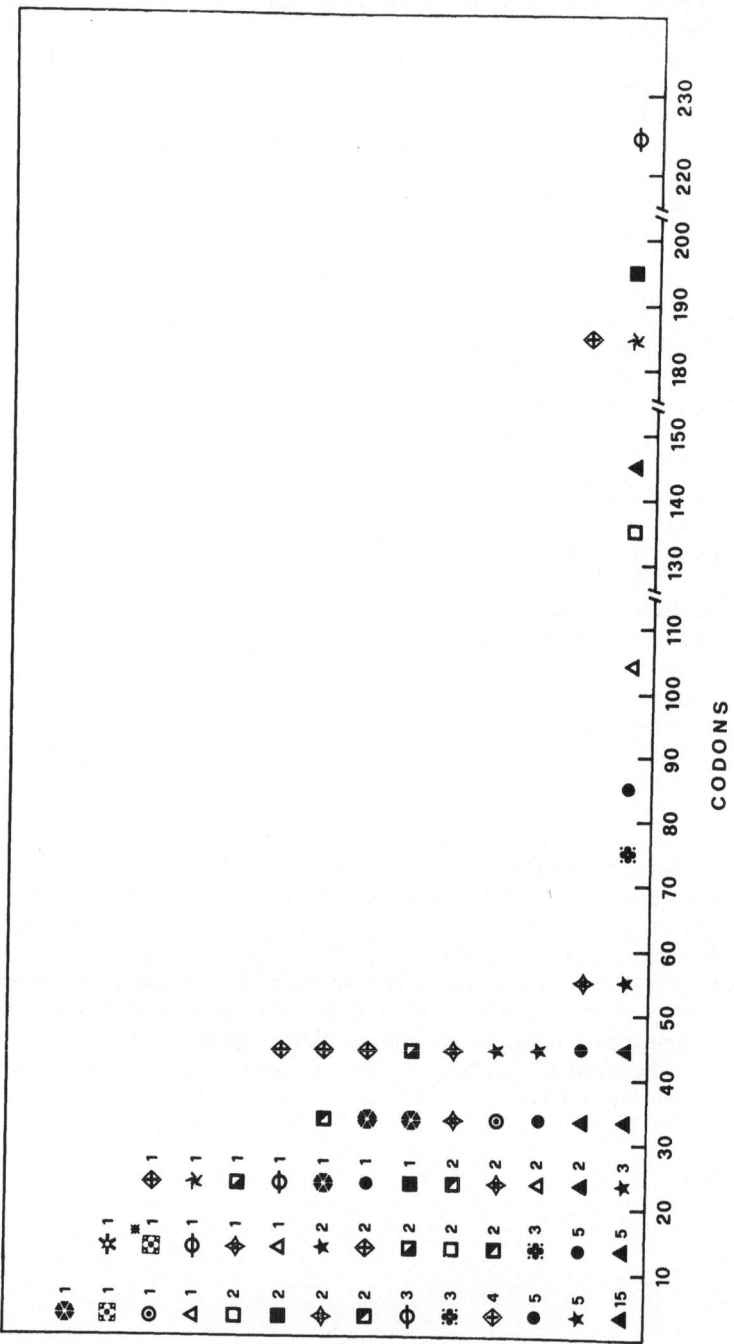

Fig. 2 - Virtual gene length distribution as found by the computer
   code in the complementary strands. Complementary strand
   means the DNA strand sequence obtained from the coding strand
   sequence of a given gene by the complementation rules (A-T
   and G-C). In the case of the ε-globin sequence, a possible
   c.i.p. gene initiation codon could exist beyond the recorded
   one, because of an indeterminate base. If this base is not
   A, the c.i.p. gene length would be 222 codons. Symbols used
   to identify a given sequence are the same as in Fig. 1.

The only observed difference is that $F_s > F_{ns}$. However, even this
finding is not surprising.

   The investigated sequences have been specifically chosen, in
fact, as DNA sequences where one DNA strand is indeed the coding
strand for a real protein, with the exception of the 16S ribosomal
RNA gene, so that the "a priori" probability $P^i(1)$ of finding one
gene at least on the ith strand is one. The same quantity for the
complementary strand may be, obviously, considerably smaller. It
can be written, "a priori":

$$F = \frac{1}{M} \sum_{i=1}^{M} \sum_{m=1}^{m^i max} m \, P_{m^i} \tag{1}$$

$$P^i(1) = \sum_{m=1}^{m^i max} P_{m^i} = \begin{array}{l} 1 \text{ (coding strand)} \\ \leq 1 \text{ (complementary strand)} \end{array} \tag{2}$$

where M is the total number of the investigated sequences (M = 12
in this paper). $\sum_{i=1}^{M}$ denotes the sum over all these sequences, and
m is a positive integer. P_{m^i} is the probability of finding m genes
all together in the ith strand, and $m^i max$ is the maximum number of
genes which can be contained "a priori" in the ith sequence. Since
m is positive, from equations (1) and (2) it can easily be shown
that, when $P_m \geq P_{m+1}$ (i.e., a reasonable hypothesis):

$$F_s \geq F_{ns} \tag{3}$$

   Since also this "a priori" statement is verified by the Fortran
code analysis, it is by no means possible at present to conclude that
the complementary strand gene distribution is statistically different
from the coding strand gene distribution. Also the fact that the
complementary distribution seems not to reach a peak at $140 \leq L \leq 150$
codons cannot be applied in order to demonstrate a meaningful difference since, as we said previously, 5 out of the 17 genes analysed
code for proteins (globin) of very close amino acid number.

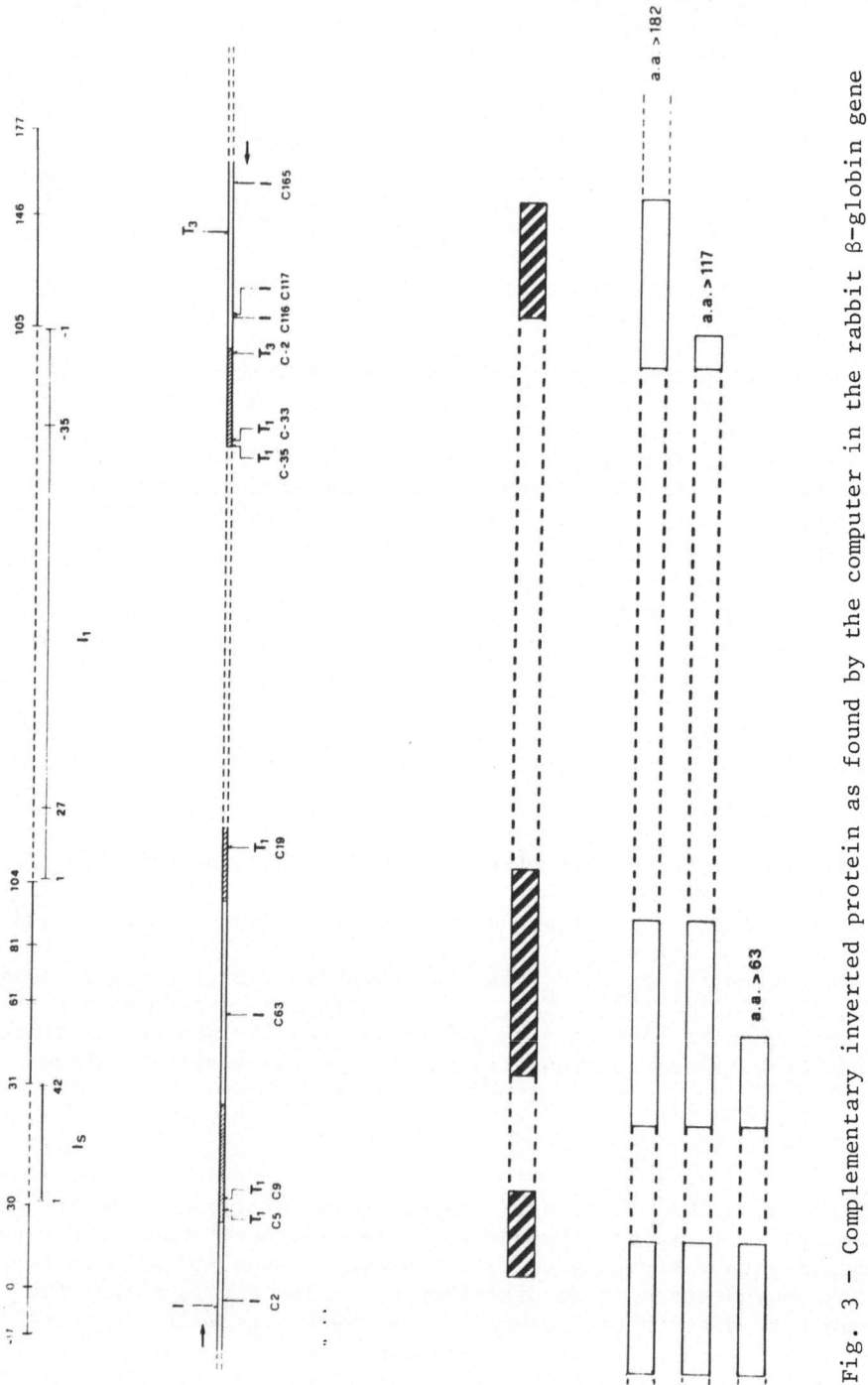

Fig. 3 – Complementary inverted protein as found by the computer in the rabbit β-globin gene sequence

The upper abscissa indicates the β-globin codon numbers. Initiation and terminator are respectively located at zero and 146. The two introns (I_s and I_1) are also included in the analysis[14]. I_s has a codon length of 42 and its sequence is fully known. I_1 sequence was determined only at the two junctions with the exon (translated parts of the β-globin gene); the known sequences have lengths of 27 and 35 codons respectively. The analysis was extended to the untranslated 5' and 3' terminus sequences of the β-globin messenger RNA molecule.

In the middle abscissa, a scheme of the DNA molecule is shown. Horizontal arrows indicate the translation direction on the DNA strands. Known sequences are drawn as a continuous line (▆▆▆▆ coding sequences, ▩▩▩▩ intron sequences) and by a dotted line (⁝⁝⁝⁝⁝⁝) when they are not known.

Initiation (I) and termination (T_1, T_2, and T_3) codons are also shown, along with the corresponding complementary (C) codon position with respect to the upper abscissa. The vertical arrows (↑) indicate the codons on the complementary strand that, with a single base substitution, might become termination codons (pre-termination codons).

At the bottom of the figure, the two proteins, β-globin (▩▩▩▩) and c.i.p. β-globin (▆▆▆▆) are shown. In the latter case, three possible c.i.p. are indicated, depending on which initiation codon is used, along with the corresponding minimum number of amino acids. Dotted lines interrupting protein sequences correspond to the intron sequences. Since terminators are found on the complementary strand of the introns, the c.i.p. sequences are also interrupted.

Excluding intron sequences from our analysis, a c.i.p. product with a codon length of more than 182, containing all amino acids except tyr, was found, according to Materials and Methods. In the figure, two other possible c.i.p. products are also shown, which are present only when other initiation codons are used (C117 and C83). Furthermore, pre-termination codons on the complementary strand are distributed in either the translated part of the β-globin gene or the 5' and 3' untranslated terminus sequences of the β-globin messenger RNA molecule. On the other hand, many termination codons are found in the complementary strand of the intron sequences[14].

These findings lead us to speculate that the farthest initiation codon should be used in determining the length of the c.i.p. gene product and that the intron sequences are intron also for the c.i.p. codon sequences. As a matter of fact, if the complementary intron sequences are taken into account, no c.i.p. genes will be found. We checked this fact also in the case of the ovalbumin gene, for which terminal parts of the intron sequence are known[6].

CONCLUDING REMARKS

On the basis of this analysis, we hypothesize the existence of proteins coded by the complementary strand of some genes. At present we are continuing our research into this subject.

REFERENCES CITED

1. L.J.Korn, C.L.Queen, and M.N.Wegman, Computer analysis of nucleic acid regulatory sequences, Proc.Natl.Acad.Sci. USA, 74:4401 (1977).
2. B.G.Forget et al., personal communication
3. F.C.Kafatos, A.Efstratiadis, B.G.Forget, and S.M.Weissman, Molecular evolution of human and rabbit β-globin mRNAs, Proc. Natl. Acad. Sci. USA, 74:5618 (1977).
4. F.Baralle, personal communication
5. H.C.Heindell, A.Liu, G.V.Paddock, G.M.Studnicka, and W.A.Salser, The primary sequence of rabbit α-globin mRNA, Cell, 15:43 (1978).
6. L.McReynolds, B.W.O'Malley, A.D.Nisbet, J.E.Fothergill, D.Givol, S.Fields, M.Robertson, and G.G.Brownlee, Sequence of chicken ovalbumine mRNA, Nature, 273:723 (1978).
7. W.Fiers, R.Contreras, F.Duerinck, G.Haegeman, D.Iserentant, J.Merregaert, W.Min Jou, F.Molemans, A.Raeymaekers, A.Van den Berghe, G.Volckaert, and M.Ysebaert, Complete nucleotide sequence of bacteriophage MS2 RNA: primary and secondary structure of the replicase gene, Nature, 260:500 (1976).
8. O.Bernard, C.Brack, M.Hirama, N.Hozumi, G.Matthyssesn, and S. Tonegawa, Nucleotide sequence of an embryonic V_λ gene, Basel Institute for Immunology, annual report 1977.
9. A.Ullrich, J.Shine, J.Chirgwin, R.Pictet, E.Tisher, A.J.Rutter, and H.M.Goodman, Rat insulin genes: construction of plasmids containing the coding sequences, Science, 196:1313 (1977).
10. J.Brosius, M.L.Palmer, P.J.Kennedy, and H.F.Noller, Complete nucleotide sequence of a 16S ribosomal RNA gene from Escherichia coli, Proc. Natl. Acad. Sci. USA, 75:4801 (1978).
11. W.Shaffner, G.Kunz, H.Daetwyler, J.Telford, H.O.Smith and M.L. Birnstiel, Genes and spacers of cloned sea urchin histone DNA analyzed by sequencing, Cell, 14:655 (1978).
12. L.A.M.Hensgens, L.A.Grivell, P.Borst, and J.L.Bos, Nucleotide sequence of the mitochondrial structural gene for subunit 9 of yeast ATPase complex, Proc. Natl. Acad. Sci. USA, 74:5618 (1977).
13. G.Macino, G.Coruzzi, F.G.Nobrega, M.Li, and A.Tzagoloff, Use of UGA terminator as a tryptophan codon in yeast mitochondria, Proc. Natl. Acad. Sci. USA, 76:3784 (1979).
14. J.van den Berg, A.van OOyen, N.Mantei, A.Schamböck, G.Grosveld, R.A.Flavell, and C.Weissman, Comparison of cloned rabbit and mouse β-globin genes showing strong evolutionary divergence of two homologous pairs of introns, Nature, 276:37 (1978).

COMPLEX FORMATION BETWEEN 9-AMINOACRIDINE DERIVATIVES AND DNA

A. Zedda

Istituto di Fisica

Università di Sassari

ABSTRACT

In order to evaluate the influence of side-chains on the ability and on the specificity of binding to DNA of 9-Aminoacridines, some physico-chemical features of complexes derived from the interaction of DNA of different sources with 9-Aminoacridine and with N-9 alkyl (A), N-9-ω-hydroxyalkyl (B) and N-9-ω-diethylaminoalkyl - 9 amino-acridines (C) have been examined.

(C) derivatives bind strongly to DNA at sites which appear to be saturated when one dye molecule is bound for every four or five nucleotides; (A) and (B) derivatives bind in a weaker way to DNA at sites which appear to be saturated when one dye molecule is bound for every nine or ten nucleotides. After the primary sites have been filled, a secondary binding process can occur in a stronger way for (A) and (B) derivatives than for (C) derivatives.

The primary binding to DNA is not influenced by the base composition of the DNA itself. The different interaction of 9-Aminoacridine derivatives with DNA is consistent with biological behaviour previously observed.

INTRODUCTION

9-Aminoacridine and a series of derivatives with N-9 aliphatic side chains unsubstituted, or ω-substituted with hydroxy and diethyl-amino groups have been recently submitted to biological essays (1-2).

The ω-diethylaminoalkyl derivatives (C) show a specific activity in preventing the degradation of DNA by pancreatic DNase I and the

induced synthesis of β-galactosidase, both involving d A- d T rich
- regions of DNA.

The alkyl and hydroxialkyl derivatives exhibit a significative
inhibitory activity on E. Coli more marked than that of diethylamino
derivatives. The observed differences in the behaviour of these
series suggested that the nature of side-chain could influence the
way of binding to DNA of the 9-Aminoacridine derivatives.

The experiments described in this report were undertaken with
the aim of characterizing the 9-Aminoacridine derivatives - DNA
complex and in order to gain information on the influence on the mode
of binding to DNA of the side-chain length and of the terminal functional group if present.

The interaction was studied by the spectrophotometric method
employed by Peacocke and Skerrett (3) to investigate the interaction
between proflavine and nucleic acids and depicted by Blake and Peacocke (4) to describe the interaction of aminoacridines with nucleic
acids. This method was particularly suitable for studies on 9-Amino-
acridine derivatives - DNA complex because of the large metachromatic
effect on the spectrum of the dyes.

MATERIALS AND METHODS

DNA from Calf Thymus (G+C = 40%), from E.Coli (G+C = 50%), from Micro-
coccus Lysodeikticus (G+C = 71%) and from Cl. Perfringens (G+C = 31%)
were purchased from the Sigma Chem. Comp.; DNA from B. Subtilis, a
gift from prof. E.Calendi, was prepared according to the method of
Marmur (5). The poly (dA) - poly (dT) and poly (dG) - poly (dC)
homopolymers were purchased from the P.L. Biochemicals Inc.. Stock
solutions of these materials were prepared in SSC-0.01 M at pH = 7.
These solutions were adjusted at DNA phosphorus concentration 1.4X
10^{-3} M (P)as determined spectrophotometrically assuming ε_{259}=6000 (6-
7) for all DNA and poly (dA) - poly (dT) and ε_{253}=7400 (7) for poly
(dG) - poly (dC).
9-Aminoacridine hydrochloride monohydrate of a purity over 98% was
purchased from the Fluka A.G. Buchs S.G.. Generous supplies of 9-
Aminoacridine derivatives were given by prof. G. Cignarella (Istituto
di Chimica Farmaceutica of Sassari University); all derivatives were
tested as the mono (A and B derivatives) of dihydrochlorides (C derivative). The formula of these aminoacridine derivatives and some
of the properties of these compounds which are relevant to this study
are given hereafter (fig. 1 and tab. 1).
9-Aminoacridine and its derivatives were dried to obtain a constant
weight before preparation of solutions in SSC-0.01 M at pH=7 and at
18°-20°C. These solutions obeyed Beer's law at least up to the concentration 5X10^{-5} M at the wavelengths corresponding to the respective
absorption peaks. The concentration of solutions were evaluated
spectrophotometrically assuming the molar extinction coefficients

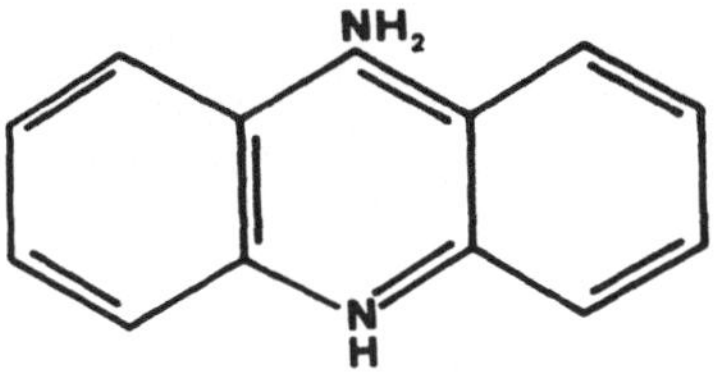

9 – Aminoacridine 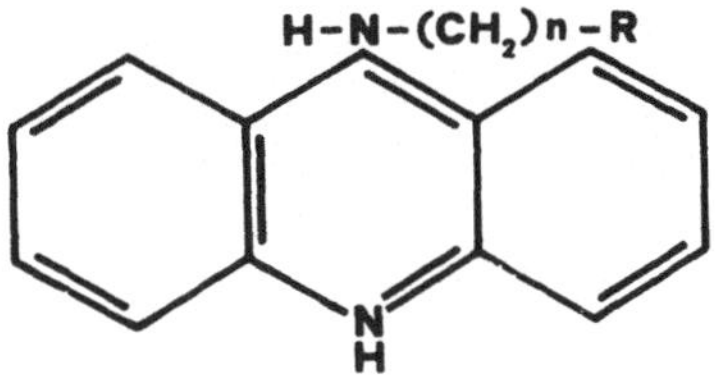   Acridine Derivatives

Fig. 1 – Chemical structure of 9-Aminoacridine and its derivatives

determined experimentally (see tab. 1).
The technique used in binding studies was that of Peacocke and
Skerrett (3). Ligand stock solution at concentration 9×10^{-5} M and
DNA stock solution at concentration 1.4×10^{-3} M(P) in SSC-0.01 M at
pH=7 were prepared at room temperature ($18°-20°C$).
A 1/1 mixture between the aforesaid solutions was prepared and diluted
with a solution of the drug alone at the same concentration as in the
mixture ($C = 4.5 \times 10^{-5}$ M); in this way measurements at the chosen
wavelengths were made on solutions containing a fixed drug concentration and decreasing nucleic acid concentrations.
All mixtures were prepared in duplicate; fresh solutions were prepared and the measurements repeated if the resulting optical densities
differed for more than three units in the third significant figure.
Therefore, the optical densities had an accuracy of 0.6%. Readings
on solutions containing DNA were made with reference to a blank
containing the same concentration of DNA in buffer.

TABLE   I

Compound	n	R	Form	M.W.	λ max in nm	$M^{-1} \varepsilon \ cm^{-1}$
A- 2	2	CH_3	mono-HCl	290.77	413	10,900
A- 3	3	CH_3	mono-HCl	304.79	413	10.700
B- 2	2	OH	mono-HCl	292.76	414	11.300
B- 3	3	OH	mono-HCl	306.22	414	11.700
B- 4	4	OH	mono-HCl	320.22	414	11.200
C- 2	2	$N(C_2H_5)_2$	di-HCl	384.32	416	9.200
C- 3	3	$N(C_2H_5)_2$	di-HCl	398.12	415	9.900
C- 4	4	$N(C_2H_5)_2$	di-HCl	412.12	414	10.100

9- Aminoacridine derivatives.

All the absorbance measurements were carried out and recorded at room
temperature (18°-20°C) in the range 470-390 nm by a Beckman mod. 25
spectrophotometer in standard quartz cuvettes with a light path of
1 cm. The range of concentration of Aminoacridine derivatives was
such to meet two requirements: to obey Beer's law and to exploit the
maximum instrument sensitivity.
Within the DNA-ligand concentration ranges used in these experiments,
no precipitation of the complex occurred, and the rate of formation
of complex appeared to be very rapid. Readings obtained a few minutes
after mixing solutions of the 9-Aminoacridine derivatives and DNA
were found to be unchanged when repeated at frequent intervals over
the following 24 hr.
In practice, solutions were allowed to stand at room temperature for
at least 10 min. before measurements were made.

RESULTS AND DISCUSSION

a) Treatment of results

When the sites on nucleic acids which bind a ligand are of a single
type and behave independently of each other, the binding process can
be described in simple mass-action terms, giving:

$$K = r/(n-r) \cdot C_f \qquad \text{or} \qquad r/C_f = Kn - Kr \qquad (1)$$

where K is the association constant, r is the number of the ligand
molecules bound per nucleotide, n is the number of binding sites per
nucleotide and C_f is the molar concentration of the free ligand,
not bound, in solution. The functions r and n are expressed in terms
of the nucleotide as the unit of nucleic acid. Concentrations of
nucleic acids are expressed in terms of molarity with respect to
nucleotide phosphorus.
According to this treatment, a plot of r/C_f versus r is a straight
line with the same slope of K and the intercept on the r-axis equal
to n. In the present experiments, values of r and C_f were calculated
from spectrophotometric measurements of the fraction α_b of 9-Amino-
acridine derivatives in a complexed form in solutions containing
known amounts of ligand and nucleic acid.
The spectral shift associated with the binding of 9-Aminoacridine
derivatives to nucleic acids and the effect of increasing concentrations of DNA on the absorption spectrum of derivatives are shown in
fig. 2.
The peak can be seen to shift progressively towards a limit which
represents the spectrum of the ligand in a fully complexed form.
All the curves pass through an isosbestic point, indicating that
they result from the contributions of two forms of 9-Aminoacridine
derivatives, free and bound, each having a characteristic absorption
spectrum.
The wavelength chosen for the measurement was the one at which the
largest change in optical density occurred on binding. Values of

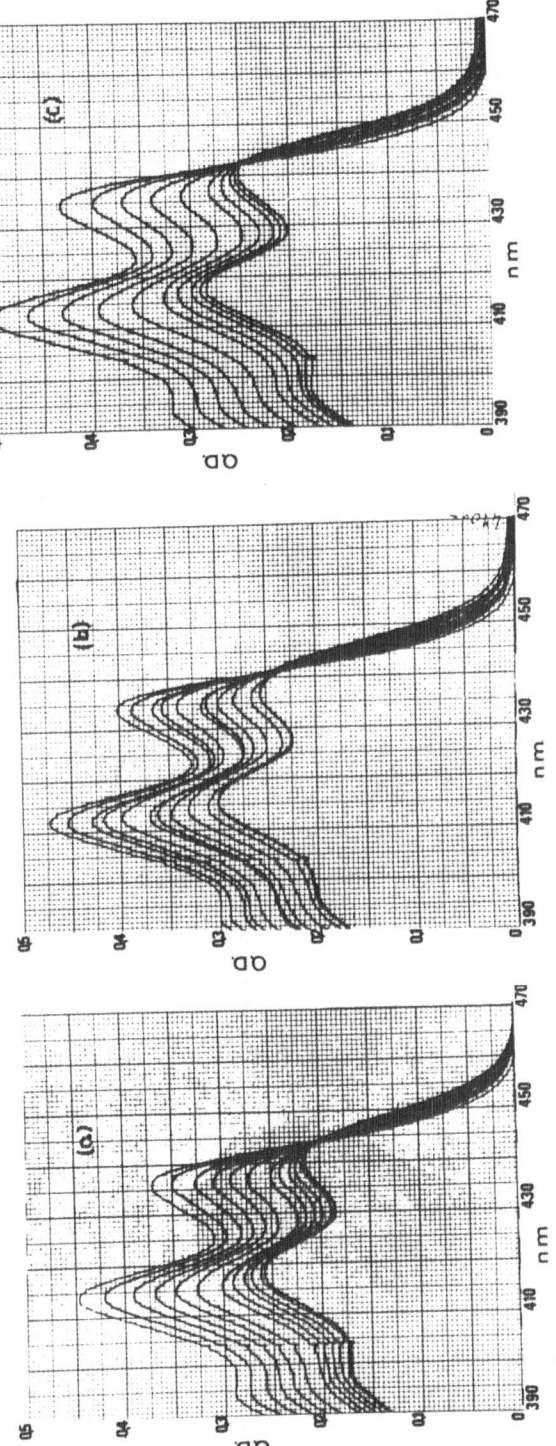

Fig. 2 – Absorption spectra resulting from the interaction between 9-Aminoacridine derivatives and DNA.

(a) C-3 and calf-thymus DNA. $C_T = 4.6 \times 10^{-5}$ M; pH=7; $\mu=0.1$ M

(b) A-3 and calf-thymus DNA. $C_T = 4.5 \times 10^{-5}$ M; pH=7; $\mu=0.01$ M

(c) C-3 and poly(dA)-poly(dT). $C_T = 4.5 \times 10^{-5}$ M; pH=7; $\mu=0.01$ M

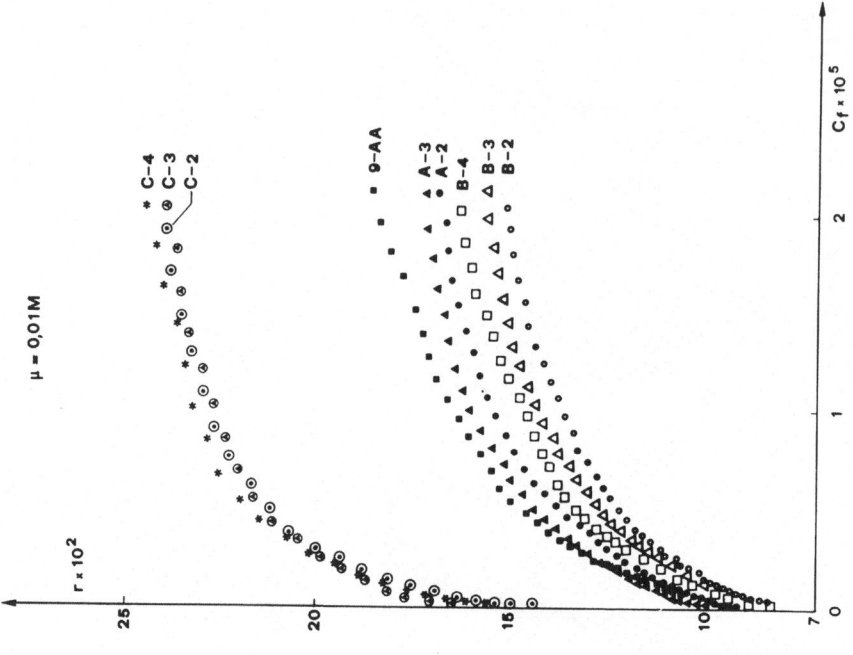

Fig. 4 – Binding of 9-Aminoacridine and its
derivatives on calf-thymus DNA.
μ=0.01 M; pH=7; temp. 20°C.

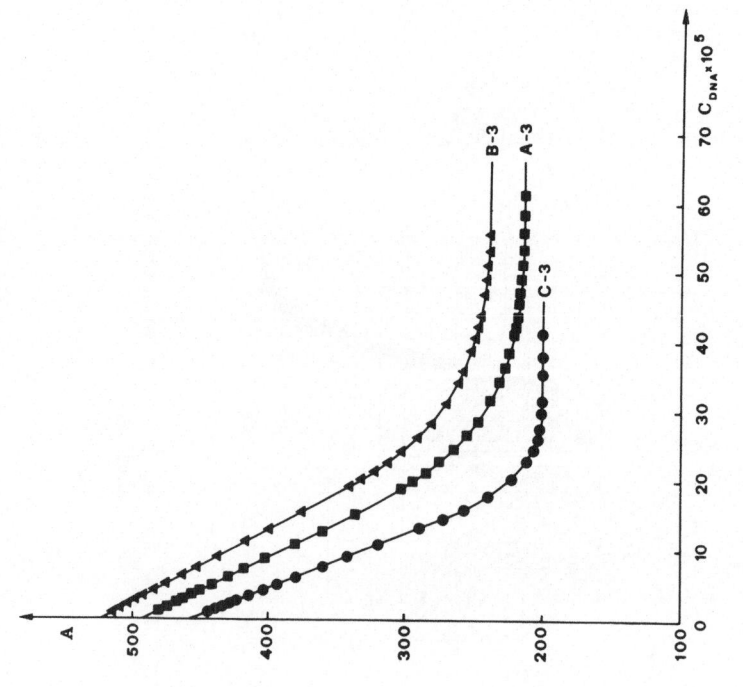

Fig. 3 – Absorbancy of A-3, B-3 and C-3
derivatives in the presence of
calf-thymus DNA. C_T=4.5x10^{-5}M;
pH=7; μ=0.01 M; light path=1cm.

TABLE II

	Cl. Perffrin. A+T = 72%	Calf Thymus A+T = 60%	B. Subtilis A+T = 58%	E. Coli A+T = 50%	Microc. Lys. A+T = 28%	poly (dA) poly (dT)	poly (dG) poly (dC)
9-AA	K=3.3x10^7 n=0.15	K=4.3x10^7 n=0.10	---	K=3.2x10^7 n=0.12	K=4.3x10^7 n=0.11	---	---
A-2	K=3.8x10^7 n=0.12	K=4.4x10^7 n=0.10	K=5.0x10^7 n=0.09	K=3.3x10^7 n=0.10	K=4.2x10^7 n=0.10	K=5.3x10^7 n=0.10	---
A-3	K=3.0x10^7 n=0.13	K=4.3x10^7 n=0.11	---	K=3.4x10^7 n=0.10	K=4.4x10^7 n=0.10	---	---
B-2	K=4.0x10^7 n=0.11	K=5.8x10^7 n=0.9	---	K=5.9x10^7 n=0.10	K=7.8x10^7 n=0.09	---	---
B-3	K=3.4x10^7 n=0.12	K=6.5x10^7 n=0.09	K=8.9x10^7 n=0.09	K=5.7x10^7 n=0.10	K=8.0x10^7 n=0.09	K=7.0x10^7 n=0.09	---
B-4	K=3.5x10^7 n=0.12	K=6.1x10^7 n=0.09	---	K=5.7x10^7 n=0.10	K=7.8x10^7 n=0.10	---	---
C-2	K=1.6x10^7 n=0.20	K=2.3x10^7 n=0.18	---	K=2.7x10^7 n=0.20	K=2.9x10^7 n=0.17	---	---
C-3	K=1.7x10^7 n=0.21	K=2.4x10^7 n=0.19	K=1.5x10^7 n=0.16	K=2.8x10^7 n=0.20	K=3.4x10^7 n=0.16	K=2.3x10^7 n=0.19	K=1.6x10^7 n=0.20
C-4	K=1.6x10^7 n=0.21	K=2.4x10^7 n=0.18	---	K=3.2x10^7 n=0.19	K=2.8x10^7 n=0.17	---	---

Parameters of binding of 9-Aminoacridine derivatives to DNA (pH=7; SSC 0.01 M; temp. 20°C).

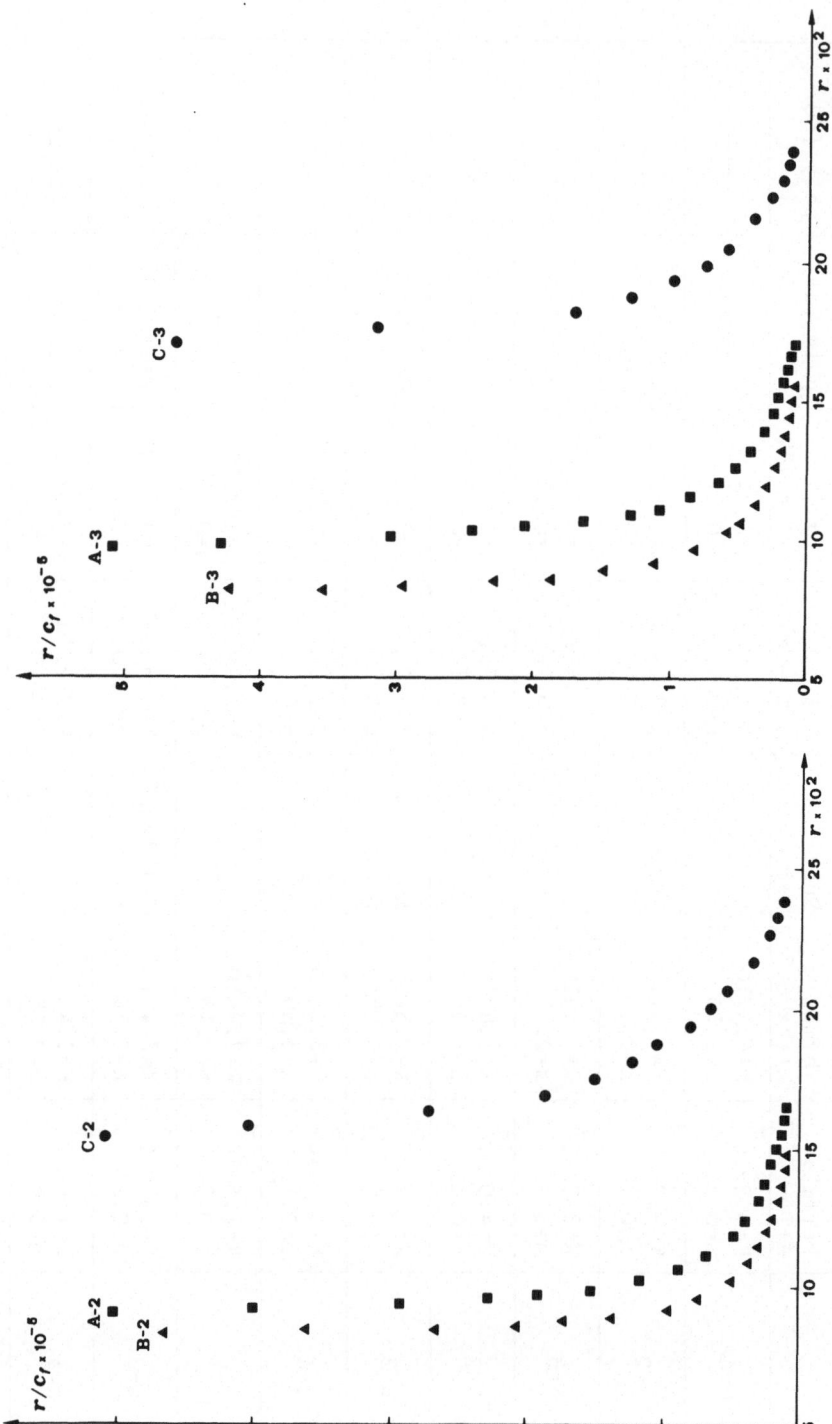

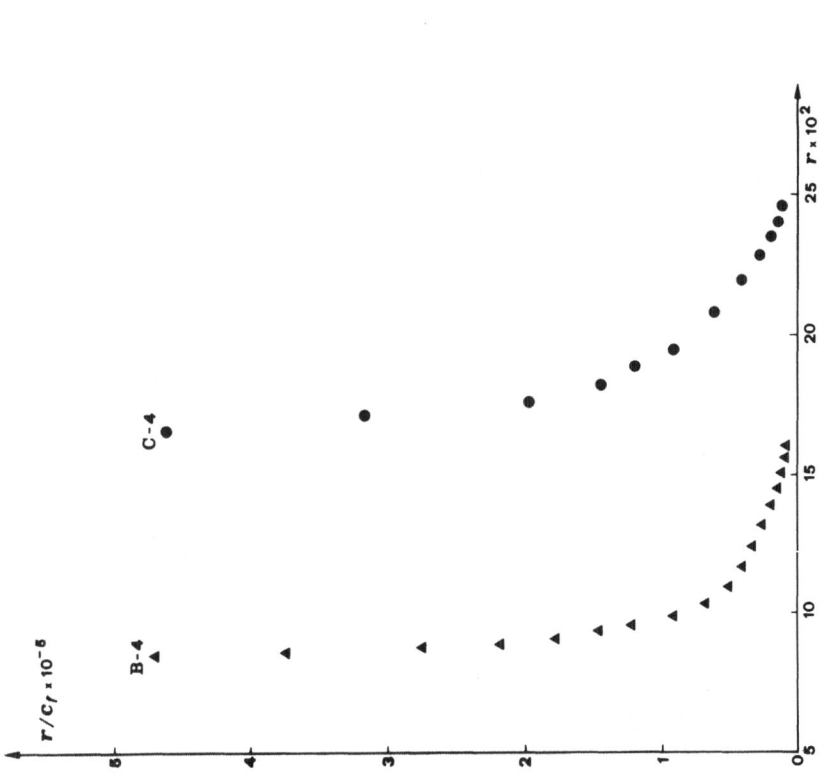

Fig. 5 – Scatchard's plots for 9-Aminoacridine derivatives in the presence of calf-thymus DNA. μ = 0.01 M; pH = 7; temp. 20°C.

α_b calculated from the optical density measurements were independent
of the wavelength over a range of 5 nm about the value chosen for
each derivative

b) Binding of 9-Aminoacridine derivatives to DNA

Absorbancy of derivatives in the presence of calf-thymus DNA are
shown in fig. 3.
Binding curves resulting from the interaction of acridines and DNA
are shown in fig. 4.
It can be seen that the binding to all types of DNA show the same
behaviour, i.e. it rose rapidly to an r value of about 0.18 for 9-
Aminoacridine, of about 0.17 for A derivatives, of about 0.16 for
B derivatives, of about 0.24 for C derivatives, and it kept constant
after this binding ratio was reached. Results show that the binding
ability order to DNA in the presence of a given concentration of
free ligand is: B-2 < B-3 < B-4 < A-2 < A-3 < 9-AA < C2=C3=C4.
Plots r/C_f versus r for tested DNA and Acridine derivatives are
included in fig. 5.
These plots are not straight lines, but the experimental points,
in the region r < 10 for A and B derivatives and in the region
r < 20 for C derivatives, can be fitted by a straight line which
provides K and n values for the complexes investigated (tab. 2).
The influence of DNA base-composition on the binding of 9-Aminoacri-
dine derivatives was investigated using DNA having A-T contents
ranging from 70% to 28%.
Clearly there were no significant differences between the values
given by the various DNA preparations, and no correlation with the
varying base-composition was evident. It is therefore improbable
that one particular type of base is specifically involved in the
binding of 9-Aminoacridine derivatives to DNA.
The significance of the curvature in these plots has generally been
explained (3) by postulating the existence of secondary sites which
the ligand can bind with a lower affinity; such secondary binding
would become significant after a certain fraction of the primary
sites had been filled, depending upon the relative affinity of the
ligand for the two types of site.
The data for 9-Aminoacridine derivatives (fig. 5) suggests that
secondary interaction with DNA does not contribute greatly to the
total level of binding until r values higher than about 0.10, for
A and B derivatives, and about 0.18, for C derivatives, are reached.
The curvature may be due whether to the secondary binding, or to a
heterogeneity among the primary binding sites (8) or even to repulsive
nearest-neighbour interactions between Acridine derivatives bound to
DNA. The present results do not permit any definite conclusions to
be drawn about these possible explanations.

DISCUSSION

The interaction between 9-Aminoacridine derivatives and DNA resembles those observed with other aminoacridines (4-9). The red--shift which occurs in the complex should be attributed to the coupling of the aminoacridine heterocyclic rings with the DNA bases. Among the series of derivatives studied, marked differences were observed in the binding to the primary sites of DNA. Thus, A and B compounds appeared to reach the saturation of the primary sites when one dye molecule was bound for every nine or ten nucleotides whereas C compounds appeared to reach the same saturation when one dye molecule was bound for every four or five nucleotides.

Since the above reported values did not change on varying the chain length in homogeneous series, one could infer that is the terminal group which determines the number of the occupied primary sites. However, the length of the side-chain though uneffective in determining the n and K values, appears to influence the r values as indicated by increase of r on increasing the chain length at equal C_f values (fig. 4).

Carrying out experiments with DNA of different sources and with poly (dA) - poly (dT) and poly (dG) - poly (dC) any significative variation could be observed in the number of primary sites occupied, thus suggesting that intercalation was not influenced by the A-T/G-C ratio.

This result was somewhat unexpected on the light of biological experiments aiming at evaluating the inhibitory activity of 9-Aminoacridine derivatives on the enzymatic degradation of calf-thymus DNA by DNase I and on the induced synthesis of β-galactosidase. Being known that these events involve A-T rich region of DNA, the finding that they were presented by C derivatives only had been interpreted as due to a preferential affinity of the later derivatives for A-T sequences. However, the biological hypothesis would not contrast with the present results if the inhibitory activity of C compounds were due to the "efficacy" of the binding to A-T sequences rather than to a preferential affinity for the same sequences.

ACKNOWLEDGEMENTS

I wish to thank Prof. G. Cignarella for the gift of the 9-Aminoacridine derivatives and for helpful advice. I also wish to thank Prof. E. Calendi for valuable discussion and suggestions in connection with this work.

REFERENCES

(1) Calendi,E., Bassu,M.C., Congiu,A.M., Fogu,G. and Muglia,M.A. (1975) - Giorn. Microbiol. 23-95.

(2) Calendi,E., Cignarella,G., Congiu,A.M. and Bassu,M.C., (1976)
     Giorn. Microbiol. 24-103.
(3) Peacocke,A.R. and Skerrett,J.H.N. (1956) - Trans, Faraday Soc.
     52-261.
(4) Blake,A. and Peacocke,A.R. (1968) - Biopolymers 6-1225
(5) Marmur,J. (1961) - J. Molec. Biol. 3-208.
(6) Chargaff,E. in "The Nucleic Acids" - vol. I, E.Chargaff and
     J.N. Davidson Ed. - Academic Press,Inc., New York 1955, pg. 307.
(7) Wells,R.D., Larsen,J.E., Grant,R.C., Shortle,B.E. and Cantor,C.R.
     (1970) - J. Mol. Biol. 54-465.
(8) Tubbs,R.K., Ditmars,W.E., and Van Winkle,Q. (1964) - J. Mol.
     Biol. 9-545.
(9) Drummond,D.S., Simpson-Gildmeister,F.W. and Peacocke,A.R.
     (1965) - Biopolymers 3-135.

VARIATIONS OF DNA PHYSICO-CHEMICAL PARAMETERS IN ITS INTERACTIONS

WITH MUTAGENIC AND/OR CARCINOGENIC COMPOUNDS (+)

Gianni Tamino and Lucia Peretta

Istituto di Biologia animale dell'Università di Padova

via Loredan, 10 - 35100 Padova

(+) Supported by a grant from the National Research Council of Italy
(Consiglio Nazionale delle Ricerche, Programma Finalizzato
"Promozione della Qualità dell'Ambiente").

Abbreviations: A, absorbance; BSS, Hank's balanced salt solution;
max, maximal wavelength; min, minimal wavelength; O.D., optical
density; poly A, polyriboadenylic acid; poly C, polyribocytidylic
acid; poly G, polyriboguanidylic acid; poly U, polyribouridylic
acid; UV, ultraviolet rays.

SUMMARY

Mugagens and/or carcinogens (genotoxics) mainly act by altering
DNA structure binding themselves to the nitrogen bases or to phosphate
residues, causing a break in the nucleic acid chain or inserting
themselves between the bases.

These structural alterations determine variations in the physico-
-chemical parameters of DNA solution, as ultraviolet light absorption,
fluorescence, thermal stability, viscosity, pH, alcaline elution etc.

The use of instruments and methods capable of detecting the
variations in the physico-chemical parameters allow us to predict
the potential genotoxicity of the chemicals under study on the basis
of their capacity to interact with DNA, and to investigate their
mechanism of action.

In this paper some physico-chemical techniques are applied to
the study of the interaction of DNA with substances whose genotoxic

action is already established or only suspected.

INTRODUCTION

Most mutagenic agents are able to cause directly or indirectly DNA alterations detectable by physico-chemical methods (1,2). On the other hand, such alterations must affect a certain number of nucleotides per DNA molecule, even though a change in one nucleotide can cause a mutation. Different physico-chemical methods exist capable of revealing at least one or two alterations every 4,000 nucleotides, which can convert "in vitro" alterations of the mutagenic type (e.g. alkylation of the bases) into modifications of another type (e.g. breakage of the DNA chain), more simple to analyse. The physico-chemical modifications can be seen not only in purified DNA treated "in vitro" with a mutagenic agent (in the presence of metabolic activation system, if that be the case), but also in DNA, purified from cells treated "in vivo", so that a more direct evaluation of the biological relevance of the observed alterations can be obtained. Thus the physico-chemical methods of DNA analysis can be useful not only to collect data about the possible mutagenic activity of a compound but also about its mechanism of action (chain breakages, cross-links, base modifications, intercalations, etc.).

On the other hand the great number of new compounds synthetized each year by chemical and pharmaceutical industries requires rapid and cheap tests to detect their possible genotoxic (namely mutagenic and/or carcinogenic) action. On the basis of the data referred by other authors and our own experience, we believe that the joint-application of some simple physico-chemical methods provides a good preliminary approach to the study of potential genotoxicity, even though a final answer can be given only by "in vivo" experiments, especially on mammals.

Table 1 gives a first partial list of such rapid and relatively cheap methods, sufficient to detect any kind of DNA alteration. Our aim has been to make a parallel use of some of these methods (UV-spectrophotometry, thermal stability spectrophotometrically determined, viscosity analysis) to study the modifications induced in DNA by known or suspected genotoxic agents, as Cr^{III} and Cr^{VI} salts, mitomycin C, and methyl-methan-sulphonate (MMS).

MATERIALS AND METHODS

Cells

Cultures of the established BHK 21 hamster cell line, clone 12, were routinely grown as monolayers in Eagle's minimal essential medium supplemented with 10% calf serum at 37°C, as detailed elsewhere (3).

TABLE 1

SOME PHYSICO-CHEMICAL METHODS FOR DETECTING CHANGES IN THE MACRO-
MOLECULAR STRUCTURE OF DNA INDUCED BY GENOTOXIC COMPOUNDS.

Effects on DNA	Method of detection
1) Chain breaks (or labile sites)	U.V.Spectrophotometry (hyperchromic effect); Thermal stability (decreased re-naturation of denatured DNA); Viscosity (lowered intrinsic viscosity); Sedimentation velocity (decreased sedimentation rate); Electron microscopy (less chain length); Alkaline elution (labelled DNA fragments pass through the filter);
2) Cross-links	Thermal stability (less hyperchromicity during denaturation and easier renaturation); Alkaline elution (after X rays treatment fewer fragments pass through the filter);
3) Stabilizing (cross-links, "sandwich" links or binding to the phosphate groups) or destabilizing actions (chelation between a base and a phosphate group) on the double helix structure	Thermal stability (decreased or increased hyperchromicity, higher or lower melting temperature and greater or lower renaturation rate, respectively)
4) Intercalation of compounds	Viscosity (increase of intrinsic viscosity); X rays diffraction (changes in crystallographic parameters); Spectrophotometry (absorption spectra of compounds interacting with DNA are modified);
5) Base alteration	U.V.Spectrophotometry (shift of DNA absorption spectra and decrease of the bases extinction).

DNA and polynucleotides

DNA was extracted from BHK 21 cells according to the method
described by Kirby (4). Synthetic polynucleotides of the Sigma Chemical Company were used. DNA and synthetic polynucleotides were
solubilized in 0.15 M NaCl.

Treatment with genotoxic compounds

Treatment "in vitro" - Potassium dichromate ($K_2Cr_2O_7$ Mallinc-
root), Chromium chloride ($CrCl_3 \cdot 6H_2O$, Merck), Mitomycin C (Sigma C.C.)
and MMS (methyl-methansulphonate, Merck) were used: 1) the chromium
compounds were solubilized in bidistilled water at the concentration
of 10^{-1} M immediately before use; 2) Mitomycin C was prepared at
100 µg/ml in bidistilled water then chemically reduced (5) and
immediately used; 3) MMS 1mg/ml was dissolved in bidistilled water
and immediately used. All the compounds so dissolved were then
brought to the desired dilution with nucleic acids to obtain a ratio
Compound used/DNA Phosphorus = 1/1 in molar concentration. The
samples were analyzed within one day.

Treatment "in vivo" - The above mentioned compounds were added
to culture medium (chromium salts) or to Hanks balanced salt solution
+ Hepes (Mit.C and MMS) in such a way to obtain the following final
concentration: Potassium dichromate 10^{-3} M, chromic chloride 10^{-2} M,
Mitomycin C 10^{-5} M and MMS 10^{-3} M. The cells were treated for one
hour at 37°C, then washed with Hanks BSS and utilized for the extraction of DNA. Controls, lacking the compound in examination,
were treated in the same way.

UV Analysis

All the readings were determined with a mod. 124 Hitachi-Perkin
Elmer double beam spectrophotometer. The solutions of nucleic acids
or their derivates were brought to a concentration between 0.3 and
0.5 O.D. at 260 nm and their spectra were recorded between 320 and
220 nm to determine λ max, λ min and the A max/A max - A min ratio,
where A is the absorbance.

Thermal stability

In the same double beam spectrophotometer, the nucleic acids
at concentrations of 30 µg/ml, ionic strength 0.15 and pH 6.5, were
denatured in quartz cuvettes with water circulating from an ultra-
thermostat (Lauda Comp.) raising the temperature from 25°C to 100°C
(10°C every 6'). Nucleic acids were then partially renatured by
slow cooling (15' every 10°C) to 50°C.

On the basis of denaturation and renaturation patterns we determined the hyperchromicity (A at 100°C/ A at 25°C), the melting

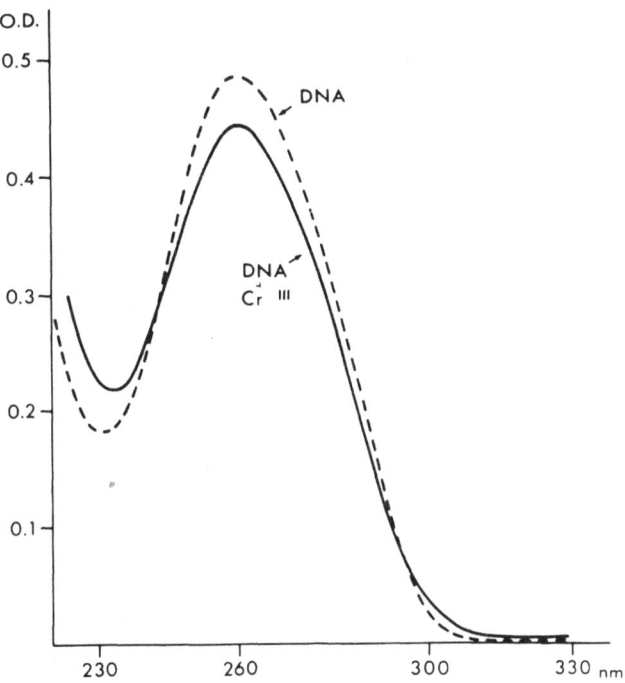

Fig. 1 - U.V. absorption spectra of control DNA and DNA after
treatment with $CrCl_3$ (Cr/P 1/1; NaCl 0.15 M).

temperature (Tm = temperature determining 50% Hyperchromicity), the
degree of renaturation (ratio between the decrease in optical density
during renaturation between 100° and 50°C and the increase in optical
density during denaturation between 25° and 100°C).

Viscosity

A Ubelhode viscometer was used with a range of shearing between
200 and 500 s^{-1} for solutions of DNA of PM $\simeq 10^6$ Daltons, at decreasing concentrations from 70 to 7 µg/ml to determine the intrinsic
viscosity (η).

TABLE   2

U.V. Absorption spectra variations of DNA after treatment with $CrCl_3$.

	Cr/P	$\dfrac{A\ \lambda_{max}}{A\ \lambda_{max} - A\ \lambda_{min}}$	λ_{max}(nm)	λ_{min}(nm)
DNA NaCl 0.5 M	0	1.76	258	227
" " " ε	1/5	1.82	258	230
" " "	1/1	1.84	258	230

RESULTS

UV analysis

        As an example of UV analysis of the interaction between DNA and the presumably genotoxic compounds we describe the experiments carried out with chromium chloride ($CrCl_3 \cdot 6H_2O$), a trivalent chromium compound, and potassium dichromate ($K_2Cr_2O_7$), an hexavalent chromium compound.

        The genotoxic compounds that bind to the bases modify their aromatic ring and therefore the UV absorption spectrum of DNA, while those that bind to phosphate residues modify the absorbance maximum (6) at the most. The data obtained for the "in vitro" treatment of DNA with chromium chloride are reported in figure 1 and table 2: O.D. decreases in correspondence to the λ max and increases in correspondence to the λ min. There is also a red shift of the λ min. Analogous behaviour was observed in DNA extracted from cells treated in culture with $CrCl_3$ compared to the control. The adding of hexavalent chromium induces no variation in the spectrum.

        Table 3 shows the effects observed treating "in vitro" synthetic polynucleotides with $CrCl_3$: poly G and, to a lesser extent, poly A behave practically as DNA, while poly U is only slightly affected, and poly C reacts in a completely different fashion, namely with an increase of λ max absorption and a decrease of λ min.

        On the basis of these data it is possible to exclude a direct link between DNA and hexavalent chromium and to infer that trivalent chromium binds to the bases (alteration of the spectrum) and to

TABLE   3

U.V. absorption spectra variations of synthetic polynucleotides
after treatment with $CrCl_3$.

	Cr/P	$\dfrac{A\ \lambda_{max}}{A\ \lambda_{max} - A\ \lambda_{min}}$	λ_{max} (nm)	λ_{min} (nm)
POLY A	0	1.57	251	229
	1/2	1.79	253	230
	1/1	1.78	254	231
POLY C	0	1.30	275	242
	1/2	1.27	274	242
	1/1	1.20	273	242
POLY G	0	1.31	251	234
	1/2	1.48	252	232
	1/1	1.63	253	231
POLY U	0	1.36	260	230
	1/2	1.39	259	230
	1/1	1.43	259	230

phosphate residues (modification of O.D. at the λ max level) and
that such an interaction depends on the primary structure of DNA
(greater effect on guanine rich sites).

Thermal stability

The study of denaturation and renaturation patterns of DNA is
an useful tool to analyze the stabilizing and destabilizing actions
of metal ions (6), the formation of cross links (7) or single strand
breaks (8).

We examined three parameters: hyperchromicity, melting temperature (Tm) and renaturation degree, in DNA treated with potassium
dichromate, chromium chloride, Mitomycin C and methylmethansulphonate
(MMS).

The data are shown in table 4, and compared to the control
values. Potassium dichromate acts differently "in vivo" and "in

TABLE   4

Variations of the thermal stability parameters compared to the controls.

COMPOUND AND TREATMENT	HYPERCHROMICITY (% variation)	ΔTm (°)	RENATURATION RATE (% variation)
$K_2Cr_2O_7$			
a) "in vivo" $10^{-3}M$	$-$ 3%	$\simeq$	$-$ 30%
b) "in vitro" Cr/P 1/1	$-$ 30%	$-10°$	$-$ 45%
$CrCl_3 \cdot 6H_2O$			
a) "in vivo" $10^{-2}M$	$+$ 5.5%	$-$ 5°	$-$ 10%
b) "in vitro" Cr/P 1/1	$+$ 30%	$-23°$	precipitate
Mitomycin C			
a) "in vivo" $10^{-5}M$	$-$ 18%	$\simeq$	$+$ 5%
b) "in vitro" M/P 1/1	$-$ 22%	$\simeq$	$+$ 30%
MMS			
a) "in vivo" $10^{-3}M$	$-$ 6%	$-$ 2°	$-$ 10%
b) "in vitro" M/P 1/1	$-$ 11%	$\simeq$	$-$ 30%

(°) ΔTm = Treated Tm $-$ Control Tm.

vitro", probably because in the presence of reducing organic molecules Cr^{VI} is converted to Cr^{III}, with parallel formation of oxidizing substances, e.g. epoxides (9), which may be the actual responsibles of the observed genotoxic effect. In any case, both the oxidizing substances produced intracellularly and the dichromate directly added to DNA cause a reduced level of renaturation and less hyperchromicity, signs of breakage along the polynucleotide chain. The decrease observed "in vitro" may be due to Cr^{III} formed by reduction of Cr^{VI}. Chromium chloride 'in vivo" acts at higher concentrations than dichromate and produces very different effects, minimally altering the level of renaturation, lowering the Tm and raising the hyperchromicity. All these effects can be referred to a destabilizing action on the DNA molecule due to Cr^{III} chelation to a base site and a negatively charged phosphate group. Cr^{III} can also interact with DNA producing "sandwiches" between guanine and guanine of the same strand or cross-links, as was observed using polynucleotides rich in G-C (10); but in our experimental conditions the action of Cr^{III} on DNA is mainly due to destabilizing bonds.

Mitomycin C is known to produce stable cross-links at high temperatures (7) in G-C rich sites of DNA, this effect was observed also in our experiments, where it accounts for the quicker renaturation of treated DNA compared to the control. Moreover, the lesser degree of hyperchromicity "in vivo" as well as "in vitro" can be attributed to G-C rich sites that do not denaturate, being "cross--linked".

Methyl-methan-sulphonate (MMS) is a monofunctional alkylating agent producing alkylation of the bases (mainly guanine and adenine) and therefore apurinic sites that are easily transformed to single strand breakage (1). Our data obtained with "in vivo" treatments indicate: 1) a modest reduction of the hyperchromicity due to limited fragmentation of the molecule (fragmented DNA causes a slight increase of optical density even at room temperature, therefore the increase of O.D. during denaturation is lower); 2) a slight lowering of the Tm, probably caused by methylated bases or apurinic sites, (i.e. weakness of hydrogen bonds that held together the double helix); 3) a slightly lowered degree of renaturation, imputable to single strand breakages.

In "in vitro" treatments Tm is unchanged, while the lowering of hyperchromicity and degree of renaturation are more marked, probably because these experimental conditions are more drastic, inducing more breakages along the polynucleotide chain, paralleled by the removal of altered bases and the immediate transformation of apurinic sites into breakages.

Viscosity

In large molecules as DNA, variations in viscosity are caused mainly by changes in molecular weight and modifications of the linear structure of the molecule. Compounds that break the polynucleotide chain lower its viscosity (11), compounds able to elongate the molecule (intercalating agents) increase it (12), and a decrease of viscosity can be caused also by substances that favor the folding of the chain thus enhancing its globular shape (6).

The compounds used by us always produced, in "in vitro" treatments, a decrease in viscosity (table 5) compared to the controls. For dichromate and MMS such a decrease can be attributed to breakages along the chain resulting in its fragmentation with reduction of molecular weight; chromium chloride, instead, exerting its action on the nitrogen bases probably makes the molecule more flexible,

TABLE   5

Per cent variations of intrinsic viscosity compared to the controls.

COMPOUND AND TREATMENT	VISCOSITY (% variation)
$K_2Cr_2O_7$	
a) "in vivo" $10^{-3}M$	- 20
b) "in vitro" Cr/P 1/1	- 26
$CrCl_3 \cdot 6H_2O$	
a) "in vivo" $10^{-2}M$	+ 9
b) "in vitro" Cr/P 1/1	- 24
Mitomycin C	
a) "in vivo" $10^{-5}M$	+ 15
b) "in vitro" M/P 1/1	- 30
MMS	
a) "in vivo" $10^{-3}M$	$\approx$
b) "in vitro" M/P 1/1	- 7

stabilizing, via the formation of cross-links, a less bulking steric
form. Finally, mitomycin C is certainly able to stabilize with
cross-links the less bulking structures produced by alkylation of
the bases and depurination of the double helix.

When the same compounds are used to treat cell cultures and the
modifications induced on cell DNA are analyzed, following its extraction, the following effects can be observed (table 5): 1) dichromate
reduces DNA viscosity through the intracellular formation of the
above mentioned oxidizing agents, which can break the chain; 2) no
alteration is detected after treatment with MMS, probably because
the brakages, inferred on the basis of the data of thermal stability,
are too few to sufficiently modify DNA viscosity; 3) chromium chloride and mitomycin C, on the contrary, increase DNA viscosity in
treated cells, so that links between DNA molecules (a known effect
of Cr^{III} at high concentrations, given the formation of precipitates
(13)) or links between DNA and proteins (already described for mitomycin C (5)) can be hypothesized.

DISCUSSION AND CONCLUSIONS

Our experiments show that the joint application of some physico-
-chemical techniques makes possible to discriminate the actions of
very different compounds (Cr^{III}, a metal ion; Cr^{VI}, a strong oxidizing agent; Mitomycin C, a bifunctional alkylating agent producing
cross-links; MMS, a monofunctional alkylating agent).

In the case of potassium dichromate we did not observe modifications of UV absorption spectra (which means that no direct bond
to DNA bases is formed), while the effect on thermal stability differed "in vivo" and "in vitro", in both cases suggesting a break
down of the molecule, supported by viscosity variations, too. These
data are in agreement with the mutagenic (14) and carcinogenic (15)
action of Cr^{VI} compounds and strengthen the hypothesis that Cr^{VI}
effect on the cells is mediated by oxidizing molecules produced by
its intracellular reduction to Cr^{III}(9).

From our experiments chromium chloride turns out to interact,
if it enters the cells (this takes place only at rather high Cr^{III}
concentrations (16)), both with the bases (UV absorption spectrum
variations) and phosphate groups (chelating effect which destabilizes
the double helix) being also able to originate cross-links between
DNA strands or molecules, as the increase of viscosity observed in
"in vivo" treatments suggests. It looks therefore possible that
Cr^{III} compounds can have a genotoxic action, when present at high
concentration or accumulated inside the cells, even if up to now
epidemiological data and studies on experimental animals do not
confirm such hypothesis.

As for mitomycin C, whose action on DNA has been widely studied
and whose mutagenic and carcinogenic activity is well known (17),
our experiments confirm its effects on the bases (greater flexibility
of DNA), the induction of cross-links (renaturation data) and DNA-
-protein links (greater viscosity after "in vivo" treatments).
For methyl-methan-sulphonate besides confirming its already known
actions (that is methylation of the bases, formation of labile apurinic sites and therefore intrastrand breakage (see data of thermal
stability and viscosity) that are at the basis of its genotoxicity
(18), we can add that in our experimental conditions the breakage
formation is more evident in direct treatment of DNA than adding
the compound to cell cultures.

In our opinion our data can be taken as an example of how these
physico-chemical methods can be usefully adopted to carry out a first
level screening on substances of suspected genotoxic action.

REFERENCES

1. B.S.Strauss, Physical-chemical methods for the detection of the
    effect of mutagens on DNA. In "Chemical Mutagens" edit. by
    A.Hollaender, vol. 1 (1971) 145.
2. P.Brookes and P.D.Laweley, Effects on DNA: Chemical Methods.
    In "Chemical Mutagens" edit. by A. Hollander, vol. 1 (1971) 121.
3. A.G.Levis, V.Bianchi, G.Tamino and B.Pegoraro, Cytotoxic effects
    of hexavalent and trivalent chromium compounds on mammalian
    cells in vitro. Br. J. Cancer. 37 (1978) 386.
4. K.Kirby, A new method for the isolation of deoxyribonucleic
    acid: Evidence on the nature of bonds between deoxyribonucleic
    acid and protein, Biochem. J. 66 (1957) 495.
5. V.N.Iyer and W.Szybalski, Mitomycin and porfiromycin: chemical
    mechanism of action and cross-linking of DNA, Science, 164
    (1964) 55.
6. I.Sissoëff, J.Grisvard and E.Guillè, Studies on metal ions-DNA
    interactions: specific behaviour of reiterative DNA sequences,
    Prog. Biophys. Molec. Biol. 31 (1976) 165.
7. V.N.Iyer and W.Sqybalski, A molecular mechanism of mitomycin
    action: linking of complementary DNA strands, Proc. Natl.
    Acad. Sci. USA, 50 (1963) 355.
8. E.P.Geiduschek, On the factors controlling the reversibility
    of DNA denaturation, J. Mol. Biol., 4 (1962) 467.
9. R.Schoental, Chromium carcinogenesis, formation of epoxy-alde-
    hydes and tanning, Br. J. Cancer, 32 (1975) 403.
10. G.Tamino, unpublished data.
11. H.Massie and B.Zimm, The use of hot phenol in preparing DNA,
    Proc. Natl. Acad. Sci. USA, 54 (1965) 1641.
12. F.Zunino, R.Gambetta, A.Di Marco, A.Velcich, A.Zaccaria, F.Qua-
    drifoglio and V.Crescenzi, The interaction of adriamycin and
    its β anomer with DNA, Bioch. Bioph. Acta, 476 (1977) 38.
13. A.Danchin, Labeling of biological macromolecules with covalent
    analogs of magnesium. II. Features of the chromic Cr(III) ion,
    Biochimie, 57 (1975) 875.
14. S.Venitt and L.S.Levy, Mutagenicity of chromates in bacteria
    and its relevance to chromate carcinogenesis, Nature (Lond.)
    250 (1974) 493.
15. C.Maltoni, Occupational carcinogenesis, Excerpta Med. Int.
    Congr. Ser., 322 (1974) 19.
16. G.Tamino, Interactions of chromium with nucleic acids of mam-
    malian cells, Atti Ass. Genet. It., 22 (1977) 69.
17. International Agency for Research on Cancer, Mitomycin C, in
    Monographs on the Evaluation of Carcinogenic Risk of Chemicals
    to Man. Vol. Lo I.A.R.C., Lyon (1976) p. 171.
18. International Agency for Research on Cancer, Methyl methansulpho-
    nate, in Monographs on the Evaluation of Carcinogenic Risk of
    Chemicals to Man. Vol. 7, I.A.R.C., Lyon (1974) p. 253.

COMPARTMENTALIZATION IN THE CONTROL OF RNA AND PROTEIN SYNTHESIS

DURING NUTRITIONAL SHIFT-DOWN IN THE YEAST Saccharomyces cerevisiae

F. Tassi, C. Donnini, C. Bruschi, P.P. Puglisi, and N. Marmiroli

Institute of Genetics, University of Parma

INTRODUCTORY REMARK

RNA and protein synthesis play a key role in the cells since they are involved in the informational flow in the living matter, and a relevant amount of available molecules and energy is channeled to the two macromolecular syntheses. Therefore it is important to analyze how the cells regulate the rate of RNA and protein synthesis under conditions of energy limitation or precursors starvation, two situations that occur rather frequently in the life cycle of the cells.

In bacteria it has been discovered that a control system exists which links the protein to RNA synthesis (for a review, see[1,2]).

The dependency of bacterial RNA synthesis on protein synthesis which has been called "stringent control" has been demonstrated to take place under conditions such as aminoacid-starvation[3], nitrogen starvation,[4,5,6], phosphate and Mg^{++} starvation[7], energy deprivation [4,8,9] and in strains mutated in amino-acyl-t-RNA synthetases[10]. In all these conditions the synthesis of r-RNA and t-RNA is inhibited; as far as the m-RNAs are concerned, some m-RNAs are inhibited, whereas the synthesis of others is enhanced[11].

During the stringent response, the 'unusual' nucleotide ppGpp is synthetized and accumulation of pppGpp[12] takes place (for a review, see[13]). Since ppGpp is in vivo and in vitro an inhibitor of bacterial RNA polymerase[14], a relationship between ppGpp and the stringent phenotype has been proposed[15]. Mutated bacteria have been isolated in which RNA synthesis escapes from protein synthesis (RC^{rel})[1,2,16]. These mutants do not synthetize ppGpp and do not accumulate pppGpp[17] and, making use of these RC^{rel} strains, it was possible to demonstrate that the synthesis of the former depends on the relA gene[15], the degradation of the latter on the spoT gene[18,19] allowing the conclusion that the stringent control system is genetically determined. On the other hand, a relaxed condition can be achieved when wild type stringent bacteria (RC^{str}) are starved for aminoacids, and exposed to inhibitors of translation[20,21]: this induced relaxation in genotypically 'stringent' bacteria has been called 'phenotypic relaxation'.

The relaxed/stringent control (rel/str) appears therefore a system for the adjustment of RNA contents under particular environmental conditions, on which depends not only the total amount of relaxed RNA synthetized, but also the specific type of RNA whose synthesis is enhanced or blocked.

The regulatory role of t-RNA, the participation of r-RNA in ensuring the proper flow through the different informational macromolecules, the fact that only some m-RNAs are synthetized under the 'rel' condition allows us to propose that the rel/str system has to be taken into account as a regulatory system which could act at different levels in the control of gene expression in eukaryotic organisms. In particular, in eukaryotic cells, the anisomorphism of the cytoplasm in the course of cell division could introduce some environmental conditions around the system for the phenotypic expression of the genetic information, that could lead to a specific balance of different types of RNA molecules through a mechanism similar to the one operating in the rel/str control system.

However, three features of the eukaryotic cells must be taken into account, as far as the study of relationship between RNA and protein synthesis is concerned:

a) the compartmentalized synthesis of both protein[22,23,24] and RNA[25,26];
b) the non-coordinated synthesis of RNA due to different and specific polymerizing enzymes[25,27] (for a review, see[28]);
c) the different pattern of sensitivity to antibiotics of mitochondrial and cytosol protein[23,24,29] and RNA synthesis[25,27].

Therefore we have compared the properties and features of both nuclear and mitochondrial RNA and protein synthesis during the cell division and in shift-down nutritional conditions to analyze:

i) the properties of the compartmental systems for rel/str control, and
ii) the properties of the environment-dependent variation of their biosynthetic interlock.

PROTEIN SYNTHESIS DURING NUTRITIONAL SHIFT-DOWN

Total and mitochondrial protein synthesis show during aminoacid starvation a similar trend and time course of inactivation; however, protein synthesis that residues in aminoacid-starved cells, shows an enhanced sensitivity to Erythromycin (ERY) in comparison with growing cells. On the other hand the aminoacid starvation of Erythromycin--resistant cytoplasmic mutants (ERY[R]) does not change the in vivo Erythromycin sensitivity of the total protein synthesis, suggesting that cytosol protein synthesis of aminoacid-starved cells acquired a property, the sensitivity to this drug, that under normal cultural conditions is restricted to the mitochondrial compartment.

A typical and well established example of nutritional shift-down associated with microdifferentiative effects (the switch off of the mitotic division and the switch on of the meiotic process) is the transfer from synthetic growth medium to acetate sporulation medium of the genetically competent heterozygous α/a diploids. In these cells, but not in a/a and in α/α diploids, premeiotic DNA synthesis, nuclear division and ascospore formation (for a review, see[30]) occur in the acetate medium. Under this condition of nutritional shift--down the global protein synthesis of α/a diploids is sensitive to the same extent (90%) to Cycloheximide (CHI) as well as to ERY, in agreement with the fact that both the inhibitors block the sporulation process. The ERY sensitivity of total protein synthesis is not observed when α/α and a/a diploids are transferred into sporulation medium, suggesting that the abnormal sensitivity of global protein synthesis to ERY should depend on the genetic constitution at the mating type locus or from the capacity of this type of cell[31], to achieve the commitment to sporulation, during the nutritional shift--down. The sporulation-deficient temperature sensitive (ts) mutant spo 11, has a global protein synthesis that is not blocked at the non-permissive temperature, allowing therefore the analysis of the ERY-sensitivity of global protein synthesis at the permissive (25°C) and non permissive (34°C) temperatures. At both temperatures, ERY blocks global protein synthesis as it occurs in SPO[+] cells, suggesting that spo 11 gene does not interfere with the acquisition of the ERY--sensitivity, although it blocks the sporulation in the cells shifted to the sporulation medium. This leads to the conclusion that the capacity or incapacity to sporulate in this condition is not the cause of the abnormal sensitivity to the drug.

For what concerne the mechanism of the acquisition of the sensitivity at ribosomal level to non-legitimate inhibitors (i.e. ERY for total protein synthesis and CHI for the mitochondrial one) during nutritional shift-down, it has been reported that in bacteria aminoacid starvation interferes with the ribosome assembling[8].

It could be assumed therefore as a working hypothesis that aminoacid starvation involves in yeasts ribosomal disassembling

which leads the particles to show the sensitivity to non-legitimate
inhibitors we have reported. The system for protein synthesis could
share therefore the sensitivity to non legitimate inhibitors, as a
consequence of this new ribosomal processing.

RNA SYNTHESIS DURING NUTRITIONAL SHIFT-DOWN

In aminoacid-starved yeasts cells, total RNA synthesis (i.e.
mitochondrial plus cytoplasmic RNA synthesis) is arrested with protein
synthesis[32,33]. This fact indicates that between yeast RNA and protein syntheses a relationship exists of the type described as 'stringent control' in bacteria. However, while the bacterial stringent
response affects t-RNAs, r-RNAs and some classes of m-RNAs, it has
been found that in yeast only r-RNAs are stringently controlled,
whereas t-RNAs synthesis[34] and m-RNAs[35] synthesis are only slightly
reduced.

As far as the effects of aminoacid starvation on the compartment-
-specific RNA synthesis are concerned, quite controversial data have
been obtained. In the yeast Saccharomyces cerevisiae strain A364A,
tyrosine starvation[36] inhibits the synthesis of mitochondrial RNA,
indicating that the molecules are stringently controlled. However,
in Saccharomyces cerevisiae strain 5428/5c, methionine starvation
gives rise to a partially different figure in mitochondrial RNA behavior with respect to the block of protein synthesis. In fact, the
amount of RNA per mg of proteins synthetized under methionine starvation in this strain, is about the same as in the control mitochondrial preparation derived from non-starved cells. However, the
sedimentation pattern in alkaline sucrose gradient of this RNA, shows
that the peak corresponding to the lighter type of mitochondrial
r-RNA is missing while a peak with intermediate density appears.

The existence of a phenotypic relaxation in aminoacid-starved
cells has been shown with the inhibitor of chain elongation CHI.
Addition, in vivo, of CHI increases total RNA in starved cells.
However, the labelled uracile incorporated is not polymerized into
r-RNAs molecules but in t-RNAs molecules, the only RNA whole synthesis
is stimulated in this condition[34].

The behaviour of another yeast strain (R7/6a) indicates that a
strain-dependance of the phenotypic relaxation exists; in fact aminoacid-starved R7/6a cells show a residual RNA synthesis that is partially sensitive to CHI, indicating that in this strain RNA synthesis
is under stringent control but that at variance with the result obtained with A364A, CHI is ineffective in inducing RNA relaxation.

The addition of ERY to the aminoacid-starved CHI-treated cells
of the strain R7/6a determines a further reduction in RNA synthesis.
The previously reported observation that protein synthesis of aminoacid starved cells had an abnormal sensitivity to ERY could help in

explaining this behavior. In fact, the inhibitory effect of ERY on
RNA synthesis could lead to a further inhibition of total protein
synthesis by the macrolide antibiotic, additional to the one due to
CHI.

Other controlled conditions of metabolic deprivation, such as
glucose and ammonia starvation, have been reported to affect RNA
synthesis.

The transfer of cells from an high to a low concentration of
carbon sources gives rise to a reduction of RNA synthesis[34,36].
However, CHI appears to reverse this effect at least in the strain
R7/6a.

Nitrogen deprivation obtained withdrawing ammonia or reducing
the concentration of the yeast extract into the growth medium determines the block of the synthesis of both t-RNA and r-RNA, but the
synthesis of t-RNAs only can be _relaxed_ by the addition of CHI[34,37].

The concomitant deprivation of nitrogen and glucose determines
a decrease in RNA synthesis which is counteracted by CHI. As concerned with the genetic control of this phenomenon, it has been observed that Erythromycin also resumes RNA synthesis in the strain
carrying the allele a but not α at the mating type locus. More
compelling evidences about the role of the mating type genes on the
interlock between protein and RNA synthesis have been found studying
the pattern of RNA synthesis in α/α, a/a and α/a diploids transferred
into the sporulation medium[31]. When glucose grown α/α, a/a and α/a
diploids are transferred to acetate sporulation medium, RNA synthesis
rate is reduced to 10% of the control within 6-8 hours: however, if
CHI is added to the cells in the sporulation medium, RNA synthesis
is stimulated in all the strains. This indicates that the _phenotypic
relaxation_ induced by CHI is independent from the genetic constitution
at the mating type locus and is unconnected with the capacity to
sporulate. However, if ERY is added to the sporulation medium, RNA
synthesis is stimulated in α/a but not in α/α and a/a strains. Therefore, it could be concluded that ERY triggers in the sporulation
medium a _phenotypic relaxation specific for the α/a genetic consti-
tution_. The role of these two types of phenotypic relaxation that
occur in the acetate medium during the sporulation process, has been
proved with the sporulation deficient ts mutant spo 11. In fact,
at the non-permissive temperature spo 11 maintains the CHI-induced
phenotypic relaxation but loses the ERY-induced stimulation of RNA
synthesis. Therefore two types of _phenotypic relaxation_' take place
during the transfer into the sporulation medium and determine the
commitment to sporulation. The first one is triggered by CHI, depends
on the specific chemicals present in the medium and is independent
from the genetic constitution of the strains at the mating type locus.
The second one is triggered by ERY, depends on the genetic constitution of the strains at the mating type locus and in addition it

appears to be connected with the capacity of the strains to sporulate, i.e. on the α/a SPO⁺ genetic constitution.

MODEL FOR REGULATION OF RNA SYNTHESIS IN COMPARTMENTALIZED SYSTEM

A - Nuclear compartment

A feature of the phenotypic relaxation of RNA's synthesis when protein synthesis is inhibited by CHI is the increase of t-RNA synthesis. Some hypotheses must be proposed therefore to explain the selection that the relaxed phenotype exerts on the synthesis of the different cellular RNAs.

A general statement must be made: the inhibition of protein synthesis is per se not a condition which leads to RNA relaxation.

Inhibitors of late chain initiation (Verrucarin 76), elongation (CHI, Cryptoleurine, Anisomycin) and termination (Trichodermin) of protein synthesis, relax RNA synthesis, whereas inhibitors of early chain initiation (Edeine A+B and MDMP (2-(4-methyl-2-6-dinitroanilino)- -N-methylpropionamide)) are ineffective as relaxing agents[34].

These data demonstrate therefore that specific targets of the inhibitors of protein synthesis are involved in the relaxation phenomenon.

Since r-RNAs synthesis is inhibited in aminoacid starved cells both in the absence and in the presence of CHI, whereas t-RNAs synthesis is stimulated in the presence of CHI, this indicates that among the molecules involved in protein synthesis, whose pool depends on the specific step inhibited by the different drugs, must be searched the regulatory signals for RNA synthesis regulation (Figure 1)

Let us consider the fact that during aminoacid starvation t-RNAs in their non amino acylated forms prevail upon the acylated t-RNAs, since protein synthesis has been driven to consume the latter. In this condition, when inhibitors of protein synthesis, whose target is in the biosynthetic steps after the formation of m-RNA-80 S ribosome complex, are added to the starved cells they prevent residual translation promoting a trickle-charging at the ribosome level that results in a sparing of charged t-RNA molecules.

A working hypothesis therefore could be assumed in which charged t-RNAs accumulated during the block of protein synthesis do not inhibit RNA polymerase specific for the t-RNA genes, which is, on the contrary, inhibited by the uncharged t-RNAs prevailing during aminoacid starvation.

According to this model, in starved cells that still possess a residual protein synthesis, uncharged t-RNAs prevail upon charged t-RNAs leading to the decrease of the main component of the RNAs pool, i.e. the r-RNA.

On the other hand, in the presence of inhibitors of protein synthesis, t-RNA molecules whose synthesis is slightly affected in aminoacid-starved cells, increase, reaching a pool-size which inhibits r-RNA synthesis. In this condition, protein synthesis inhibition is connected to RNA synthesis through the inhibitory effect on r-RNA synthesis of the accumulated uncharged t-RNA molecules.

B - Mitochondrial compartment

In the mitochondrial particle the regulation of RNA synthesis seems to take place through effector molecules which differ from those hypothesized as active in the regulation of nuclear RNA. In fact unusual nucleotides seem to have a regulatory role as it occurs in prokaryotic cells, whose polymerizing system closely resembles the mitochondrial one at least in terms of the action mechanisms of several antibiotics and drugs.

Aminoacid starved cells do not produce ppGpp and the same results were obtained in an amino-acyl t-RNA synthetase thermosensitive mutant grown at the non perimissive temperature[38]; on the other hand, in vitro experiments with mitochondrial ribosomes and the stringent factor of E.coli indicated that the production of ppGpp[13] could take place in the subcellular particle. It could be hypothesized that a basic difference between the cytosol and the mitochondrial compartments in the regulation of RNA synthesis is therefore that the mitochondrial particle only is endowed with the capacity to synthetize this nucleotide. The observations that inhibition of MPS with Tetracyclin (TET) blocks the production of ppGpp[39] and that the inhibition by Chloramphenicol (CAP) of MPS does not affect mitochondrial r-RNA synthesis[36] (as it occurs in CAP-treated bacteria) are in agreement with this hypothesis.

It could be suggested that mitochondrial synthesis of all types of stable RNAs is coordinately regulated and that this property depends on the fact that a single form of mitochondrial RNA polymerase is deputed to the synthesis of the different types of mitochondrial RNAs.

CONCLUDING REMARKS

In yeasts, RNA and protein synthesis are functionally linked, but the compartmentalized nature of the eukaryotic cells imposes some compartment-specific features to the RNA/protein interlock which occurs during nutritional shift-down.

In particular it appears that, in the yeast cells, in one of the

PROTEIN SYNTHESIS

RNA: SYNTHESIS: PROFILES EXAMPLE

(a)

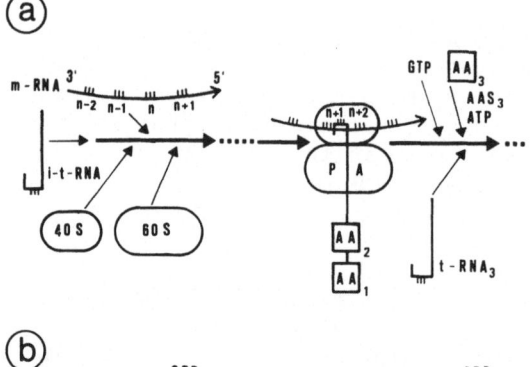
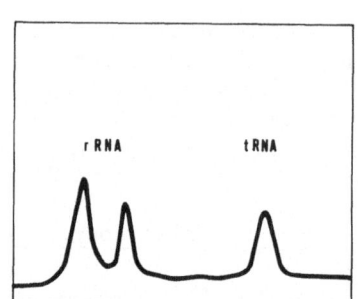

(b)

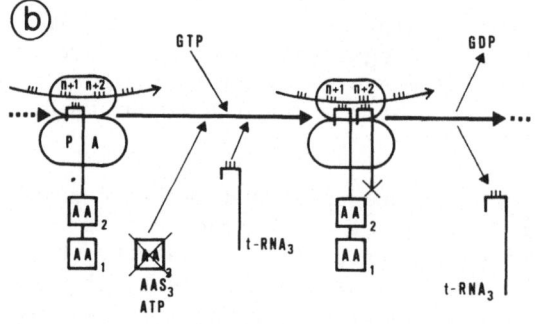
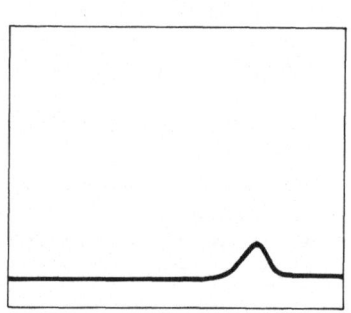

(c)

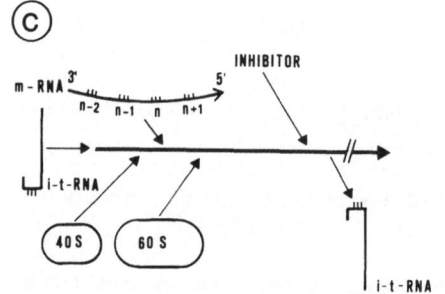
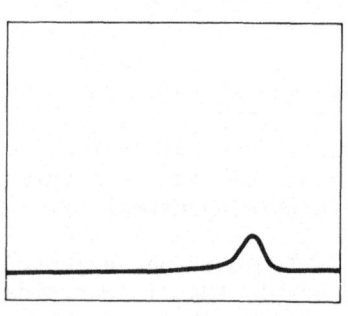

(d)

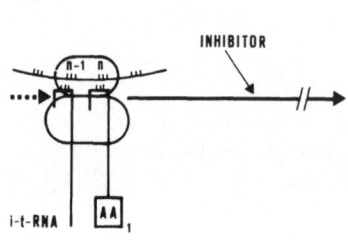
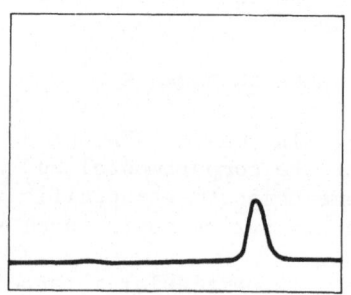

Fig. 1 <u>a-b-c-d</u>. In figure is illustrated as example what can occur
  to r-RNA and t-RNA synthesis when the cells are:
  a) in growth conditions; b) starved for an aminoacid;
  c) treated with an inhibitor of early initiation and
  d) treated with inhibitors of initiation, elongation and
      termination.
  The symbols utilized indicate: AA = aminoacid; AAS = amino-
  acyl synthetase; ATP = adenosinetriphosphate; GDP = guano-
  sinediphosphate; GTP = guanosinetriphosphate; 40 S and 60 S =
  the two ribosomal subunits of eukaryotic 80 S ribosome; P and
  A = peptidyl and aminoacyl sites on 60 S ribosomal subunit.
  The height of the RNAs peaks indicates the amount of RNA
  synthetized, the width among the various peaks the sedimentation values.

a - the different steps of cytoplasmic protein synthesis normally
    lead to peptide bound formation and to the releasing of uncharged
    t-RNA. It has been assumed that initiation does not need formil-
    -t-RNA, but a generic initiator i-t-RNA and that aminoacylation
    of t-RNA occurs after the formation of m-RNA ribosome t-RNA
    complex[42,43].

b - the absence of a given aminoacid (in the case indicated as AA3)
    arrests protein synthesis and the acylation of t-RNA 3 that al-
    ready have reached the A site on the ribosome. This results in
    a liberation of uncharged t-RNA that accumulates in the cyto-
    plasm and blocks r-RNA and partially t-RNA synthesis.

c - treatment with an inhibitor of early initiation prevents the
    formation of the t-RNA-m-RNA-ribosome complex and the acylation
    of t-RNA: therefore increases the number of uncharged t-RNAs
    present in the cytoplasm. Also this condition affects r-RNA
    synthesis and partially t-RNA.

d - the treatment with an inhibitor of initiation (but the same
    should occur blocking elongation and termination) cannot prevent
    the acylation of t-RNA that occurs a step before. However r-RNA
    synthesis remains inhibited whereas t-RNA synthesis returns to
    the normal values.

two compartments (the cytosol) the synthesis of rRNA, tRNA and mRNA
is uncoordinated, wheras in the other one (the mitochondria) all the
species of RNA are coordinatively synthetized.

The regulation of RNA synthesis that takes place in sporulating
cells represents a paradigmatic example of the role that the connection between protein and RNA synthesis has in order to finalize
the cell physiology to specific environmental modifications. In
fact, the sporulation-specific genetic program encoded in SPO genes
seems someway involved in this type of control: a modification in
the sporulating capacity reflects in fact the regulation of the
synthesis of RNA which is linked to protein synthesis. This fact
demonstrates that epigenetic factors can have a role in the expression
of the nuclear genetic information.

Mitochondria could be the connection through which environmental
effectors not active on the nucleus or on the cytosol could influence
the regulation of total RNA synthesis.

In fact it has been shown that during the nutritional shift-down
accompanying sporulation in yeasts, regulation of total RNA synthesis
rests on the interlock between the mitochondrial and cytoplasmic
protein synthesis.

Since intracellular and extracellular chemical changes occur
during the growth, it could be suggested that the nucleo-mitochondrial
interlock we have discussed before could be of some importance also
in modulating the growth-rate to the changing composition of the
surrounding milieu through the connection that exists between RNA
synthesis and protein synthesis activity[40].

This methodological approach could be undertaken not only to
analyze the differences existing between compartments of the same
cell but also to enter into the detail of the differences existing
in basic regulatory mechanisms in different species. A similar
approach to the taxonomy based on the physiogenetic analysis of
regulatory mechanisms has been reported[41]. The combination of the
two types of analysis could provide a series of molecular details
whose importance appears of increasing interest for taxonomy as
well as for cellular modelling.

REFERENCES

1. G.Edlin and P.Broda, Physiology and genetics of the 'ribonucleic
    acid control' locus in Escherichia coli, Bacteriol. Rev. 32:
    206 (1968).
2. A.N.Ryan and E.Borek, The relaxed control phenomenon, in "Progr.
    Nucl.Ac.Res.and Mol.Biol.", J.N.Davidson and W.E.Cohn, eds.,
    Academic Press, New York and London (1971).

3. A.B.Pardee and L.S.Prestidge, The dependence of nucleic acid synthesis on the presence of aminoacids in Escherichia coli, J. Bacteriol. 71:677 (1956).

4. F.Neidhart, Properties of a bacterial mutant lacking aminoacid control of RNA synthesis, Biochem. Biophys. Acta 68:365 (1963).

5. J.D.Irr, Control of nucleotide metabolism and ribosomal RNA synthesis during Nitrogen starvation of E.coli, Journal of Bacteriol. 110:554 (1972).

6. B.H.Fakher and D.Schlessinger, Synthesis and break-down of ribonucleic acid in E.coli starving for nitrogen, Biochem. Biophys. Acta 119:183 (1966).

7. H.Maruyama and D.Mizuno, Changes in base sequence of ribosomal RNA during degradation induced by phosphate and magnesium starvation, Biochem. Biophys. Acta 199:166 (1970).

8. R.Kaplan and D.Apirion, The fate of ribosomes in E.coli cells starved for a carbon source, J. Biol. Chem., 250:1854 (1975).

9. K.C.Westover and L.A.Jacobson, Translation and functional inactivation of m-RNA after energy source shift-down, J. Biol. Chem., 249:6279 (1974).

10. A.Bock, L.E.Faiman and F.C.Neidhart, Biochemical and genetic characterization of a mutant of E.coli with a temperature sensitive valyl ribonucleic acid synthetase, J. Bacteriol. 92:1076 (1966).

11. G.Reiness, H.L.Yang, G.Zubay and M.Cashel, Effects of guanosine tetraphosphate on cell free synthesis of E.coli ribosomal RNA and other gene products, Proc.Natl.Acad. U.S.A. 72:2881 (1975).

12. M.Cashel and J.Gallant, Two compounds implicated in the function of the RC gene in E.coli, Nature 221:838 (1969).

13. M.Cashel, Regulation of bacterial ppGpp and pppGpp, Ann. Rev. Microbiol. 29:301 (1975).

14. A.Travers, Modulation of RNA polymerase specificity by ppGpp, Molec.gen.Genet. 147:225 (1976).

15. M.Cashel and B.Kalbacher, The control of ribonucleic synthesis in Escherichia coli V. Characterization of a nucleotide associated with the stringent response. J. Biol. Chem. 245:2309 (1970).

16. J.Gallant and R.A.Lazzarini, The regulation of rRNA synthesis and degradation in bacteria, in "Protein Synthesis a series of advances", E.H.McConkey, ed., M.Dekker, New York (1976).

17. J.Sy and E.Lipmann, Identification of the synthesis of guanosine tetraphosphate (MSI) as insertion of a pgrophosphoryl group into the 3'-position in Guanosine 5'-diphosphate, Proc. Nat. Acad. Sci. 70:306 (1973).

18. T.Laffler and J.Gallant, Spo T, a new genetic locus involved in the stringent response in E.coli, Cell 1:27 (1974).

19. C.Kari, J.Török and A.Travers, ppGpp cycle in E.coli, Molec. gen.Genet. 150:249 (1977).

20. F.Gros and F.Gros, Role des aminoacides dans la synthèse des acides ribonucleiques chez E.coli, Biochem. Biophys. Acta 22:200 (1956).

21. A.Muto, A.Kimura and S.Osawa, Effect of some antibiotics on the stringent control of RNA synthesis in E.coli, Molec.gen.Genet. 139:321 (1975).

22. M.R.Siegel and M.D.Sisler, Site of action of Cycloheximide in cells of Saccharomyces pastorianus. III Further studies on the mechanism of resistance in Saccharomyces species, Biochem. Biophys. Acta 103:558 (1965).

23. G.D.Clark-Walker and A.W.Linnane, In vivo differentiation of yeast cytoplasmic and mitochondrial protein synthesis with antibiotics, Biochem. Biophys. Res. Commun. 25:8 (1966).

24. A.J.Lamb, G.D.Clark-Walker and A.W.Linnane, The biogenesis of mitochondria 4. The differentiation of mitochondrial and cytoplasmic synthesizing systems in vitro by antibiotics, Biochem. Biophys. Acta 161:415 (1968).

25. R.Adman, L.D.Schultz and B.D.Hall, Transcription in yeast: separation and properties of multiple RNA polymerases, Proc. nat. Acad. Sci. (Wash) 69:1702 (1972).

26. E.Wintersberger and G.Viehhauser, Function of mitochondrial DNA in yeast, Nature 220:699 (1968).

27. E.DiMauro, C.P.Hollemberg and B.D.Hall, Transcription in yeast: A factor that stimulates yeast RNA polymerases, Proc. Natl. Acad. Sci. 69:2818 (1972).

28. E.K.F.Bautz, Regulation of RNA synthesis, in "Progr. Nucl. Ac. Res. and Mol. Biol.", J.N.Davidson and W.E.Cohn, eds., Academic Press, New York and London (1972).

29. A.H.Scragg, A mitochondrial DNA-directed RNA polymerase from yeast mitochondria, in "The biogenesis of mitochondria", A.M. Kroon and C.Saccone, eds., Academic Press, New York and London

30. B.S.Baker, A.T.C.Carpenter, M.S.Esposito, R.E.Esposito, L.Sandler, The genetic control of meiosis, Ann. Rev. Genet. 10:53 (1976).

31. N.Marmiroli, F.Tassi, L.Bianchi, A.A.Algeri, P.P.Puglisi and M.S.Esposito, Decompartmentalization of erythromycin sensitivity during sporulation of Saccharomyces cerevisiae: Genetic control by chromosomal and mitochondrial genes, Submitted to J. Bacteriol. (1979).

32. M.Yeas and G.Brawerman, Interrelationships between nucleic acid and protein biosynthesis in microorganisms, Arch. Biochem. Biophys. 68:118 (1957).

33. R.M.Roth and C.Dampier, Dependence of ribonucleic acid synthesis on continuous protein synthesis in yeast, J.Bacteriol. 109: 773 (1972).

34. S.G.Oliver and C.S.McLaughlin, The regulation of RNA synthesis in yeast, I: Starvation experiments, Molec.gen.Genet. 154:145 (1977).

35. R.W.Shulman, C.E.Sripati and J.R.Warner, Noncoordinated transcription in the absence of protein synthesis in Yeast, J. Biol. Chem. 252:1344 (1977).

36. D.B.Ray and R.A.Butow, Regulation of mitochondrial ribosomal RNA synthesis in Yeast. I: In search of a relaxation of

stringency, Molec.gen.Genet. 173:227 (1979).
37. F.Foury and A.Goffeau, Stimulation of yeast RNA synthesis by cycloheximide and 3',5'-cyclic AMP, Nature New Biol. 245:44 (1973).
38. R.Kudrna and G.Edlin, Nucleotide pools and regulation of ribonucleic acid synthesis in yeast, J.Bacteriol. 121:740 (1975).
39. C.C.Pao, J.Paietta and J.A.Gallant, Synthesis of guanosine tetraphosphate (magic spot L) in Saccharomyces cerevisiae, Biochem. Biophys. Res. Comm. 74:314 (1977).
40. C.Waldron and F.Lacroute, Effect of growth rate on the amounts of ribosomal and transfer ribonucleic acids in yeast, J.Bact. 122:855 (1975).
41. A.A.Algeri, F.Tassi, I.Ferrero and P.P.Puglisi, Different phenotypes for the lactose utilization system in Kluyveromyces and Saccharomyces species, Antonie van Leeuwenhoek 44:177 (1978).
42. J.D.Irvin, B.Hardesty: Binding of aminoacyl transfer ribonucleic acid synthetases to ribosomes from rabbit reticulocytes. Biochem. 11:1915 (1972).
43. J.S.Tscherne, I.B.Weinstein, K.W.Lanks, N.B.Gerstein, C.R.Cantor, Phenylalanyl-tRNA synthetase activity associated with rat liver ribosomes and microsomes. Biochem. 12:3859 (1973).

ACKNOWLEDGEMENTS

The original data reviewed in this paper constitute a part of the program "Functional Taxonomy of Yeasts" supported by a grant of the Ministero della Sanità – Divisione Generale dei Servizi Veterinari.

CONTRIBUTOR INDEX

SUBJECT INDEX